Topics in Environmental Physiology and Medicine

Edited by Karl E. Schaefer

Herbert L. König
Albert P. Krueger
Siegnot Lang
Walter Sönning

Biologic Effects of Environmental Electromagnetism

Translation by Terry C. Telger

With 142 Illustrations

Springer-Verlag
New York Heidelberg Berlin

Herbert L. König
Institute for the Foundation of Electrotechnics
Munich Technical University
Munich, West Germany

Albert P. Krueger
School of Public Health
University of California
Berkeley, California 94720 U.S.A.

Siegnot Lang (deceased)

Walter Sönning
Emil Dittlerstrasse 4
D-8000 Munich, West Germany

Library of Congress Cataloging in Publication Data

Main entry under title:
Biologic effects of environmental electromagnetism.
(Topics in environmental physiology and medicine)
Bibliography: p.
1. Electromagnetism—Physiological effect. I. König, Herbert L., 1925– II. Series. [DNLM: 1. Electromagnetics. 2. Electrophysiology. 3. Ions. QT 34 K77u]

QP82.2.E43B54 574.19'17 81-8898
AACR2

Softcover reprint of the hardcover 1st edition 1981

9 8 7 6 5 4 3 2 1

ISBN-13: 978-1-4612-5861-2 e-ISBN-13: 978-1-4612-5859-9
DOI: 10.1007/978-1-4612-5859-9

Contents

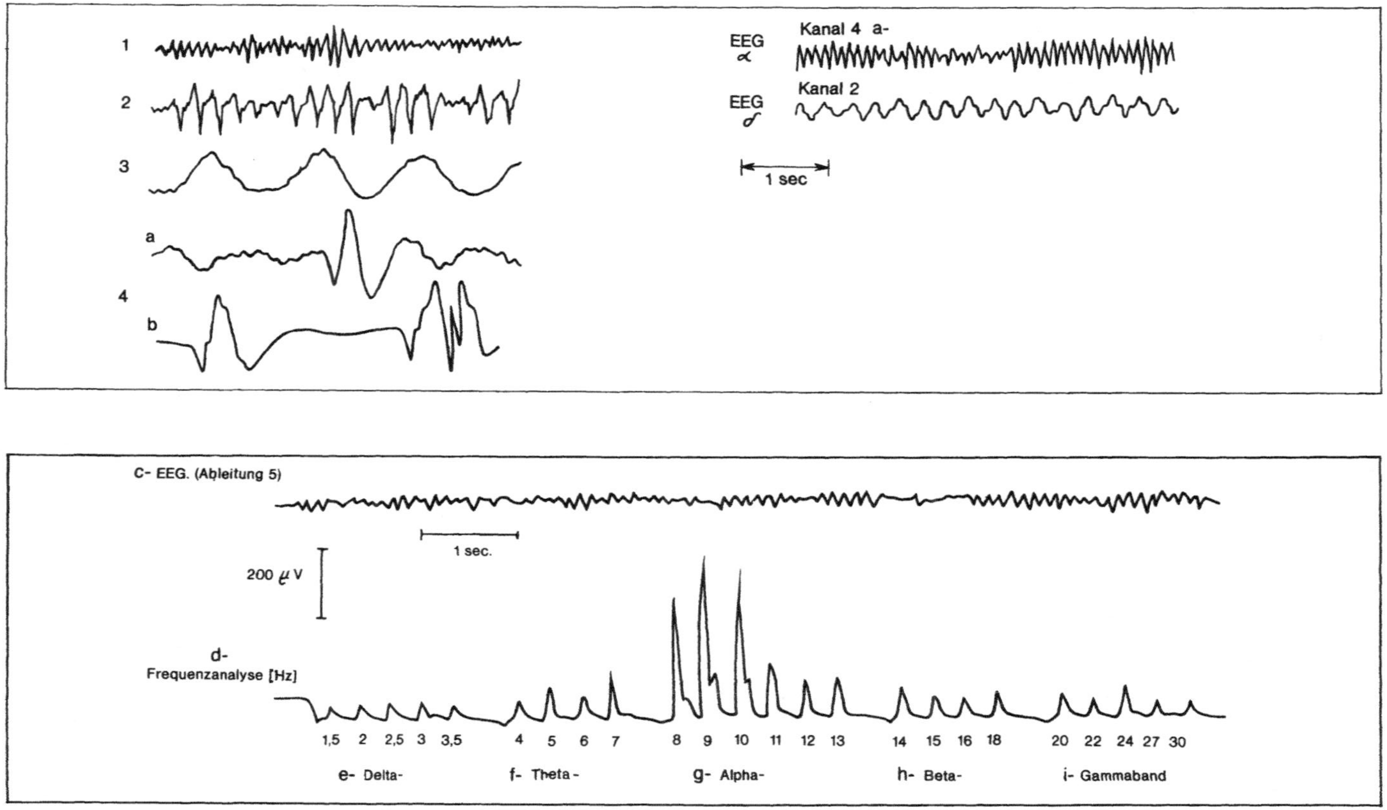

Figure 1-10. ***a*** Natural electromagnetic processes in the environment (1–4) compared with human EEG records. Striking similarities are seen between Schumann resonance vibrations (1) and the EEG alpha rhythm, and between local electric field fluctuations (2) and the EEG delta rhythm. *b* Sample EEG record with eyes closed. The top curve shows the EEG as a function of time. The bottom curve is a frequency analysis of the EEG. The height of the peaks is a relative measure of the frequency with which signals at the indicated frequencies occur. The range between 8 and 11 Hz, called the alpha band, is predominant. After Schneider (1964).

Key: a-Channel 4; b-Channel 2; c-EEG (lead 5); d-Frequency analysis (Hz); e-Delta band; f-Theta band; g-Alpha band; h-Beta band; i-Gamma band excerpt from such records, it would be difficult even for an expert to determine whether EEG currents or naturally occurring electric field fluctuations were involved.

There is also a remarkable similarity in the conditions under which each group of signals typically occurs. For example, type I signals are observed primarily during fair-weather periods, under quiescent weather conditions. These conditions correspond closely to the mental state associated with the appearance of alpha waves in the EEG. A similar comparison can be made with respect to type II signals in the ELF range. Irregular occurrence, relatively high amplitudes, a strong harmonic content in the fundamental wave, and a labile fundamental frequency which is in every case lower than the type I frequency are all typical of such signals. On a local scale, they are usually indicative of strong weather disturbances and represent the antithesis of type I signals: the disturbed, unsettled, anomalous, unbalanced. In this respect, these field phenomena bear a striking resemblance to the delta rhythm in the EEG.

A small circle of physicians is even conducting studies on the extent to which a very definite frequency-specific signal activity or function can be assigned to the human organism in the frequency range around 1–10 Hz.

The electric control processes which are associated with the activity of the heart and are recorded in the electrocardiogram (ECG) also fall within the ELF range.

Similar comparisons can be made with respect to processes in the higher VLF range, as in the control of muscular reactions. In order to move a muscle, the brain sends forth control impulses that are relayed to the muscle via a complex system of neurons and stimulate it to make the desired movement. If these commands are to be executed with sufficient speed, the electric control impulse which carries the command must be transmitted at a correspondingly high velocity. This is ensured by the use of an electric current as the signal transmitting agency. The signal information, or the command to the muscle, must also be conveyed from the arriving signal to the muscle in a very short time. This requires that the control impulse contain a considerably higher-frequency spectrum than is present in the ELF range. This is, in fact, the case, for muscular control is accomplished by means of electric impulses whose frequency spectrum lies in the VLF range, about 1–10 kHz and higher. A natural, external stimulus is also present in this higher-frequency signal range: the VLF atmospherics produced by lightning strokes exhibit exactly the same properties in terms of information content as do such processes within the body.

But what of the role played by natural electromagnetic processes in the frequency ranges directly above the VLF and directly below the ELF? As mentioned earlier, the intensity of natural electromagnetic phenomena, and thus their role as environmental stimuli, decreases steadily from 10 kHz to higher frequencies. Moreover, electromagnetic energies are increasingly transformed into heat as their frequency increases, which further diminishes their capacity to bear information. As their frequency decreases, on the other hand, the intensity of natural electric and magnetic fields rises to a value which is described by the natural electrostatic field (fair-weather field), or to the intensity of the static terrestrial magnetic field (see Fig. 2-5). But such quasi-static fields have little in the way of information content. Viewed with respect to time, information means a change in a pre-existing state, which is simply not present under static conditions. Hence, static fields and direct currents are unsuitable for conveying information of a complex nature.

Becker (1963a) pointed out connections between biologic signals and natural micropulsations in the frequency range from 0.01 to several hundred hertz, stating that these pulsations were particularly marked in the range from 8 to 16 Hz based on then-available data on ELF atmospherics). His intention was to show a connection between the alpha rhythm, and thus the activity of the human brain, and this type of electromagnetic phenomenon.

Herron (1965) reports on research results pertaining to micropulsations of 0.029–0.031 Hz. His results are similar to those of Schumann et al. (1966) and Siebert (1964 a,b). Herron associates these natural processes with the familiar ultraslow fluctuations of potential in the EEG.

Finally, there is much evidence which points to an internal utilization of electromagnetic energies by organisms. According to Presman (1970), there are sufficient indications and experimental results to confirm this. Examples include the distribution of an electric surface potential in man and animals, the electric brain currents recorded in the EEG, the processes controlling cardiac function as recorded in the ECG, and muscular control processes in the VLF range.

To summarize, only the frequency range which is described by the ELF and VLF regions and lies between 0.1 Hz and 10 kHz (including high-frequency processes which are correspondingly modulated) is suitable for biologic control processes. All control and regulatory processes in biologic systems take place within this range of frequencies, and it is in this frequency range that naturally occurring electromagnetic phenomena play a significant role. It is very likely that life was significantly influenced in its evolution by the low-frequency electromagnetic component of the atmosphere.

2

Electrophysical Forces of Natural Origin

Natural Electromagnetic Fields

Scientific investigations since the late 1950s have shown that electromagnetic processes of natural origin, covering the full spectrum from static processes to the extremely high frequency range, exist in our immediate environment. These electromagnetic energies can be classified with regard to their frequency, and thus their position in the frequency spectrum, or described in terms of the wavelength they would possess if they were to propagate as waves in empty space. An overview of the relationship between wavelength and frequency, and the various frequency ranges included in the electromagnetic spectrum are illustrated in Fig. 2-1. It is customary to assign processes with frequencies below 1 Hz to the ultralow frequency (ULF) range. This is followed by the extremely low frequency (ELF) range (ending at 300 Hz), and the very low frequency (VLF) range (ending at 300 kHz). High-frequency radiation includes the familiar long-wave, medium-wave, short-wave, ultrashort-wave and microwave ranges. Next in the spectrum are infrared radiation, visible light, ultraviolet (UV) radiation, X-radiation, and α-, β-, and γ-radiation. At the top of the spectrum is cosmic radiation, which has the highest frequencies known. The location of the optical windows in the atmosphere, along with comparative energy data, is also presented in Fig. 2–1.

Some of the details of what we have learned about the geophysical and electrophysical properties of natural electromagnetic fields are outlined below.

Quasi-static Fields, Electric Air Current

In 1752 Franklin demonstrated that lightning is an electric discharge, similar to the sparks between the electrodes in an electric machine, and that therefore electricity must be present in the atmosphere. The associated general atmospheric electric field (fair-weather electricity) comprises a negatively charged earth and a positively charged atmosphere (Fig. 2-2). This may be explained by the phenomenon of ground lightning during thunderstorms compensating the negative charge in the clouds, leaving positive ions located at altitudes (up to the ionosphere) that are insulated from the ground by the intervening air. These ions are free to diffuse upward in the thinner air

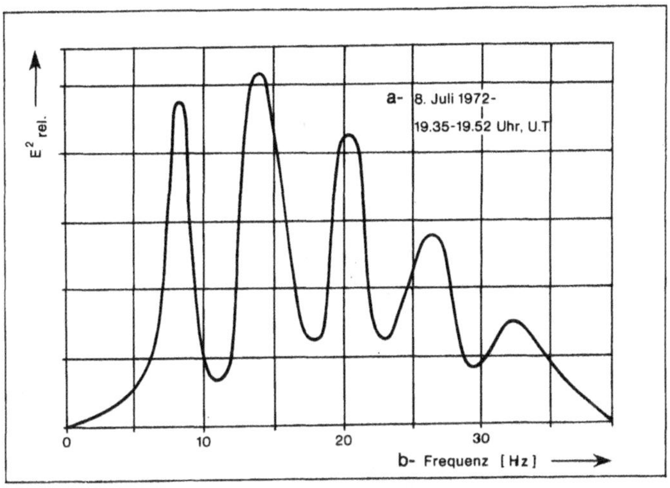

Figure 2-8. Relative power spectrum of the vertical electric field (E^2_{rel}) of natural signals in the ELF range. Lightning excites so-called Schumann resonance oscillations—electromagnetic radiation in the form of "standing waves"—in the spherical earth-ionosphere cavity resonator. After Toomey and Polk (1970).

numerical ratio exists between the electric and magnetic components; this is the "wave impedance," $E/H = W = 376\ \Omega$. This ratio holds in ELF atmospherics if these signals are measured sufficiently distant from their point of origin.

But the formation of standing waves leads to quite specific field configurations for the various modes, as shown in Fig. 2-9 for the electric field. A certain ELF signal, then, distributed over the entire globe, produces at all points a predictable relationship between the electric and magnetic field components and their modes. If these relationships are determined by measurement, conclusions can be drawn regarding the geographic location of the signal origin. Much research has been devoted in recent years to utilizing this principle as a means of monitoring global thunderstorm activity from a few isolated stations (Keefe et al., 1973; Toomey and Polk, 1970; Ogawa et al., 1969; Sao and Jindoh, 1974).

Occasionally the Schumann resonances

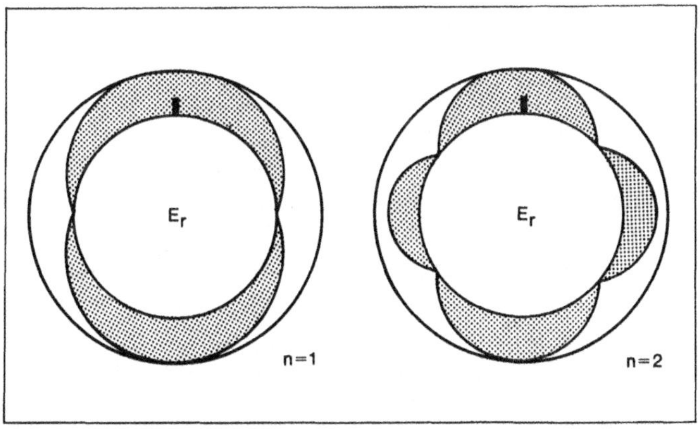

Figure 2-9. Field configuration of the electrical component of the Schumann resonance oscillations occurring as "standing waves" for the modes $n = 1$ and $n = 2$. After Toomey and Polk (1970). Key: E_r, earth.

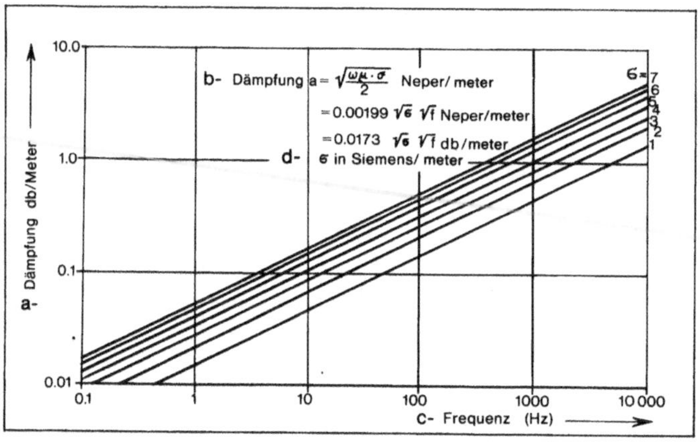

Figure 2-10. Attenuation of waves in seawater as a function of frequency for various water conductivity values. After Soderberg and Finkle (1970).
Key: a-Attenuation (db/m); b-Attenuation a = . . . neper/meter; c-Frequency (Hz); d-In mho/meter.

are masked by other irregular signals (noise) above the third resonance (at about 20 Hz), but in experimentally obtained spectra they are never observed above the fifth resonance (about 32 Hz). This is mainly because the resonant quality of the earth–ionosphere cavity is too low at higher frequencies, and so the resonance peak in the spectrum is too flat. The quality of the resonance, *Q*, tells us something about the structure of the ionosphere (Handa et al., 1971; Keefe et al., 1973; Ogawa and Tanaka, 1970; Ogawa and Murakami, 1973), as well as about certain sunspot phenomena (Polk, 1973).

As a result of disparities in the properties of the ionosphere—Tran and Polk (1973), for example, investigated its conductivity with respect to signals in the ELF and VLF ranges—different diurnal variations are found in mean received signal intensities as a function of frequency. While the signal intensity is greatest at night in VLF atmospherics, it is greatest during the day in the ELF range. The transition occurs at about 120 Hz (Fig. 2-11).

More precise observations and theoretical analyses (König, 1959; Chapman and Jones, 1964; Keefe et al., 1973; Rycroft, 1965; Ogawa, 1973; Polk, 1973) of atmospherics in the Schumann resonance region have shown that, due to their resonant and wave characteristics, the signals have a practically sinusoidal time course for the fixed observer (in Fig. 2-12, signal type I), whose envelope may be steep (in signals consisting of few oscillations) or flat (in signals which extend over several oscillations).

In recordings of wave trains, the frequency often shows a tendency to decrease toward the end of the signal. This decrease may be up to 30% of the initial value. The degree of this decrease varies greatly from one signal to another. However, it has been determined that wave trains which attain a very large amplitude, and thus have a steep envelope and relatively short duration, usually undergo very marked changes in their frequency. Longer-lived signals with a smaller amplitude and flat envelope show practically no frequency change. The signals are greatly favored by fair weather and as a rule can be easily recorded under such conditions. It is probable that the signals in question, which are not measured in bad weather, are simply masked by other phenomena of greater intensity during storm periods. As stated earlier, signals of this type are considerably less intense at night than during the day. The ratio of day to night intensities ranges from about 3:1 to 10:1 (Fig. 2-13).

The measurement of signals with

frequencies below 100 Hz is extremely difficult. It is necessary in such cases to employ a suitable filtering technique to suppress interference from man-made signals (e. g., 50-Hz or 60-Hz power current, 16 $^{2}/_{3}$-Hz railway frequency, 25-Hz telephone ring) whose intensity usually far exceeds that of natural fields (König, 1959; König and Behringer, 1970; Keefe et al., 1973; Ogawa, 1973; Sao et al., 1973).

These low frequencies pose special antenna problems in the case of both electric (Clayton et al., 1973) and magnetic fields (König, 1968b). Nevertheless, a close coupling has been established between ELF phenomena associated with wave-propagation phenomena on the one hand, and with magnetospheric or geomagnetic field fluctuations on the other (Polk, 1973). This is clearly shown by earth current and geomagnetic field measurements by Fournier (1960a,b), who reports on oscillations with periods of 30–0.025 s (40 Hz). Magnetic field oscillations with a period of 4.5 and intensity on the order of 10^{-12} T (milligamma) were especially marked. Besides these purely geomagnetic phenomena, the magnetic components of the Schumann resonance oscillations in the 10 Hz range were also evident from the earth current measurements.

VLF Atmospherics. Electromagnetic phenomena of natural origin in the VLF range, which are caused by lightning, show all the properties of electromagnetic waves after propagating a relatively short distance. Thus, signals with a frequency of 10 kHz have a wavelength in free space (30 km) which approaches the spatial magnitude of the lightning flashes which produced them.

The VLF electromagnetic waves triggered by electric discharge phenomena in the atmosphere undergo a change of

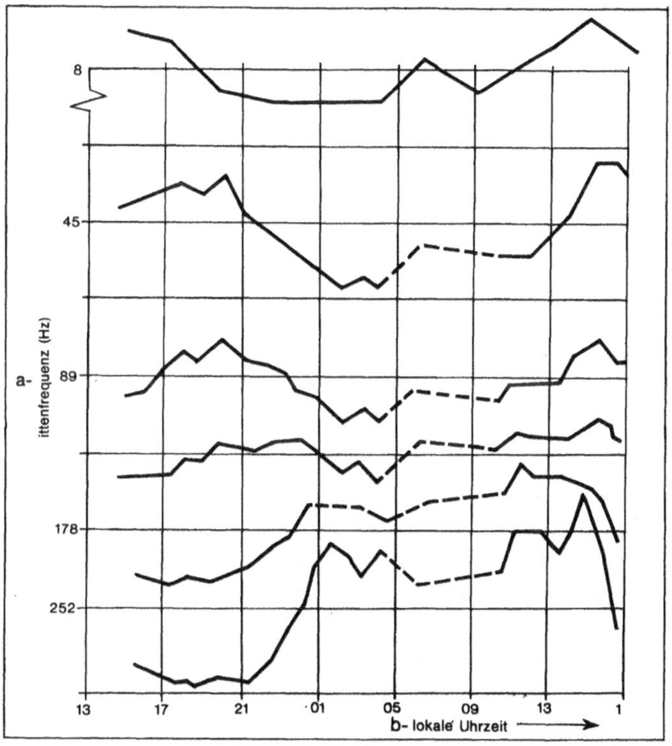

Figure 2-11. Comparison of daily variations of natural signal amplitudes measured in various frequency bands. After Holzer et al. (1957), with additions.
Key: a-Center frequency (Hz); b-Local time

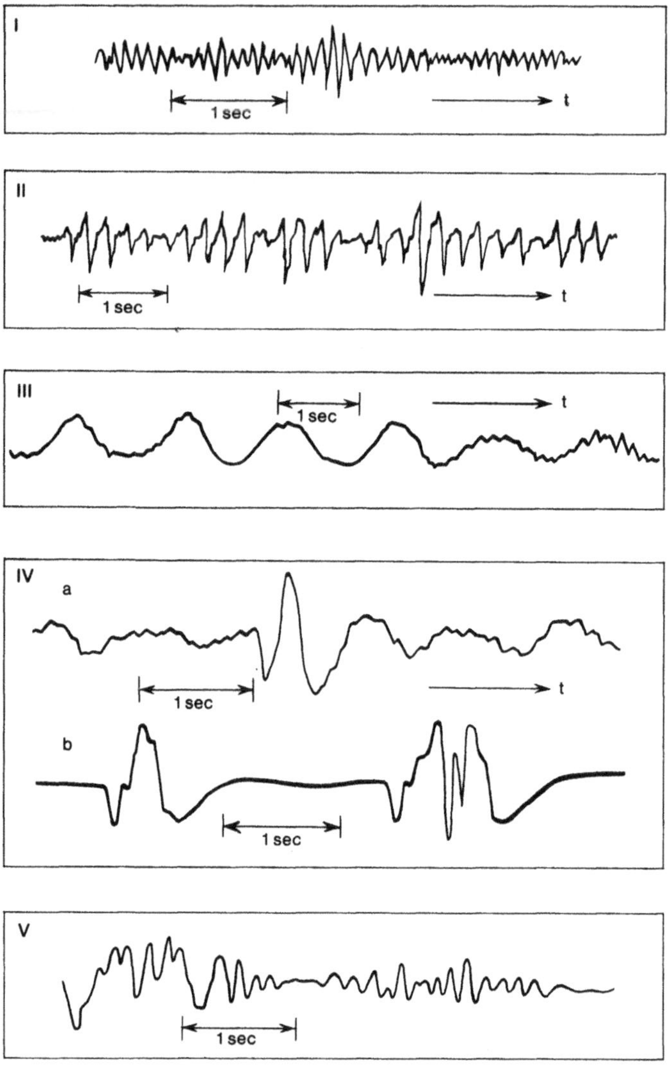

Figure 2-12. Electromagnetic processes of natural origin in the ELF range. **I** Schumann resonance oscillation, approx. 8 Hz. **II** Local electric field fluctuation, 3–6 Hz. **III** Local electric field fluctuation, approx. 0.7 Hz. **IV** Electric field during thunderstorm: (*a*) thunderstorm not yet visible on horizon, (*b*) thunderstorm visible on horizon (amplitude greatly reduced). **V** Sunrise effect, electric field.

waveform for the fixed observer as they propagate farther from their point of origin in the earth–ionosphere waveguide (Fig. 2-14). Rhoads and Garner (1967) devised a measure for estimating the attenuation of propagating VLF signals: between about 200 km and 2000 km, the signal amplitude diminishes by about 15–18 db/1000 km. At greater distances the attenuation of 15-kHz signals is approximately 4 db/1000 km (Fig. 2-15) or 2 db/1000 km (Fig. 2-16). The attenuation as a function of frequency is plotted over a greater range in Fig. 2-16 (Watt and Maxwell, 1957; Bernstein et al., 1974). By comparison, Fig. 2-10 (Soderberg and Finkle, 1970) shows a mean attenuation rate in seawater of about 2 db/m for a 10-kHz signal.

It is also well known (Al'pert and Fligel, 1970; Storey, 1953) that a considerable portion of the energy of low-frequency waves, especially those emitted by lightning strokes, penetrates out through the ionospheric layers and into space. In the

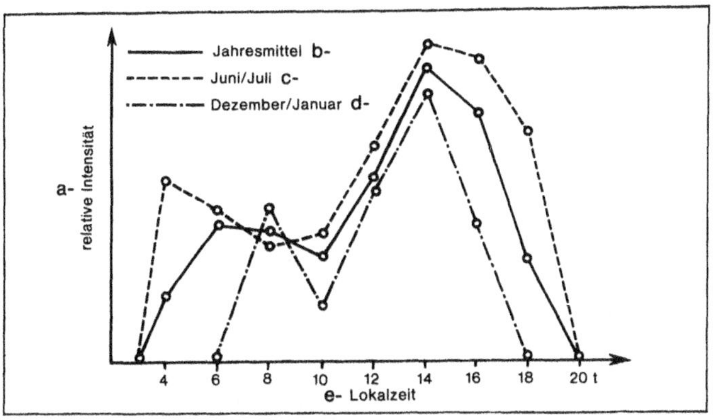

Figure 2-13. Daily variation of relatively strong ELF atmospherics (8-Hz Schumann resonance). Key: a-Relative intensity; b-Yearly average; c-June/July; d-December/January; e-Local time

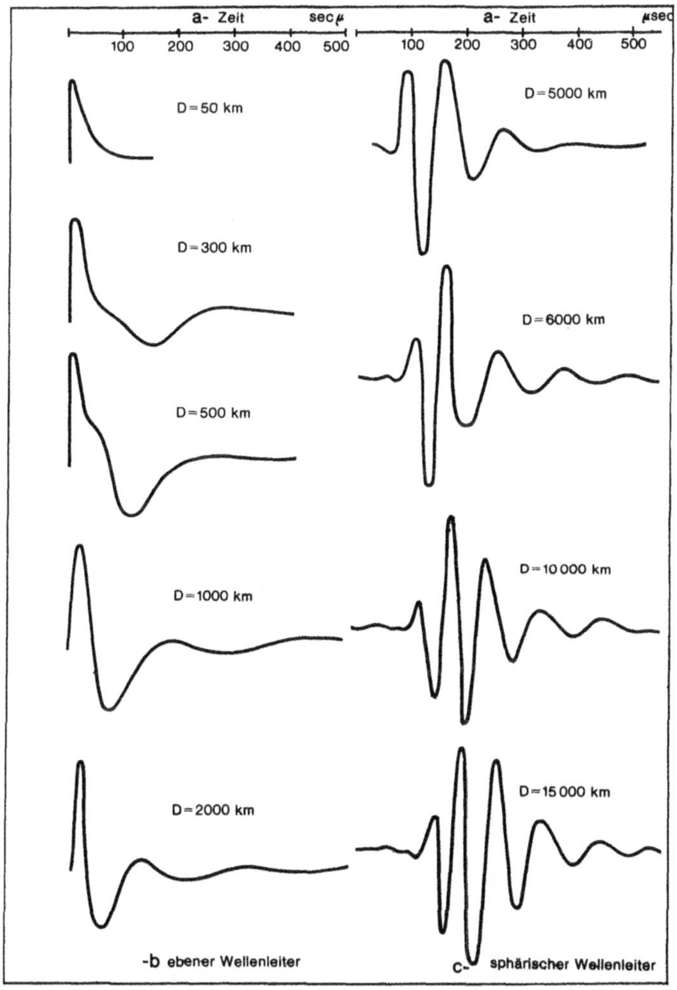

Figure 2-14. Waveforms of VLF atmospheric signals at various distances D from the point of origin (lightning) between 50 and 15,000 km for plane and spherical waveguides. After Al'pert and Fligel (1970).
Key: a-Time; b-Plane waveguide; c-Spherical waveguide

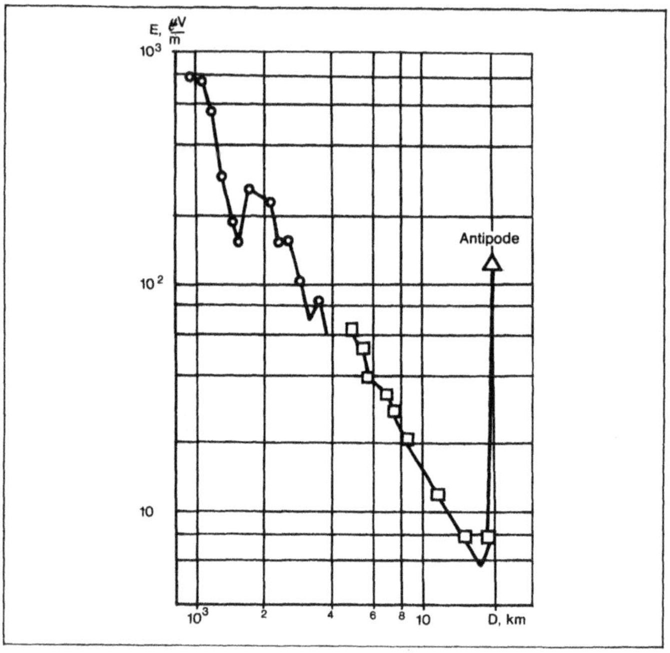

Figure 2-15. Result of field-intensity measurements at signal frequencies of 16.6 kHz (circles), 17.5 kHz (squares) and 14.7 kHz (triangles) over very great distances *D* with the special case of the magnetic antipode. After Watt and Groghan (1964).

frequency range of 1–10 kHz and above, this leads to the formation of "whistlers," which propagate along the lines of force of the earth's magnetic field between two points on the earth's surface that are connected by corresponding magnetic field lines (Fig. 2-7). This also accounts for the high signal level at the "magnetic antipode" (Fig. 2-15).

There is another important phenomenon

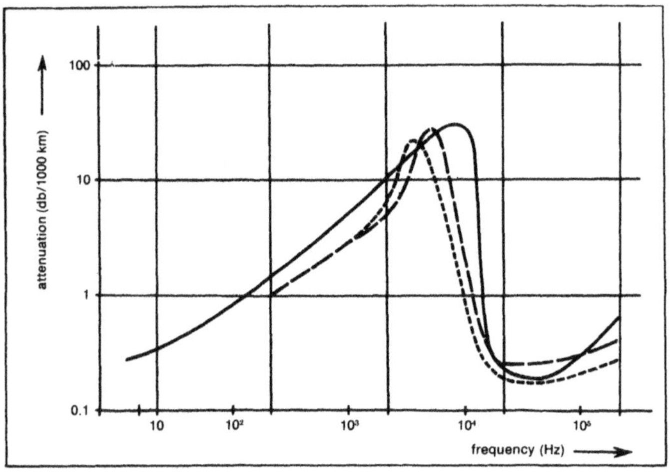

Figure 2-16. Mean attenuation under various conditions in the ELF range (below 1 kHz), after Bernstein et al., and in the VLF range (above 1 kHz), for atmospherics as a function of frequency, at distances d ≥ 1000 km; ———— during the day; — — — at night/over sea; -------- at night/over land. After Watt & Maxwell (1957).

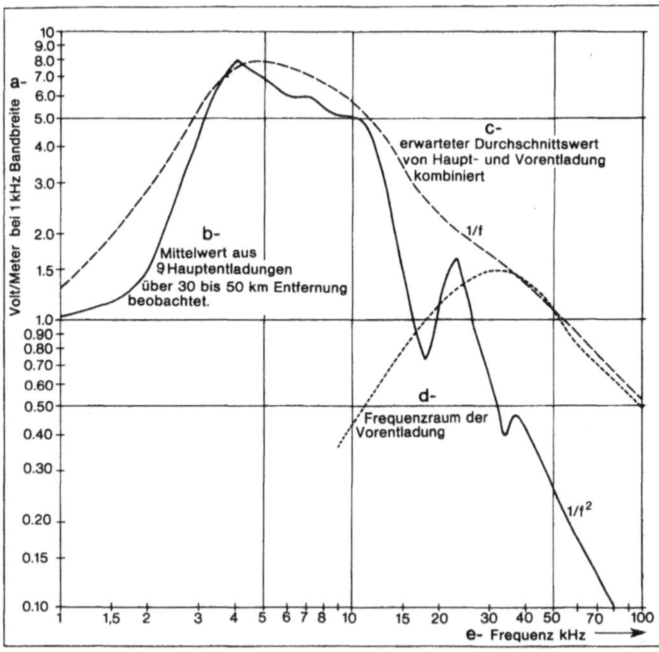

Figure 2-17. Frequency spectrum of the radiative component of a lightning stroke; field intensity at a distance of about 1.6 km from the point of origin. After Watt and Maxwell (1957).
Key: a-Volt/meter at 1 kHz bandwidth; b-Mean value for 9 main discharges; c-Expected average value for main and predischarge combined; d-Frequency spectrum predischarge; e-Frequency (kHz)

associated with the natural geomagnetic field. It has been shown in a number of experiments that the reciprocity principle is violated within the earth–ionosphere waveguide. Thus, a VLF wave propagating over a great enough distance from east to west undergoes a greater attenuation than it would if it propagated from west to east (Watt and Groghan, 1964; Taylor, 1960a,b; Hanselmann, 1964; Martin, 1961; Crombie, 1958, 1963).

Near-field observations of lightning discharges have revealed very high field intensities (Fig. 2-17). It is not surprising that these atmospherics can be recorded as discrete signals whose intensities far exceed the ordinary general signal level. Electric field intensities are plotted as a function of frequency in Fig. 2-18.

In correlational studies on the biologic effects of such naturally occurring VLF signals, it is customary to record the number of VLF atmospheric pulses which are incident per unit of time (second). Also, special instruments are used to divide the atmospherics into groups according to intensity. They can be classified further by making the recordings in different frequency bands.

Since VLF signals have shorter wavelengths than ELF signals, they can be measured with more modest equipment. By utilizing DF systems and data on the propagation of VLF atmospherics, a bearing can be taken on thunderstorm centers. Investigations by Heydt (1971) have produced very promising results in this regard for the African and American continents.

Experiments with Models

One method of investigating the physical properties of atmospherics under laboratory conditions is to construct scale models of the earth and the ionosphere and perform experiments with the aid of microwaves. Polk (1969) used this method to study the effect of the day–night boundary in the ionosphere on the propagation of ELF at-

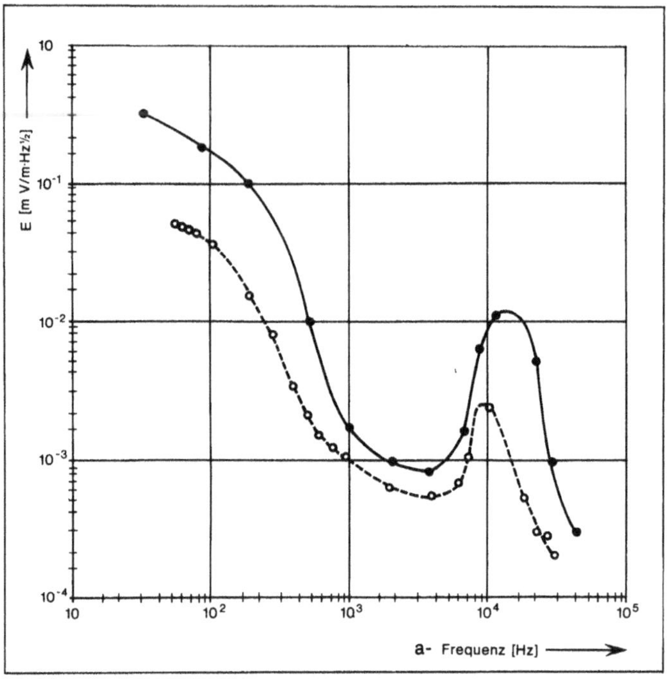

Figure 2-18. Frequency spectrum (intensity of vertical electric field E) of ELF and VLF atmospherics (open circles: all day, at distance of 6000 km; filled circles: Alaska, summer, 16:00 to 20:00). After Galejs (1970).
Key: a-Frequency (Hz)

mospherics. In connection with environmental influences of physiologic interest, Ludwig et al. (1968) investigated VLF atmospherics to study physiologic influences of an electric climate. They also utilized a microwave model in order to obtain more detailed information on the influence of conductivity jumps in the earth–ionosphere waveguide and on the effects of mountainous terrain (orography) on wave propagation over great distances. The model experiments showed that with each conductivity jump, a region with standing waves develops, giving rise to characteristic field and reception changes.

Connection between ELF and VLF Signals

As will be discussed later in more detail, studies have shown that electromagnetic fields in both the ELF and VLF ranges exert effects on living forms. This raises the question, what are the physico-technical reasons that such effects are observed in both ranges?

Biologic effects of VLF electromagnetic fields are reported primarily in connection with signals of at least intermediate intensity, and thus with atmospherics which propagate over distances of around 1000 km and can be detected as individual events. As mentioned earlier, the number of atmospheric pulses (1 wave train = 1 pulse) per unit time is recorded, and the resulting "pulse repetition frequency" is used as the parameter. Correlations between biologic factors and the occurrence of VLF atmospherics are found principally when the pulse repetition frequencies of these phenomena are in the ELF range, that is about 1–50 Hz. If we further take into account the nonlinear electric properties which are typical of biologic systems, we find that every VLF pulse creates a rectifying effect in such systems, producing a pulse of one polarity which (according to

X-rays and short-wave UV radiation are absorbed in the upper layers of the atmosphere, where they lead to the dissociation of nitrogen and oxygen molecules and to the ionization of these molecules and the free atoms formed by dissociation. "Absorption layers" develop, associated with the selective absorption bands of the various dissociation and ionization processes. The UV component emitted by the photosphere penetrates the atmosphere almost unhindered to heights of 30 km, where its short-wave component contributes to the formation of the ozone layer. Ultraviolet radiation with a wavelength of 0.3 μm can penetrate the atmosphere virtually unhindered.

Cosmic Rays

Cosmic rays basically consist of electrons, mesons, protons, neutrons and photons which enter the earth's upper atmosphere from deep space. The neutrons and photons are not deflected by the earth's magnetic field and can penetrate the atmosphere even at low energies. The neutrons decay into electrons and protons, which are captured by the earth's magnetic field and are found primarily in the terrestrial radiation belts. H. Schulz (1970) reports that the earth's atmosphere attenuates cosmic rays to about the same degree as 10 m of water or 90 cm of lead. It is noteworthy that nearly all particles are absorbed in the upper layers of the atmosphere by a selective filtration effect, and only high-energy photons and a few electrons are left to reach the earth's surface. As a result, only about 30 fast electrons/m^2s are found at the earth's surface despite the powerful current of material which enters the atmosphere.

Earth's Surface As an Infrared Radiation Source

The earth's surface and the sea's surface, like the atmosphere, emit an infrared radiation which depends primarily on their surface temperature. Like atmospheric radiation, these infrared rays have wavelengths ranging from 6 to 60 μm. Their intensity is also comparable, amounting to about 0.4 kW/m^2 on cloudy days and about 0.5 kW/m^2 on cloudless days.

The natural radioactivity of the ground and the radioactive isotopes in the atmosphere lead to the formation of 1–2 ions/cm^3s in our living environment. By coincidence, the intensity of this radiation corresponds closely to that of cosmic rays (10^{-13}W/m^2 or 0.1 roentgen/year). Although this ionization rate is low, it is of sufficient magnitude to ensure a continuous supply of ions.

Infrared Emissions from the Atmosphere to the Biosphere

The infrared emissions from water, carbon dioxide and ozone molecules in the atmosphere to the biosphere constitute what is called the atmospheric counter-radiation. It spans the aforementioned range of wavelengths from 6 to 60 μm (see Fig. 1-4). The intensity of the atmospheric counter-radiation is 0.3 kW/m^2 on the average, approximately equals the radiation emitted into space. Besides varying with the temperature of the lower atmospheric layers, this value also depends on cloud conditions: on clear winter nights it may fall to 0.16 kW/m^2, while on overcast summer nights with low-hanging clouds it may increase to 0.45 kW/m^2.

An interaction thus exists between infrared radiation from the ground and that from the atmosphere. For the latitude of middle Europe, it is known that only about 25% of the radiation impinging upon the earth's surface during the course of a year is attributable directly to solar radiation; the remaining 75% is due to long-wave infrared radiation from the atmosphere. Up to 84% of the radiant energy thus imparted to the earth's surface is radiated back again in the same form. Of the remainder, about 14% is expended in evaporation, and only 2% is expended in heating the ground and air.

Interaction of Radiations

In the range of wavelengths between UV and infrared radiation, an interaction exists among extraterrestrial solar radiation, long-wave emissions from the earth's surface, and atmospheric counter-radiation. The relative wavelengths and intensities of these radiations are illustrated in Fig. 1-4.

Natural Radioactivity, Ground Radiation

Natural radioactivity originates in the atmosphere and in the earth's crust. Cosmic rays, during their absorption in the atmosphere, trigger the formation of isotopes with infinitely long half-lives (period in which total activity is diminished by half), which make possible a continuous diffusion of radiation from the atmosphere into the living environment.

The radiation emitted from the ground is produced primarily by the radioactive substances present there (uranium, thorium, actinium). These substances in turn trigger the release of radioactive radium, thorium and actinium emanations, both by direct irradiation (alpha, beta, and gamma rays) and by ionization. Direct measurements have shown a mean ionization rate near the ground of 2–10 ion pairs/cm^3s, which is consistent with measurements of the content of radioactive elements in the upper ground layer. The ionization rate may be more than 20 times the normal value in some places, because of a high radium content in the ground.

Alpha rays originate in a very thin surface layer of the ground. Because their range is only a few centimeters, their contribution to ground radiation is negligible. Beta rays originate at greater depths, have a greater range, and produce an ionization of about one ion pair/cm^3s near the ground to 0.1 ion pair/cm^3/s at a height of 10 m. Even more important are the gamma rays, because they come from still greater depths, and so from a larger reservoir of radioactive substances. Their ionizing power is estimated to be 3 ion pairs/cm^3s near the ground, 1.5 ion pairs/cm^3s at a height of 150 m, and 0.3 ion pairs/cm^3s at a height of 1 km. The contribution made by potassium radioactivity is probably greater than that of the "classical" radioactive elements because of the high potassium content of the earth's crust.

Radium, thorium, and actinium release a gaseous product into the atmosphere, especially when the air pressure falls, which undergoes radioactive decay and leads to a measurable level of natural radioactivity in the air. According to available data, the unusually high fluctuations in natural radioactivity can be interpreted as follows. When the air pressure falls, more ground emanations are released than during a pressure rise. Rain, moreover, washes natural radioactivity out of the air. The vertical air currents generated by solar radiation also act to remove natural radioactivity from the air layers near the ground. At night, on the other hand, the natural radioactivity in the biosphere is replenished by new emanations from the earth's crust. Various values are given in Fig. 2-23 for the radon content of the air. The unit 1 curie (Ci) corresponds to the activity of a radioactive nuclide with 3.700×10^{10} disintegrations/s, while the radiation dose unit 1 R (roentgen) indicates the production of 2.0822×10^{9} ion pairs/cm^3 in air at 0° C temperature and 1013 millibars air pressure.

Summary

Radiation which may be of significance with respect to biologic processes can be classified according to its origin as (1) radiation originating in space, and (2) radiation originating in the atmosphere and the earth's surface. Table 2-1 outlines these forms of radiation.

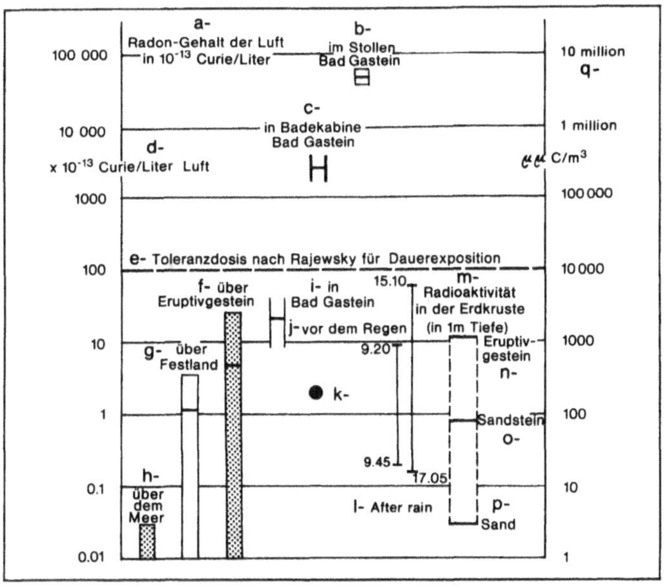

Figure 2-23. Radon content of the air in 10^{-13} Ci/l over the ocean and over land; the height of the columns indicates the fluctuation of the radon content (heavy line: mean value); right: washout of radon from the air during rainfall; far right: radioactivity of the earth's crust at depth of 1 m. After R. Schulze (1970).
Key: a-Radon content of air in 10^{-13} Ci/l; b-In tunnel. Bad Gastein. Austria; c-In dressing room, Bad Gastein. Austria; d-Ci/l air; e-Tolerance dose for prolonged exposure, according to Rajewsky (1956); f-Over igneous rock; g-Over mainland; h-Over the ocean; i-In Bad Gastein; j-Before rain; k-Time; l-After rain; m-Radioactivity of the earth's crust (at depth of 1 m); n-Igenous rock; o-Sandstone; p-Sand; q-10 million

Air Ionization, Aerosols

The degree of ionization of the air is a major determinant of its conductivity; it also strongly influences the vertical electric current in the air and the electric fields there. Thus radioactivity, which is primarily responsible for the ionization of the air (see also Fig. A-7 in Appendix A), indirectly influences electrostatic conditions, the vertical atmospheric electric current, and low-frequency electric fields. An awareness of this influence is especially important in the study of biologic processes in connection with electromagnetic phenomena for which local effects are suspected or known.

Air ions are typically classified as small or large atmospheric ions (Fig. 2-24). Small ions consist of up to 10 molecules with a positive or negative elementary electric charge. The density of small ions near the ground is 100–1000/cm³. Long-term observations indicate a declining trend: from 1900 to 1970, the particle density decreased from 850 to 500 small ions/cm³.

Large ions are positively or negatively charged aerosol particles. Their concentration is normally 10^3–10^4/cm³ but may reach 10^5/cm³ in the air around large cities and industrial areas. The term aerosol refers to a group of colloids which is characterized by a mist-like dispersion of solid or liquid substances in a gas. If the particles are too small, that is, of atomic or molecular dimensions, they simply mix with the gas in the ordinary manner. If the particles are too large, "coarsely dispersed systems" are formed, such as damp fog, rain or snow. Particles with a diameter in excess of 10 μm sink through the air at a rate of 0.5–5 cm/s. Thus aerosols include all gaseous, liquid or solid substances that are suspended in a gas as physicochemical entities 1–10

Table 2-1. *Radiations of significance to biological processes.*

Radiations from space	Radiations from the atmosphere and earth's surface
1. Corpuscular radiation Almost completely absorbed 2. Cosmic radiation 0.1 r/yr Slight biologic effects 3. γ radiation Almost completely absorbed 4. X-radiation Completely absorbed 5. Ultraviolet radiation 6. Visible light, window I 7a. Infrared radiation, short-wave On 5, 6 & 7a: λ: 0.3–2 μm 10^3 W/m^2 CO_2 assimilation evaporation heating selective biologic reactions 7b. Infrared radiation, long-wave Completely absorbed 8a. High-frequency radiation, ultrashort wave; short wave; window II λ: 10^{-2}–10^2 m 10^{-9} W/m^2 No particular biologic effects 8b. High-frequency radiation, long wave; medium wave Completely absorbed 9. Lowest-frequency radiation ULF radiation from the sun (ELF radiation from quasars not yet explored)	1. Infrared radiation λ: 6–60 μm 10^2 W/m^2 Life-sustaining effect Compensates for radiation deficiency (greenhouse effect) 2. High-frequency radiation λ: 1–100 km Pulses up to 10 V/m Transmitter of processes in the upper atmosphere to man 3. Low-frequency radiation ELF Schumann resonance ULF radiation from the ionosphere and magnetosphere 4a. Natural radioactivity (resulting from radium content of the earth's crust) 0.1 r/year Tolerance acquired No significant biologic effects Biologic concentration slight 4b. Artificial radioactivity (resulting from nuclear detonations) Radiative intensity low at present Danger of biologic concentration

μm in diameter, or as clusters of 10^3 to 10^{18} atoms. Electro-aerosols—particles, suspended in a gas, that have been given a unipolar electric charge—have found applications in medicine.

In connection with the phenomena discussed here, we should mention the phenomenon of ionized rain, or electrically charged raindrops. The meteorologic or electrophysical processes which account for the electric charge that occurs on rain, especially after prolonged dry periods, remain somewhat obscure. Nevertheless, it is possible under certain circumstances to detect electric charges on individual raindrops by the use of special measuring apparatus. An antenna struck by raindrops can detect an electric impulse for each individual drop. It is obvious that variations in rain density will lead to certain fluctuations in local electric fields. However, this phenomenon is not always observed. If a uniform, prolonged rainfall occurs, practically no electric charge can be detected on individual raindrops after a certain period of time.

The so-called Lenard, or waterfall, effect is observed whenever water is spattered into small particles by the action of a waterfall, pounding surf or even a bathroom shower. Lenard discovered that a charge separation takes place under such conditions. The larger drops acquire one charge—positive for distilled water, negative for salt solution—while the smaller

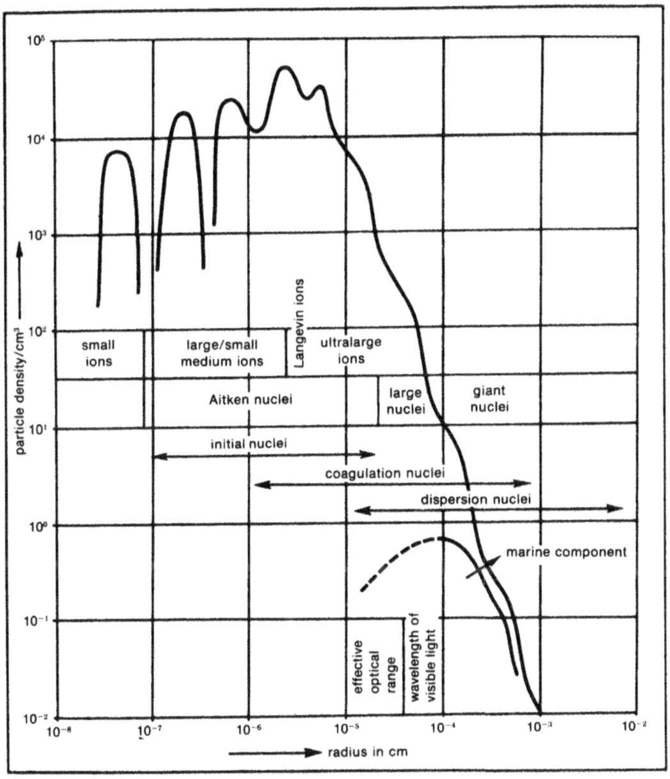

Figure 2-24. Survey of natural aerosols in air with a high nucleus concentration, with corresponding details. After Israel (1957).

droplets and the air acquire the opposite charge. It appears, however, that the Lenard effect plays only a minor role in thunderstorm electricity.

Vertical Electric Current in the Atmosphere

Electric charge carriers, or air ions, give the air a weak electric conductivity. The positive and negative small ions, with a concentration of 100–1000/cm³, produce a conductivity of $0.5\text{–}5 \times 10^{-14}$ S/m (mho/m) near the ground and about 0.3×10^{-14} S/m above the sea. Air ions are formed almost exclusively by the ionizing action of cosmic radiation and radiation from substances in the air and ground. In the upper troposphere and stratosphere, of course, cosmic radiation is the sole ionizing agency. Intermediate and large ions contribute little to the conductivity of the air. However, they do carry space charges which drift with the wind and cause time variations in the field intensity at a given location. Due to the electric conductivity of the air, both space charges and charged solid or liquid surfaces are discharged with half-value times of 100–1000 s. The concentration and mobility of the small ions, and therefore the electric conductivity of the air, increase with altitude. For example, the conductivity at 1, 5 and 10 km above the ground is 2.5, 10 and 25×10^{-14} S/m, respectively. At a height of 15 km, about 5000 small ions/cm³ produce a conductivity of 100×10^{-4} S/m.

From the field intensity E and the conductivity σ of the air, we can compute a vertical conduction current density j (from $\vec{j} = \sigma \vec{E}$) of approximately 10^{-12} A/m². This can be confirmed by measurements with a collector plate on the ground surface. This current density, which is more or less altitude-dependent over fair-weather regions, would correspond to a current on the order of 1800–2500 A toward the entire earth surface.

The formation of strong space charges in precipitation clouds, and especially in thunderclouds, has not yet been adequately explained.

Earth Currents

As mentioned earlier in connection with electromagnetic processes in the ULF and ELF ranges, electric currents of natural origin, which are causally related to processes in the ionosphere and magnetosphere and to the earth's magnetic field, flow in the earth. The strength of these currents depends on time variations of magnetic disturbances. Earth currents produce a voltage drop across the earth's surface which, measured over some distance, amounts to several volts. In the auroral zone, however, such as in northern Norway, voltages of almost 1000 V are commonly observed over distances of up to 100 km. These lead to such strong currents that the drop-indicator switchboards formerly used in telegraph offices frequently heated up or burned out. Transatlantic cables from Europe to America sometimes developed end-point voltages of 1500 V. These earth currents are not distributed uniformly over the earth's surface, of course, and so are a source of localized electromagnetic effects.

3

Man-made Electromagnetic Fields and Ionizing Processes

The increasing use of electricity accompanying the growth of industry and technology and the rapid expansion of the radio and television industries have created many new and varied sources of electromagnetic fields. The introduction of modern, highly insulating building materials has also led to the formation of electric fields by the build-up of electrostatic charges.

Static Processes

The annoying electrostatic charges which build up on clothing of synthetic material and on shoes with insulating soles can lead to spark discharges over a distance of several centimeters. The troublesome effects accompanying spark-overs from the hand can be avoided by holding a metallic object, such as a key, tightly in the hand and letting the discharge take place through this object. The build-up of charge on highly insulating materials, curtains, carpets and similar objects can also exert a permanent influence on the air–ion balance in occupied rooms. A burning candle or cigarette smoke also has this effect. The formation or breakdown of electric fields in closed rooms must in any case be related to such circumstances.

Low-frequency Field Fluctuations

Static conditions are disturbed by changes in geometric conditions, which play an important role where electrostatic charges are concerned. These charges may result from a change in position of the objects serving as charge carriers, certain changes in the environment of the charges which in effect transfigure the field space, or from field intensity changes connected with the inflow or outflow of charges. These time-dependent processes do not normally occur at an extremely high rate, but on a time scale of minutes, seconds or perhaps fractions of a second (neglecting the extreme case of a spark discharge). They are then regarded as electric field fluctuations lying in the ULF or ELF range.

Fluctuations of relatively strong man-made magnetic fields are less common. However, in large cities electric transportation systems that are powered by direct current (e.g., subway, streetcars) set up powerful magnetic fields which are subject to the same fluctuations as the current which produces them.

Of less significance are magnetic field variations caused by the movement of field-altering bodies, such as automobiles, in their vicinity.

Railway Current

In Germany and other European countries, electric railway systems are powered by current with a frequency of $16^2/_3$ Hz, usually conducted by overhead lines and rails. Due to the high operating voltage—15,000 V—that the contact wire has with respect to the rails and thus the ground, the electric field produced by the overhead line extends relatively far beyond the limits of the track. This effect is amplified if an entire network of overhead lines is present, as is the case near train stations. The lines must be separated by many hundreds of meters before the electric fields they generated are reduced to values below 1 mV/m.

Much more critical are the magnetic fields associated with the operating current of railway systems. Since the rails are used as current-return lines and are not insulated from the ground, a portion of the current flows underground parallel to the rails. These currents, called stray currents, flow in paths below ground which are determined primarily by the ground conductivity, as well as by the momentary operating conditions of the train. Our own measurements have shown the great distances over which stray currents can propagate from the rails. In one instance, magnetic field intensities measured at a perpendicular distance of 2 km from the rails were about 1000 times greater than those of corresponding natural phenomena determined at the 10-Hz Schumann resonances. Even at a measuring point located at least 20 km from the nearest railway line, the magnetic field intensities were found to be significantly greater than the comparable Schumann resonance signals.

Railway systems are also powered by high-tension wires which carry electric currents with a frequency of $16^2/_3$ Hz. There is no problem with stray currents in this case, because both the supply and return lines are suspended from high-tension poles and so are well isolated. The individual circumstances determine whether the electric field in the vicinity of such high-tension lines is relatively intense due to the higher transmission voltage, leading to appreciable electric field intensities at some distance from the lines. This problem will be discussed in more detail in connection with the supply of electric power (see "Electric Power Supply" below).

Telephone Bell Current

The current which is used to activate the telephone bell in telephone lines lies in the same frequency range as railway current. It has a frequency on the order of 25 Hz. Of course the voltage and current of this bell signal are considerably weaker than those used in railway systems. But because telephone lines are channeled into living areas, the distance between the cables and persons living in the vicinity of the telephone may be quite small. If a telephone connection is installed next to a bed, the electric and magnetic fields originating in the telephone cables must also be taken into account under some circumstances. There are also signals in the voice-frequency range, of course (about 300 Hz–3 kHz), and possibly stray fields picked up by the lines from outside sources (external lines or switching pulses).

Electric Power Supply

The supplying of households and businesses with light and power current at a frequency of 60 Hz (or 50 Hz in Europe) represents a special problem. A far-ranging network of power supply lines ensures that

High-frequency Fields

As stated earlier, radios and television sets radiate electric and magnetic energies at a frequency of 60 Hz (50 Hz). Owing to their mode of operation, however, these devices also emit stili higher-frequency electromagnetic fields ranging from 625 Hz or 16.25 kHz (image formation) to about 40 MHz (intermediate frequency) depending on the TV system employed. Added to this are the direct transmission frequencies of VHF and television transmitters in the range of about 100–500 MHz. The result is that such devices emit electromagnetic radiation over a very broad range of frequencies. It has also been found that radiation in the X-ray region of the spectrum is emitted from the picture tube itself. General Electric, for example, was compelled to recall a series of about 100,000 television sets in 1967 because tests had revealed an unacceptably high level of X-ray emission.

According to regulations of the U.S. Department of Health, the radiation dose rate measured at any point outside a television set may not exceed a value of 0.5 mR/hr. The Union of German Electrical Engineers tolerates the same maximum dose up to a distance of at most 50 mm from the cabinet, a standard which was also adopted in the Radiation Protection Orders of March 1973, issued by the West German Labor Ministry. More recent investigations of measurable dose rates have shown that the industry has apparently succeeded in solving this problem, for no unsafe radiation levels could be detected.

Electromagnetic waves that carry radio and television signals penetrate readily into occupied rooms (most readily through window openings), as is demonstrated by portable sets and similar devices with indoor antennas. Using a small, portable transistor set, it can be shown quite simply that the incoming signal strength, and thus the reception conditions, varies greatly at different points in a room. Reflections of these relatively short-wave electromagnetic waves (~0.5–3 m wavelength) from the walls and especially from mirrors and other highly reflective surfaces lead to the formation of standing waves in rooms. As a result, the fields present in the room cancel in some places and add in other places, leading to local variations of reception within the same room. When an indoor antenna is used, reception from a weak transmitter can be improved by finding the most favorable location for the antenna. A knowledge of the spatial variations (maxima and minima) of the electric and magnetic fields can also be of help in this regard. The proper orientation of the antenna and the possible reception of reflected signals (the cause of "ghosts" on the TV screen) are other important factors.

In summary, it can be said that man is exposed to a mixture of many natural and artificial electromagnetic processes in his environment (Fig. 3-2).

Shielding of Electromagnetic Processes

The shielding of electromagnetic energies is a relatively complicated physical process. It depends upon whether the shielding of electric, magnetic or electromagnetic fields is involved. Moreover, the electric properties of the shielding material, as well as its structural form (plate or grid), play an important role. Both lead to a frequency-dependence of the shielding effect.

The penetrating ability of electromagnetic waves varies with the conductivity of the medium. For example, electromagnetic waves cannot penetrate far into metals. This fact is utilized in the Faraday cage: a space surrounded on all sides by metal remains free of electromagnetic fields (Fig. 3-3). By contrast, a considerable depth of penetration is achieved when nonmetals serve as shielding materials. The shielding effect, or depth of penetration, is a function of (1) the frequency of the electromagnetic wave ("skin effect"), (2) the specific conductivity of the material, and (3) the permeability of the material. The depth of penetration refers to the depth in the mate-

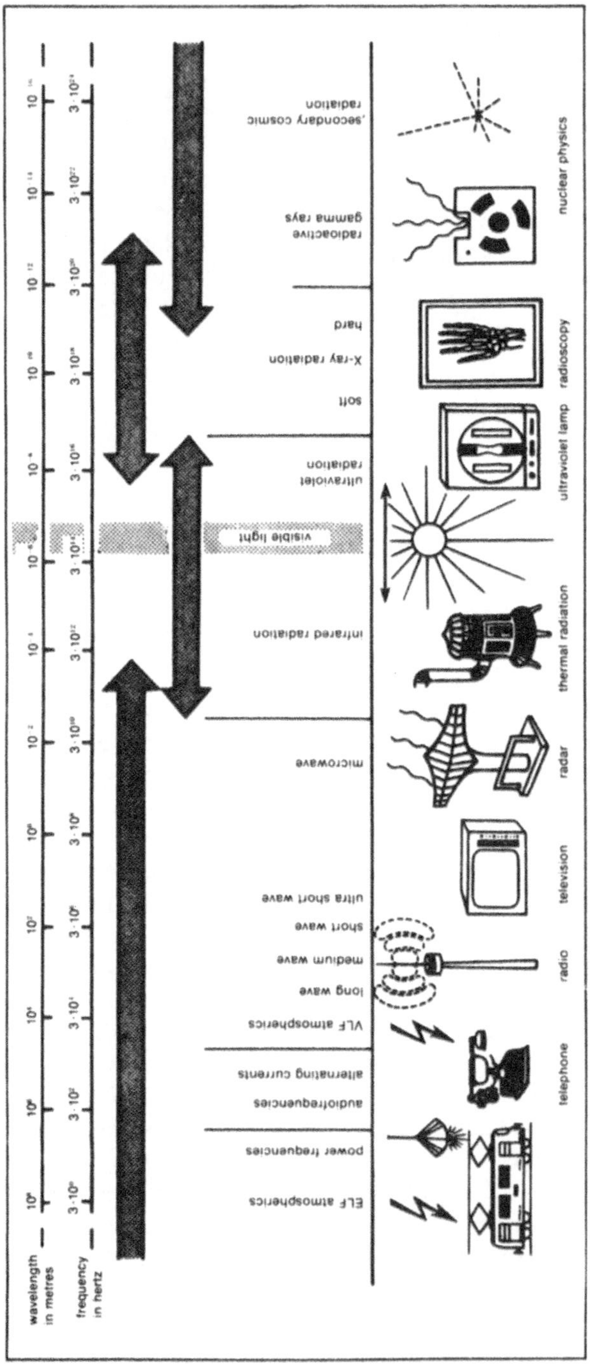

Figure 3-2. The various forms and applications of electromagnetic energies over a broad range of frequencies; the principal ranges are designated.

sunspot activity has varied in a 22-year cycle for the last 265 years;

the air pressure in Edinburgh varied in a 22-year cycle in the 19th century, according to measurements made over a period of 112 years;

the thickness of growth rings in trees has varied in a 54-year cycle during the last 1070 years;

sunspot activity has also exhibited a 54-year cycle during the last 266 years.

So far a total of 19 cycles which apparently influence certain events on earth have been identified. These are 4.0, 5.9, 6.0, 8.0, 9.0, 9.2, 9.5, 9.6, 9.8, 11.2, 12.0, 12.6, 16 2/3, 17 1/3, 17.7, 18.2, 22, 54, and 164 years. Even this very small sampling from available numerical data shows that the course of all events in our environment is partly dependent on periodic processes, or cycles, although the causal connections remain largely obscure.

Besides these relatively short-term cycles, there are other cycles which are measured on an astronomical time scale. The associated characteristic variations have no immediate effect on man, of course, although there may be some links between them and evolutionary processes. Of interest in this context are climatic cycles, whose existence during the past 35,000 years has recently been discovered. Mörner (1973) reports the following:

> There appear to be three main climatic cycles, as shown in [Fig. 3-5]. One is struck by the fact that the two cycles with the shortest periods are not constant, but fluctuate between 230 years and 1000 years in one case, and between 1000 years and 3600 years in the other. In technical terms, one might say that the climate-related temperature variations on earth undergo a frequency modulation with time. The climatic cycle of longest duration has a period of 21,000 years. It is apparent that our period of observation is still too brief for us to discern the existence of similar cycles.

Finally, recent studies by Dorland and Brinker (1973) show that the human "mood barometer" is also subject to a periodic rhythm. From a long-term observation of 1500 consecutive days, these researchers found clear evidence of cycles from 2.4 to 57.8 days in human mood, ranging from happy to extremely unhappy.

Technique of Measuring Electric and Magnetic Fields

In the previous sections we discussed in detail the various types of electric and magnetic fields which occur in our environment. It is now time to deal briefly with the problems involved in the detection and measurement of such fields. In our discussion of fields, the frequency of the fields was used as a systematically ordered element, for it allowed us, at least in purely abstract physical terms, to introduce a certain order into the variety of phenomena discussed. Moreover, frequency provides a framework within which the biological effects of electric and magnetic fields can be more meaningfully discussed.

But the frequency range of a field also gives us an indication of the technology and instrumentation needed to obtain information via the field, especially if it is relatively weak. For example, a considerable technical effort is needed to detect static electric fields or magnetic fields if they are in the range of μV/m or 10^{-7} T (mG), respectively. This effort is reduced only slightly if processes with frequencies up to about 20–30 Hz are involved. The next higher frequency range facilitates measurement considerably, and even low-intensity fields in this range are not difficult to detect. The frequency range employed for radio and television in particular makes it possible to obtain useful information at low field intensities, even with a modest antenna system. At higher frequencies, in the range above 1000 MHz, the technology again becomes more complex. This is due less to antenna

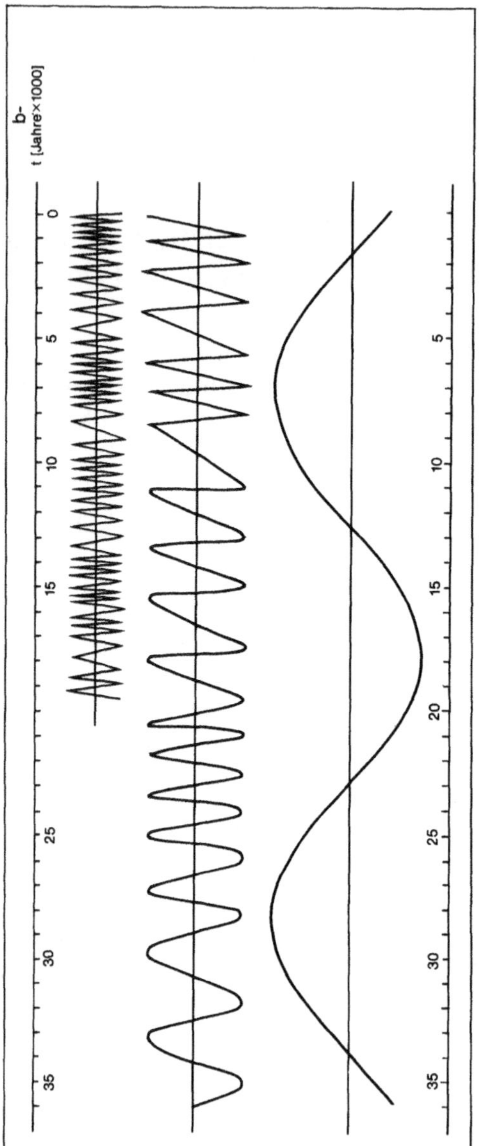

Figure 3-5. Climatic cycles. Time scale in 10^3 years. Warm is shown above the line, cold below. Cycle A has a period of 230–1000 years, cycle B 1000–3600 years, and cycle C approximately 21,000 years. After Mörner (1973). Key: a-Climatic cycles; b-t (years × 1000)

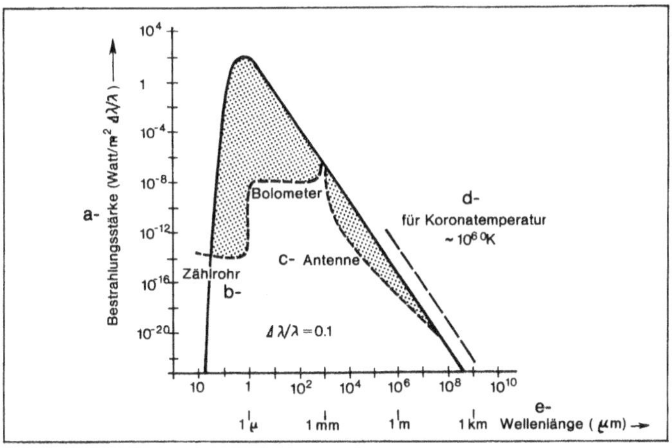

Figure 3-6. Spectral distribution of a black-body radiator at 6000 K, for $\Delta\lambda/\lambda = 0.1$, in W/m^2; roughly equal to the radiative intensity of the sun at the boundary of the earth's atmosphere. Broken line: lower limit of measurability; shaded area: measurable range. After R. Schulze (1970).
Key: a-Intensity of radiation (W/m^2) ($\Delta\lambda/\lambda$); b-Counter tube; c-Antenna; d-for corona temperature $\sim 10^6$ K; e-Wavelength

problems (the antenna size decreases with increasing frequency) than to the increasing complexity of the amplification system. Particular difficulties are involved in the measurement of electromagnetic processes in the infra-red and adjacent long-wave region. For light waves and higher frequency ranges, equipment requirements again become more modest (Fig. 3-6).

The terms measurement and measuring technique as used here refer to the detection of an arbitrary, and thus often broadband, signal of a certain intensity. But measurements can also be made selectively, that is, limited to a very specific, perhaps very narrow, range of frequencies. This leads to additional complications besides those mentioned above, which may make it difficult to satisfy all requirements. Continuous progress is being made in this area, however, and new outlooks and possibilities are emerging which were inconceivable even a few years ago.

4

Biologic Activity of Static and Low-frequency Fields

Researchers have long been performing experiments, studies and calculations in an effort to learn more about the biologic activity of electromagnetic fields. If we consider the entire frequency spectrum from static fields to fields of extremely high frequencies, we begin to appreciate the technical difficulties involved in the treatment of such questions. Moreover, biologic systems are highly complex and, in many respects, difficult to analyze, for the researcher must deal not with processes which can be reproducibly measured and observed, but with variable modes of behavior and relatively slight alterations in the processes of interest. This necessitates a statistical treatment of the subject.

Experimental Possibilities

In experiments which are meant to show whether certain observable biologic processes are caused by electromagnetic fields, it is possible to work with artificially produced fields. If results are to be absolutely reliable, however, all extraneous influences must be excluded. This is seldom possible in practice.

The attempt to keep even the simplest environmental conditions (room temperature, humidity, lighting, drafts, etc) constant over a certain period of time is an extremely laborious task. Added to this are such factors as irritation of the subject caused by noises and floor vibrations. The subject should also be shielded or protected against all electromagnetic influences which lie outside the controlled conditions. This is extremely difficult over the entire frequency range, especially in the case of extraneous static magnetic fields and high-frequency electromagnetic fields. It may even be necessary in such cases to conduct the experiment in underground bunkers with walls fabricated from the necessary shielding materials. Ionization of the air is yet another factor to be considered, for it could exert effects on electric conditions in the air. In practice, however, the influence of air ionization caused by cosmic radiation cannot be eliminated, because these rays readily penetrate all ordinary shielding and continuously produce new charged air particles in the vicinity of the experimental subject.

So the production of a so-called zero field is not really possible in practice. For this reason a second approach is usually taken in such experiments. The experimental subjects are exposed to different experimental conditions either simultaneously or in succession. A skillful comparison of the results can yield meaningful relative data.

The experimenter must also be aware of the initial state of his subject. Even in comparative investigations, nonlinear subject responses cannot be ruled out.

A residual stimulus level can also influence the experimental result. It is evident that the presence of a stray 50-Hz electric or magnetic field in a laboratory area in connection with an artificially produced field of approximately equal intensity which does not have the same frequency but lies in the same frequency range, can lead to false conclusions. Extreme caution is also advised in the statistical evaluation of the experimental results in order to avoid incorrect conclusions.

These brief remarks give some idea of the difficulties which the researcher must overcome in this area.

Electrostatic Fields, Electric Air Current

Because of its relatively high intensity, the natural electrostatic field has often been used in experiments and investigations of biologic effects. The simplest approach to this problem—making observations and measurements of this field over a 24-hr period followed by a statistical comparison with certain biologic processes—holds little prospect of success. The experimental result will be adulterated by the presence of too many other parameters which have a similar time course, and whose influence on the progress of the experiment cannot be excluded. Thus any biologic effects of the electrostatic field cannot be reliably correlated with the natural field.

To obtain meaningful results, the experiments must be performed with artificially produced fields. A large number of such experiments have been conducted during the past 25 years, on both man and other living forms. For example, Busch (1961, 1963) reports on growth experiments with bacterial cultures under the influence of a 50–400 kV/m field, in which increasing field intensity resulted in a slowing of the mortality rate, and so in the inhibition of cell autocatalysis. The author attributes this effect to the depolarizing and repolarizing action of the electric field on living cells.

In another experiment, cultures exposed to an electrostatic field maintained a survival rate of about 65%–70%, while smears on a zero-field plate showed only relatively light, poorly staining cell shadows.

Hahn (1963) exposed smears of bacterial cultures on blood plates to an electric field of 6.7 kV/m. The colonies thus treated did not grow, a kind of "sterilizing effect" having occurred due to dissociation of the bacteria through ion flow. The results of the bacteriologic studies prompted Hahn to recommend the installation of relatively intense electrostatic fields in occupied rooms, for this would reproduce the electroclimatic conditions found in nature. However, Hahn neglected to consider that the natural static field does not have a constant value over a 24-hr period, as does the artificial field, but rather undergoes sizable fluctuations in intensity during the day.

A similar effect is described by Becker and Kraus (1964). Bacterial colonies in a constant electric field of about 13 kV/m exhibited less growth than colonies outside the field. This would indicate sterilization through a bactericidal action of the field.

A study by Wehner (1964) illustrates the problems that can arise in biologic experiments of this kind. Agar plates with *E. coli* smears were introduced into a 55 kV/m electric field. The colonies that developed on the treated plates were about 2.6 times larger than those that developed on the control plates. The field had a growth-promoting effect in these experiments. The author attributes this to an acceleration of metabolism due to polarization by the field.

Finally, Sale and Hamilton (1967) exposed bacterial and yeast preparations to an electric field with the relatively high intensity of 2500–3000 kV/m. The field exposure in this case was pulsed and had a killing effect, which the authors attributed to damage to the cell membrane by the field's action.

Observations by Hicks (1957) also show that growth processes in trees are accelerated in the presence of high-intensity electromagnetic fields, and retarded under artificial shielding.

Backmann and Reichmanis (1973) studied the growth rate of barley seedlings under vertical electrostatic high-tension fields of 50–400 kV/m and the gaseous byproducts of such a high electric field stress. To produce this field, they used a meshed electrode (mesh size 0.6 cm) measuring 25 × 75 cm, charged to + 15 kV with respect to ground. It was mounted in the form of a slanted roof over a container with the barley seedlings, thereby exposing the plants to varying electric field intensities due to the steadily increasing electrode spacing. However, this arrangement did not create ideal conditions for static field experimentation. Any inadvertant change in the distance between electrode and ground would cause a change in the electric field in the environment of the plants. For example, a change in spacing of only 0.1% at the point of the smallest electrode spacing (8 cm) means a variation of 0.08 mm. Even the slightest ground vibrations or air movements could cause such a change unless special precautions were taken. But these variations in spacing also impress a low-frequency alternating field onto the electric field whose amplitude (voltage-wise) is on the order of 15 V_{pp}, and whose frequency results from mechanical vibration processes, placing it in the ELF or VLF range. Of course, the results obtained with such an arrangement are not diminished in their general significance, at least as far as the effect of electric fields on barley seedling growth is concerned, but it is impossible in this instance to determine with certainty the respective contributions of the electrostatic field and superimposed alternating field to the effects observed.

Nevertheless, the experiments showed that direct exposure to fields below 200 kV/m generally intensifies growth, while higher field intensities tend to retard growth. In all fields up to 750 kV/m, gaseous byproducts (CO_2, O_2, N_2) led to an initial increase in growth. This gain was later lost, and overall growth patterns returned to normal after 150 hr.

The effects on growth produced by an electrostatic field polarized opposite to the natural (negative-ground) field and by an alternating field were analyzed by Murr (1966). He found that the growth of corn and bean plants apparently depends in large measure on the magnitude of the current actively flowing through the plant. Although plant growth in the electric field remained unaffected by the nature of the field or current (static or dynamic), it was generally retarded by leakage currents over 10^{-5} A and stimulated by currents of 10^{-8} A. Current strengths of 10^{-16} A or less had no measurable effect on plant growth.

Wilhelmi (1962) investigated the effect of thunderstorms on the radial growth of woodland trees, prompted by similar biometeorologic problems. Although numerous internal and external factors are involved in the growth of a tree, and no single element can be isolated as the biotropic factor, thunderstorms were presumed to have some effect on growth rhythms. Measurements confirmed that strong electric discharges do produce a physiologic effect in trees. This effect has two components. One component involves a disturbance of sap flow, resulting in a gradual growth-altering process. The probable agency of this effect is bound nitrogen, which is formed under the influence of electric discharges, carried into the soil with precipitation, and absorbed by the roots of trees. Though the electrically induced formation of nitrogen is an immediate event, the absorption of nitrogen is dependent upon the uptake of nutrients by the tree. The second component is an immediate physiologic response which can be observed in the daily radial growth of trees and is probably due to an electric stimulating effect which is expressed in an immediate growth response. Wilhelmi (1962) confirms this on the basis of electric air current recordings, although his data do not indicate the role played by ELF and VLF processes.

Experiments were also done on the effects of the electrostatic field on animal physiology. The results reported by Altmann (1959, 1962, 1969b) must be considered of particular interest in this regard, for he clearly observed an increase in oxygen consumption and metabolic rate in various animal species under the influence of the field. For example, bees showed a substantially higher oxygen consumption in the static field than control insects in a Faraday cage. The higher metabolic rate was accompanied by an increase in food consumption (sugar solution). A higher mortality rate was also observed.

Altmann (1969a) also experimented with guinea pigs exposed to both static (420 V/m) and alternating electric fields. A clear increase in metabolic rate was found in animals exposed to the static, constant field, while the animals exposed to the alternating field (2–10 Hz) showed no changes relative to the Faraday controls.

Altmann (1968) showed, on the basis of oxygen consumption measurements in goldfish, that aquatic organisms are not immune to the effects of electric fields. Field-exposed fish exhibited a higher oxygen consumption than specimens protected by a Faraday shield. A similar observation was made for oxygen consumption in frogs. This increase was observed only as long as the animals were exposed to the field. Oxygen consumption quickly normalized after a brief recovery period. The stimulating action of the direct current could be enhanced considerably by interrupting it at intervals.

Similar results on oxygen consumption were obtained by colorimetric measurements of the extinction of the amino acid–Ninhydrin color complex as a function of duration of exposure for migratory locusts, water frogs, and the white house mouse, supplemented by activity measurements in the white house mouse and parakeet. In every case it was shown that biologic parameters could be influenced by exposure to electrostatic fields. Haine (1961) demonstrated similar effects on the activity of small insects.

Experiments by Dowse and Palmer (1969) showed a change in the daily activity rhythms of mice exposed to electrostatic fields. The effects of electrostatic fields on rats were explored by Mayasi and Terry (1969). In two age groups of male and female rats, they used three different electric field intensities (zero field, 1600 V/m and 16,000 V/m) as well as two noise intensities (30 db and 90 db) in order to test the learning ability of animals during swimming. The criteria were the time taken and number of errors made in the attempt to leave the water. Evaluation of the results by an analysis of variance showed:

1. Adult rats which had previously been exposed to electric fields made far fewer errors than young rats, regardless of sex and noise conditions.
2. All rats swam more rapidly after exposure to electric fields than untreated controls.
3. Following exposure to noise, female rats took a significantly longer time to accomplish the task than male rats under the same conditions and without regard for age.
4. When the rats were exposed simultaneously to electric fields and noise, the males took less time to perform the task than the females, regardless of age.

Strampfer and Geyer (1974) obtained preliminary results on the effect of an electrostatic field and a 10-Hz pulsed field on the psychic behavior of human subjects. For a period of 21 days, 8 men under psychiatric treatment were exposed for a total of 14 hr/day to a static field of approximately 800 V/m, upon which a rectangular field of 250 V/m was impressed. The behavior of the test subjects was compared with that of a control group. In reaction tests, subjects exposed to the field showed a smaller incidence of false and delayed reactions, and evaluation of an adjective list showed that they felt less confused, fearful and passive.

Möse et al. (1970, 1972) published various studies on the effects of the electrostatic field on the sensitivity of the smooth muscles to stimulating drugs. Their investigations showed a clear decrease in the excitability of the ileum in guinea pigs and the uterus in rats exposed to the field for several days in response to histamine, acetylcholine, bradykinin and serotonin. Möse and Fischer (1972) also report on experiments with white mice exposed to a static field of approximately 24 kV/m. Based on control comparisons, some statistically significant increases were found: running activity increased 55%; food consumption increased 19%; drinking water consumption increased 15%. The body temperature also rose by 0.3° C. Hepatic oxygen uptake in the animals showed a much smaller scatter than in the controls. In mice exposed to the static field for 4 $^1/_2$ months, an initial normal litter series was followed by an inability to bear further offspring.

Möse et al. (1973) dealt with immunologic responses in the electrostatic field compared with Faraday conditions. For this purpose, mice were pretreated with ovine erythrocytes and exposed to fields of various intensities (40 V/m–24 kV/m). The degree of immunization was determined by the direct plaque technique. The highest plaque formation values were found in the spleens of animals exposed to the static field. The same was found for spleen weight, splenocyte count and hemagglutination titer. Interestingly, the field had the greatest effect on plaque values at intensities between 1 kV/m and 5 kV/m. But even at 200 V/m, a considerable increase of immunization was observed in comparison with the controls. Based on these findings, it is suggested that the natural electrostatic field of the earth is an important factor in promoting and maintaining the immune system.

Fischer (1973) discussed the biologic significance of the electrostatic field in general terms. Previous experimental results, confirmed and expanded by research in Graz, Austria, indicate that cellular respiration, and thus oxidative metabolism as a whole, can be strongly influenced by atmospheric electric processes. The implications of this, especially with regard to the climatization of occupied rooms, are discussed.

According to Kemmer (1967), studies on the biologic effects of direct electric current on aquatic animals are of general importance with regard to both direct and alternating currents, because alternating fields can also produce displacement currents in organisms, even if there is no direct galvanic contact with any external energy source. Kemmer reports that all previously studied vertebrates exhibit positive galvanotaxis (the tendency of an organism in water carrying a direct current to position itself in a certain direction relative to the field lines). A series of experiments was conducted to isolate the cause of this effect:

1. If the tactile organs of the frog are blocked by anesthesia, the galvanotactic response remains unaltered. Hence, the current penetrates through the body surface at all points and does not depend upon superficial receptors.
2. The brain also plays no apparent role, because the galvanotactic response is unaffected by its blockage.
3. If its spinal cord is drilled out or destroyed by electrocoagulation, the frog still performs its convulsive swimming movement toward the anode despite the absence of the normal motor reflexes controlled by the spinal cord.
4. The peripheral motor nerve fibers can be blocked by the injection of curare (blockage of the transmitter ace-

tylcholine at the motor end-plate). Even then, the frog still swims a short distance toward the anode with each current pulse at the same current strength which elicits the galvanotactic response in the intact animal.

5. It was thus shown that the electric current directly excites the musculature, which performs a movement imposed upon it by this current. Galvanotaxis, then, is the result of an electrical stimulus which, like the action potential of the motor nerve fiber, acts directly on the muscle (via the chemical transmitter acetylcholine).

It was assumed that the frog always turned toward the anode, and that the current thus exerts a polar effect. However, it can be shown in the curarized animal that this assumption is no longer true. If the frog is placed into the experimental vessel at an angle to the field lines under constant current density conditions, it will exhibit a swimming response only in the direction of the anode. But if the current density is increased, the animal swims suddenly toward the cathode with each current pulse, which apparently contradicts all previous experiments. Kemmer (1967) believes that the explanation for this lies in the different distribution of ions inside and outside the membrane of each cell, including the muscle fiber, which results in the presence of a 40–80 mV electric resting potential at the membrane. This leads to different effects for currents of different polarization. If the current density is increased sufficiently, the current will become active against this resting potential, thereby disrupting the specific orientation of the animal with respect to the current direction.

There is no doubt that electrostatic fields influence the ionization of the air, but this has almost never been taken into account in experiments reported in the literature. As a result, it is difficult to determine whether reported biologic effects are triggered by the direct action of the electrostatic field or by associated changes in air ionization.

This problem also applies to some well-known experiments with humans. Frey (1949), for example, reports on experiments with an inhalation mask whose inspiration hose was fitted with a copper ring 10 cm before the nasal opening. This ring was charged to a voltage of +411 V, −411 V or −90 V, which affected the ion content of the inspired air. But the resultant electrostatic field can also exert certain direct effects, at least on nearby portions of the head, which are quite independent of the inspired air.

A positive charge (411 V) on the copper ring led to an increase in systolic blood pressure, heartbeat, and cardiac output, with a simultaneous decrease in the peripheral flow resistance of vessels. A negative charge (90 V) can have the same effect, but it sometimes produces an opposite reaction initially before the aforementioned symptoms supervene.

Kritzinger (1957) reports improved performance in the Pauli test (working simple arithmetic problems over a long period of time) under the influence of an electrostatic field. In his experiments, subjects were exposed to a field intensity of 1800 V/m in the region of the head. Compared with a normal performance increment of about 10% relative to initial performance, the field-exposed subjects showed an increment of 18%–25%. Kritzinger also produced a calming effect on the respiration of subjects by the use of 1 kV/m fields in connection with other experiments. Although the Kritzinger experiments might not withstand the closest scrutiny and his results should be accepted with some reservation (see "Architectural Biology") they correlate very well with other available material. Daniel (1965) also reports a calming of respiration and pulse (reduction of pulse rate by about 10%) in experimental subjects as a result of exposure to 1 kV/m electrostatic fields. Varga (1968a) confirms this decrease in pulse rate as a function of field intensity in the range from 1 to 100 kV/m.

Reinders (1964) claims, based on experiments with electrode voltages of about 10 kV, that electrostatic DC fields are, beyond a certain point, capable of "influencing certain vital functions in such a way that weather phenomena which have an adverse effect on the health of many persons can be largely compensated. Prolonged and regular exposure even appears to reduce weather sensitivity in susceptible persons, thereby improving their overall state of health."

Lueders (1968) also showed by statistical and objective studies that electrostatic fields exert an effect on human resistance and the human autonomic nervous system. He made use of data obtained by K.-H. Schulz (1965) in measurements of the electric conductivity of the skin under the influence of an electrostatic field.

Schulz (1967) conducted additional experiments with electrostatic fields. His subject was seated on a plate, insulated from the ground, which was charged with either a positive or negative direct voltage of up to 500 V. The effects of the field were again monitored by measuring the various conductivity values, which in most cases decreased steadily with the field. The conductivity values measured during the course of several days showed a pronounced downward shift, while the subject clinically showed an improvement or normalization of disordered functions. Schulz sees in this a simple and suitable means of determining the effect of static electricity on the human body and its purposeful therapeutic use. He found that apparently not every person reacts in this way, however, and that there appear to be various reaction types. The initial state of the subject, which can vary widely with external factors, also plays a role.

Callot et al. (1973) reported on interesting observations of electrostatic field effects in persons who spent the winter on Kerguelen (southern hemisphere). Using a horizontal antenna attached to a radioactive probe with radium D (^{220}Ra) or americium-241, the electrostatic field was measured at a height of about 1 m above the ground, and then the mean recorded intensity values for the months of April through December were correlated with the total number of medical visits made by the people living on Kerguelen.

A striking association was found between the number of medical visits and the intensity of the electrostatic field. Another connection was revealed by the Schneider test, which is essentially a cardiac output or fatigue test to which the nonathlete is especially sensitive. An inverse proportionality was found between the atmospheric electric field intensity and the Schneider test. As the field intensity increased, the Schneider test values declined, meaning that the functional capacity of the test subjects became increasingly poorer. The blood sedimentation rate was also inversely proportional to the electrostatic field intensity. Higher sedimentation rates were always associated with very low atmospheric electric field intensities. A certain correlation could also be established between field intensity and the number of blood corpuscles present.

On the whole, these results indicate that a connection exists between certain pathophysiological reactions and the value of the electric field gradient in the immediate human environment. It is necessary, however, to conduct more precise investigations in order to obtain better information in this area of biophysics.

One word of criticism must be added at this point with regard to experimental design. The necessary improvements and refinements should not be limited strictly to the biologic testing methods, but should also include certain technical refinements as well. With the technique employed, it is not entirely clear whether it was always the electric field which the authors cited as the presumed parameter. While it is true that the radioactive probe attached to the antenna ionized the surrounding air and thus facilitated measurement of the electrostatic field, it must be considered that the movement of air in the immediate vicinity of the antenna was indirectly recorded along with the electrostatic field.

It cannot be claimed that the report presented here on the biologic activity of electrostatic fields is complete, of course. In any case, a survey of the literature shows a considerable diversity of results both in psychological tests and in physiological measurements. Certain systematic connections have been established between the action of electrostatic fields and the behavior or subjective state of man, but they are few in number. Although the biologic effects of the electrostatic field are confirmed by most authors, a comparison shows that many results are contradictory. The most probable reasons for this have been mentioned: inadequate technical resources, incomplete knowledge in the area of electrophysics, incomplete knowledge of the potential biologic activity of ELF and VLF fields, quite apart from purely biologic aspects. As a result, it is rare for a researcher to allow for the possible occurrence of ELF field fluctuations, for example, due to mechanical vibration of the electrodes (variation of spacing).

However, one must give special recognition to the work of H. Schulz (1970), who was able to show on a scientific basis a means of clarifying at least one aspect of this somewhat confusing situation. Every expert in the field of bioclimatic research knows from experience that individuals fall into various types according to their reaction to changes in environmental parameters (e.g., "vagotonics" or "sympatheticotonics," or "W types" and "K types"). This is reported by Curry (1946), Hänsche (1963), Hartmann (1967), Reinders (1964), K.-H. Schulz (1967), Schuldt (1974b), and many others. H. Schulz knew that similar principles were employed in psychopharmacological studies. He therefore made use of this well-proven personality-specific working model in his investigations of the psychological effects of one bioclimatic parameter, the electrostatic field. Schulz maintained that, for procedural reasons, the study should be limited to a few well-established dimensions of the human personality, because homogeneous groups can be selected with certainty only if clearly defined personality traits are used as criteria. Therefore, he used only two dimensions in his selection of subjects: *extroversion* and *psychic lability* (neuroticism), as determined operationally by questionnaire (Eysenck, 1959).

Schulz (1970) began his experiments with a total of 60 subjects. They were divided into four groups of 15 each for his preliminary experiment, which was preparatory to his main experiment. Three of the four groups were exposed to three different electric field intensities: 1 k V/m, 2 k V/m and 3 k V/m (measured at the level of the head). The fourth group was exposed to no electric field and served as the control. The experiments took place in a partly climatized chamber whose walls, floor and ceiling were covered with metal foil, which provided good shielding against external alternating electric fields of natural or artificial origin. The room temperature varied between 20° and 23° C, the relative humidity between 50% and 60%. The ionization strength of the experimental chamber could not be monitored due to a lack of instrumentation. The experiments themselves were conducted on an individual basis by the double-blind method. Neither the subject nor the experimenter were informed as to the conditions in effect. The preliminary experiment yielded the following results:

1. The reaction time to a simple acoustic signal was apparently improved at a field intensity of 1 kV/m compared with zero-field conditions and higher intensities of 2 kV/m and 3 kV/m. Thus the field effect showed an intensity dependence, which was confirmed by subsequent findings.

2. Long-term performance was measured by the Pauli test, in which the subjects continuously added together single-digit numbers and wrote down the sum. The result of the test, shown in Fig. 4-1, led Schulz to conclude that the performance of the groups in the Pauli test is systematically influenced by the intensity of the static electric field; specifically, a low field inten-

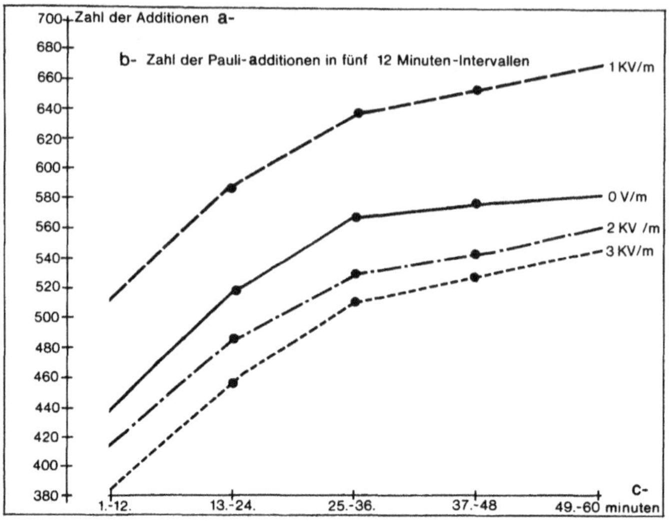

Figure 4-1. Mean number of Pauli additions for four groups in different electrostatic fields during five 12-min intervals. After K.-H. Schulz (1970).
Key: a-Number of additions; b-Number of Pauli additions in five 12-minute intervals; c) Minutes

sity (1 kV/m) tends to improve performance, while higher intensities (3 kV/m) tend to impair it.

3. In another test the subject's task was to determine the "flicker fusion frequency." Analysis of the test results showed no effect of the static electric field on test-related fatigue in the subjects.

4. In the speed "dotting test" the subject was given a pencil and paper and told to make as many dots on the paper as possible in 15 s, with no regard for the placement of the dots. With the increase in the electrostatic field a trend toward an increase in the number of dots was observed, as shown in Fig. 4-2.

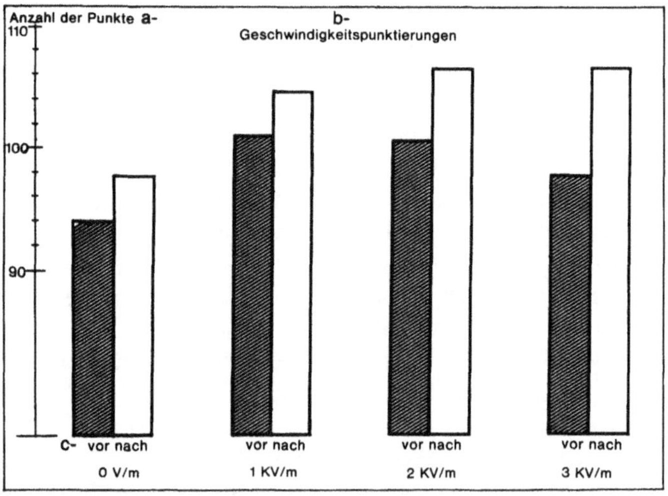

Figure 4-2. Mean number of dots made in speed dotting test before and after a dual task under various electrostatic field conditions. After K.-H. Schulz (1970).
Key: a-Number of dots; b-Speed dotting test; c-Before After

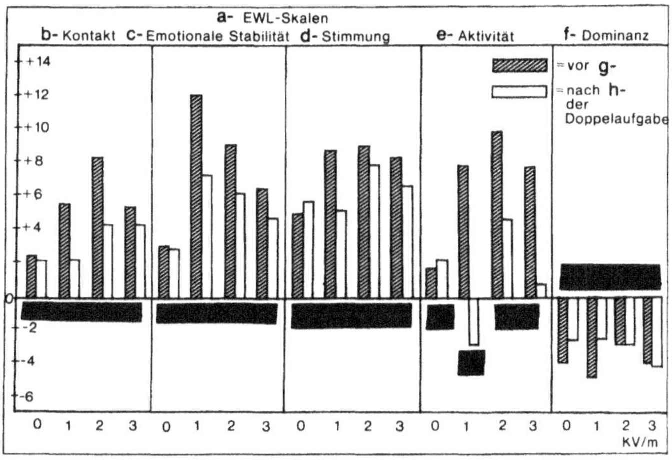

Figure 4-3. Self-assessment of four groups subjected to various electrostatic field conditions, before and after a dual task. After K.-H. Schulz (1970).
Key: a-Adjective list scales; b-Contact; c-Emotional stability; d-Mood; e-Activity; f-Dominance; g-Before; h-After the dual task

5. Each subject's self-evaluation of performance showed a covariation with field exposure conditions.

6. An adjective list developed by Jahnke (1961) was used to evaluate the momentary subjective state of the test subjects. Analysis of the data obtained by this method indicated a clear effect of the time factor on all five scales (Fig. 4-3), as well as distinct group differences for three scales (readiness for contact, emotional stability, and activity), and a significant interaction between field exposure and the time factor for two cases (contact and activity).

7. Besides the adjective list, the subjects were presented with five evaluation scales to determine their subjective state. These five-point scales, designed for the experiment, measured freshness, concentration, work enjoyment, tension, and talkativeness. This test represents a direct scaling of the corresponding dimensions by the subjects themselves. It was found that self-evaluation according to these scales was also time dependent, with the exception of the "tension" scale. The subjects in the field-exposed groups (1, 2 and 3 kV/m) felt both fresher and more talkative on first measurement than did the subjects in the control group. There was also a significant average difference between the 2- and 3-kV/m groups and the control group with regard to the concentration scale.

Based on the results of this preliminary experiment, which are of considerable importance in themselves, Schulz (1970) conducted his main experiment, designed to yield more personality-specific findings. For this purpose the subjects were divided on the basis of psychological tests into four groups: extroverted/stable (ES), extroverted/labile (EL), introverted/stable (IS), and introverted/labile (IL). Only one field intensity, 2 kV/m, was employed in the main experiment. The following results were obtained:

1. In the case of the adjective-list scales, the subjects also reacted sensitively to the experimental treatment in the main experiment; there were personality-specific reaction modes to the experimental treatment in all five scales (Fig. 4-4). The effect of extroversion on the reaction of the subjects to the static electric field is equidirectional in four of the five scales (contact, emotional stability, mood, and activity). These group-specific reaction modes were found to be statistically significant for the scales indicated. Moreover, extroverted subjects proved to have a greater readiness for con-

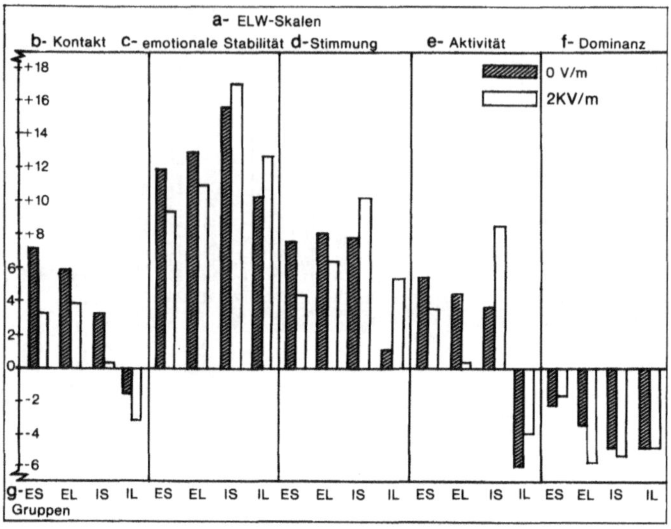

Figure 4-4. Self-assessment of four personality groups (ES, extrovert-stable; EL, extrovert-labile; IS, introvert-stable; IL, introvert-labile) on five scales of an adjective list under control and experimental conditions (electrostatic field of 2 kV/m). After K.-H. Schulz (1970).
Key: a-Adjective list scales; b-Contact; c-Emotional stability; d-Mood; e-Activity; f-Dominance; g-Groups

tact than introverted subjects in both the control and main sessions.

2. The pulse rate, as an index of physiologic activity, was recorded by random sample at the beginning, during the middle and toward the end of the two-hour experiment. As Fig. 4-5 shows, the physiologic response of the four personality groups to the experimental treatment varied. While the EL group reacted to field exposure with a slowing of the pulse rate, the IL group showed a quickening of the pulse.

3. The excretion of epinephrine and norepinephrine (Fig. 4-6) in the urine appeared to be altered by two hours' exposure to the static electric field, although this effect could not be verified statistically.

In discussing his results, Schulz stated that the hypothesis that a static electric field (2 kV/m) exerts a general effect on the subjective well-being of man is supported by the fact that self-evaluation on most of the adjective scales was influenced by the experimental treatment. Personality-specific differences could also be found in the effect of the field. Introverted subjects achieved higher values on the "contact," "mood," "activity," and "emotional stability" scales by the end of the two-hour experimental session than did the control subjects and thus showed signs of a mild "euphoria," as was also observed in the first adjective-list tests in the preliminary experiment. Extroverted subjects, on the other hand, reported a slight impairment of their subjective well-being at the end of the 2-hr session.

Schulz summarizes the results of his preliminary and main experiments as follows:

1. Static electric fields have an effect on the subject well-being of test subjects.
2. The effect has a relatively rapid onset and is probably complicated by the peripheral parameters of the experimental situation.
3. Personality traits must be taken into account when judging the effect of the field.
4. It cannot be shown with certainty that objectively measurable physical performance is affected.
5. Individual differences are present in the degree to which the subjects are irritated

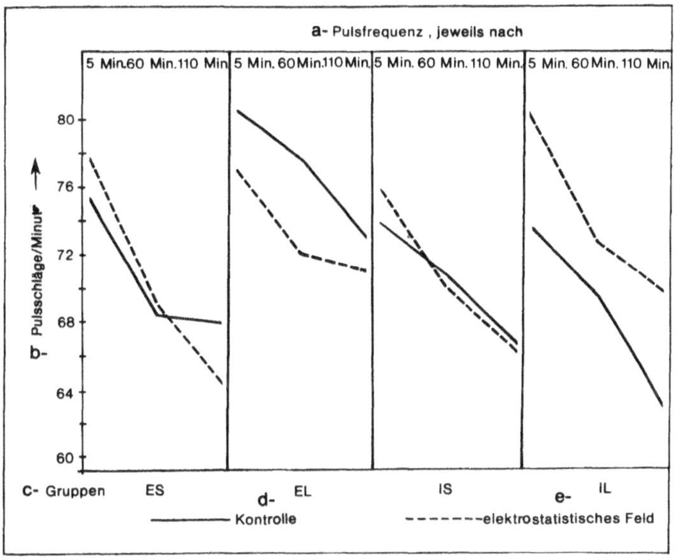

Figure 4-5. Mean cardiac pulse rate of four personality groups (ES, extrovert-stable; EL, extrovert-labile; IS, introvert-stable; IL, introvert-labile) under control and experimental conditions (electrostatic field of 2 kV/m) at three points in time. After K.-H. Schulz (1970).
Key: a-Pulse rate after: b-Pulse, beats/min; c-Groups; d-Control; e-Electrostatic field

by the fields, as indicated by pulse rate data. Catecholamine secretion does not change significantly on field exposure.

In a follow-up experiment, Schulz (1970) investigated the question, Can the effect of static electric fields on the test subjects again be detected on repeated exposure, or were the subjects' reactions unique and nonreproducible? He also wished to learn whether the effect of the field, which apparently has a very rapid onset, terminates

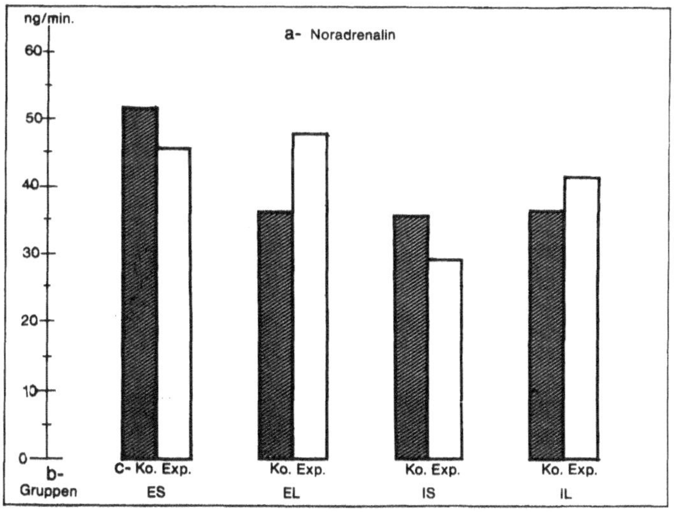

Figure 4-6. Mean norepinephrine secretion in four personality groups (ES, EL, IS, IL) under control and experimental conditions (electrostatic field of 2 kV/m). After K.-H. Schulz (1970).
Key: a-Norepinephrine; b-Groups; c-Control Experimental

as soon as the field is switched off, or whether after-effects persist. He found the following:

1. The adjective-list evaluation yielded no useful results, probably due to an insufficient random sampling.

2. Pulse-rate measurements again confirmed the spontaneous action of the field on this circulatory parameter. He also found that a certain interaction exists between the experimental treatment and psychic stability; the pulse rate of labile subjects rose during field exposure. In the stable subjects, a difference was found between the extroverts and introverts: while the ES group showed a fall of pulse rate during field exposure, the IS group reacted with a slight increase in pulse rate, like the labile groups. In the ES group the pulse rate clearly slowed under control conditions, while it remained at roughly the same level under experimental conditions. In the IL group, on the other hand, the pulse rate was relatively constant under control conditions and fell considerably under experimental conditions. The results of the pulse-rate measurements were also of interest with regard to the question of the onset of the field effect, however; for even more clearly than the main experiment, the follow-up experiment showed differences between the control measurements and measurements within the field, especially in the first measurement five minutes after the static electric field was started.

3. The flicker fusion frequency test was confirmed as a suitable means of investigating the field effect. It revealed significant differences between the two personality factors, while the treatment factor played only a subordinate role.

4. The dotting test showed the most distinct field-related effects in the follow-up experiment. The test subjects made more dots under the influence of the static field than under control conditions. The overall performance of the stable subjects changed little on exposure to the field, while the labile subjects responded to field exposure with an improvement of performance in this simple psychomotor test. The EL and IS groups each made more dots during the second measurement than during the first. For the other two groups, ES and IL, the opposite observation was made.

Schulz (1970) concluded his report by stressing the need for personality-specific planning in the majority of tests employed. Studies by Bisa and Weidemann (1955) on the effect of electro-aerosol therapy on the flicker fusion frequency confirm this. A comparison of the control and experimental conditions in all patients showed no significant threshold changes in optical resolution. However, when the patients were divided into homogeneous subgroups according to their reactions in various physiologic measurements, typical effects of unipolar positively- or negatively charged electro-aerosols on the fusion frequency could be determined.

Tromp (1963) pointed to connections between air pollution (smog) and death rates (see also Fig. 1-8). Fisher et al. (1977) investigated the significance of the natural atmospheric electric field in this regard:

> The atmospheric electric potential (E) was evaluated statistically with smoke and sulfur dioxide as a predictor of excess mortality during the London air pollution episodes of the 1950s. E, as a surrogate of respirable aerosols, in combination with SO_2 proved to be a better predictor of excess episode mortality than smoke.

Static Magnetic Fields

Barnothy (1964, 1969) presents an excellent survey of studies dealing with the biologic effects of magnetic fields, although he confines his attention to static magnetic fields. (Time-varying magnetic fields induce an electric current in biologic systems, so that the effects of these currents generally mask the biologic effects of the magnetic field itself.) The various values of the biologic effects observed by Barnothy

Table 4-1. *Values at which various biologic objects and phenomena respond to magnetic fields or magnetic field gradients. After Barnothy (1969).*

Field-sensitive phenomena (kA/m)		Field-gradient-sensitive phenomena (kA/m²)	
Snakes	0.12	Dowsing reflex	10^{-5}
Magnetophosphenes	16	Cast-off of tumors	4,800
Central nervous system (rabbit)	64	Cessation of bacterial growth (logarithmic phase)	18,400
Plant growth (wheat)	80	Magnetotropism	40,000
Embryo resorption (mouse)	240	Lethal effect (mouse)	40,000
Developmental retardation	320	Lethal effect (fly)	48,000
Hematologic changes, wound healing, pathologic changes, altered enzyme activity	400	Cessation of tumor growth	80,000
Altered oxygen consumption and degeneration of cancer cells	640		
Cessation of bacterial growth (static phase)	1,125		
Oxygen uptake in potatoes, survival of leukemic mice	1,440		

(1969) are given in Table 4-1, arranged according to field intensity or the gradient of the field intensity (local variation of magnetic field intensity).

Barnothy found that the growth of mice is substantially retarded if a magnetic induction of about 0.6 T (6000 G) with an average gradient of 1 T/m is installed in their cages. Besides the absolute field intensity, the gradient of the field also appears to play an important role, for in another experiment the growth of mice exposed to a field with an induction of 0.42 T and average gradient of 0.5 T/m, and to an induction of 0.36 T and average gradient of 6.5 T/m was compared. The growth of the mice was retarded much more strongly in the more homogeneous field than in the relatively inhomogeneous field (Fig. 4-7). In experiments with transplanted tumors, the bodies of mice in a magnetic field of about 0.3–0.4 T were able to cast off the tumor, in contrast with control animals. Similar experiments are reported by Gross (1960), who showed statistically that the life expectancy of tumor-bearing mice would be prolonged

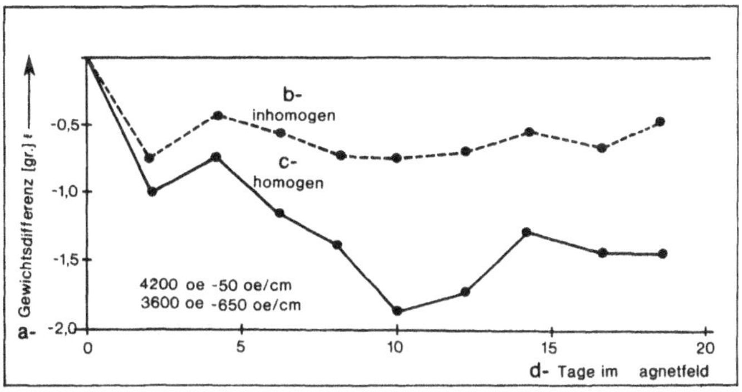

Figure 4-7. Growth retardation (weight difference between experimental and control animals) in mice exposed to a homogeneous and inhomogeneous magnetic field. The action of the static homogeneous field is stronger than that of the inhomogeneous field. The most marked effect is observed in both cases after 10 days' treatment. After Barnothy (1969).
Key: a-Weight difference (g); b-Inhomogeneous; c-Homogeneous d-Days in magnetic field

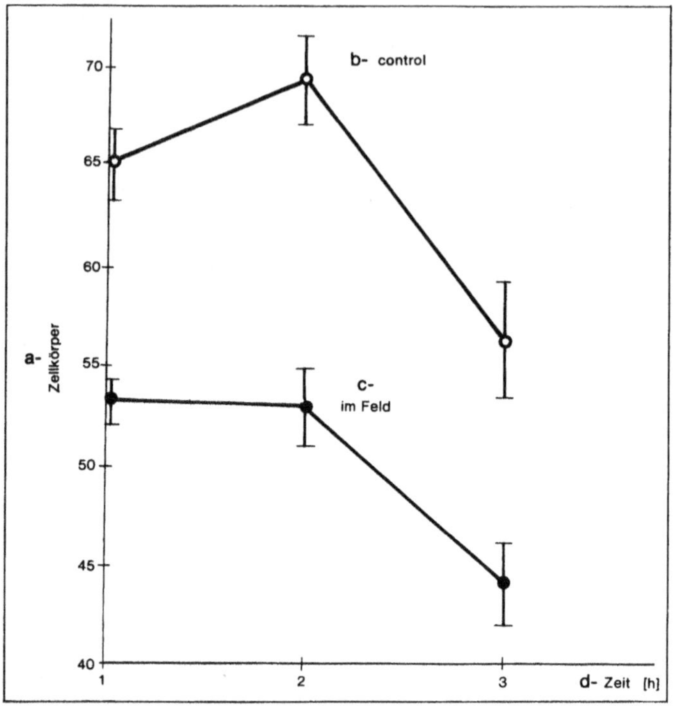

Figure 4-8. The effect of a static magnetic field (about 600 kA/m, or 0.73 T, in air) on cell bodies of a tumor suspension. The cross-bars indicate the scatter. After D'Souza (1969).
Key: a-Cell bodies; b-Control; c-In the field; d-Time (hr)

by exposing the mice to a magnetic field before implantation of the tumor. This effect presumably has to do with a leukocytosis caused by exposure to the magnetic field. Unfortunately no successful experiments are yet known which confirm these positive findings with regard to the treatment of tumors in man.

Souza et al. (1969) performed experiments with tumor suspensions. The suspensions were exposed to a magnetic field at 600 kA/m for 1–3 hr and compared with corresponding controls. As Fig. 4-8 shows, the researchers found a statistically significant reduction in the number of cell bodies in the suspensions exposed to the magnetic field.

Cook et al. (1969) reported on the effects of magnetic fields on cellular respiration. They found that at field intensities of 640 A/m or higher, respiration. was correspondingly diminished. Fields of lower intensity had no effect, while those of higher intensity (up to 800 kA/m) produced no significantly greater effect than the 640 A/m field.

Barnothy (1969) also conducted exhaustive studies on the effects of magnetic fields on the growth and development of organisms. Experiments with 680 mice yielded the following results:

1. In all cases, exposure to the magnetic field produced a retardation of growth compared with the control animals.
2. The field with a smaller gradient (640 kA/m^2 or 0.42 T/m at 336 kA/m) had a greater effect than a field with a larger gradient (5200 kA/m^2 or 0.36 T/m at 29 kA/m).
3. The mean difference in body weight between the two groups of animals (treated and untreated) was greater in young animals than in adults, although the relative weight reduction was more pronounced in the adult animals.

4. The treated group exhibited greater individual differences than the control group, indicating that the action of the field is different in different individuals.
5. On the second day of field exposure, a marked weight loss was observed regardless of the nature of the field.

A high-altitude atomic test on July 9, 1962, apparently influenced electric and magnetic conditions on the earth's surface. Simultaneous worldwide measurements of all known electric and magnetic parameters showed considerable disturbances in many areas, which were apparently caused by this explosion. The question of whether such nuclear tests can also produce biologic effects was investigated by Levengood (1967b), who experimented with two groups of flies in separate containers. Mounted to the outside of one container was a permanent magnet which produced a magnetic field of about 0.035 T (350 G) within the container. The flies in the second container served as controls.

The curves tracing the production of offspring by the two separate cultures each show a discontinuity associated with the nuclear test. However, a smaller deviation from the mean value of four generations took place in the magnetically treated culture than in the control culture. It is presumed that the magnetic field deflected a portion of the blast-generated radiation from the experimental container. Another possibility is that the magnetic field produced a protective effect which was transmitted from the original population to the first generation of offspring, which then grew up outside the field.

The static magnetic field has also proved capable of producing genetic effects. The animal most often used in such experiments is the fly *Drosophila*. The field intensities generally range from 60 kA/m to 560 kA/m (Barnothy, 1969). Interestingly, experiments with higher intensities (11,200 kA/m) led, in the case of *Drosophila*, to no discernible genetic effects. Apparently there is an optimal range of intensities for genetic effects.

Electromagnetic field effects have also been demonstrated in the growth of plants. The effect of a static magnetic field of 100 kA/m on the roots and plants of barley seedlings is shown in Fig. 4-9, which clearly indicates an acceleration of growth.

Based on available results from the ex-

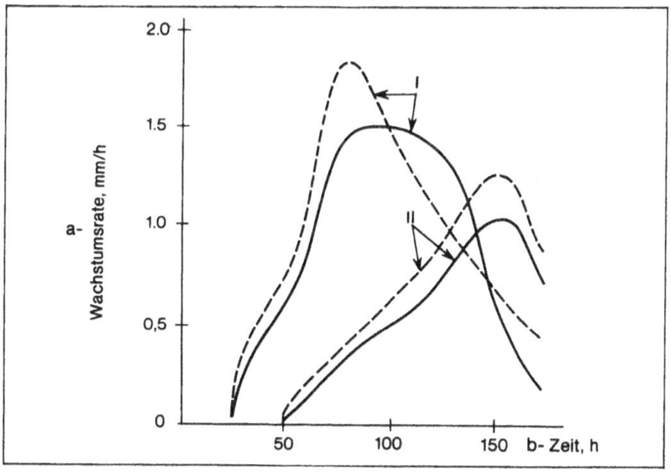

Figure 4-9. Effect of a static magnetic field of 100 kA/m (0.12 T, in air) on the roots (I) and plant body (II) of barley seedlings. Broken line: plants in the magnetic field; solid line: control. After Merikle et al. (1964)
Key: a-Growth rate (mm/hr); b-Time (hr)

periments of Novitskii et al. (1965, 1966 a,b,c), Presman (1970) draws the following conclusions:

1. The stimulating effect of weak magnetic fields (1.6–6.4 kA/m) on plant growth is especially marked in the first 2 or 3 days of seedling growth.
2. Field intensities of 6.4 kA/m lead to accelerated leaf movements and alterations in the electric resistance of the plant substance.
3. Strong magnetic fields (360 kA/m for 1 hr) influence the magnetic properties of plants.

In his study on the biologic activity of magnetic fields, Edmiston (1972) used mustard seeds, whose germination, measured in 4-day periods, allowed observation of the effect of a static field (field intensity 27 kA/m, corresponding to a magnetic induction in air of about 0.034 T or 340 G). Additional experiments with seedlings arranged on a rotating platform in a magnetic field, as well as follow-up experiments under weaker static field conditions (185 A/m) showed that seedlings that are rotated in a magnetic field exhibit more rapid growth than the rotating or stationary controls outside the field. Stationary seedlings in a magnetic field grow more slowly than the controls.

Astonishing results are reported in connection with the terrestrial magnetic field. Pittmann (1967), for example, noticed that winter wheat plants in an experimental field grew better in rows planted in an east–west direction than in rows planted in a north–south direction. He was surprised to learn in subsequent research that almost all root systems grow from north to south, parallel to the earth's magnetic field. If wheat seeds are sown in furrows oriented in this direction, the germinating plants must apparently compete with each other for nutrients in the soil. But if they grow in furrows running from east to west, they do not penetrate the root systems of other plants, and each plant has its own zone from which to draw nutrients, resulting in better and more rapid growth. Other experiments with grain and other seed corn exposed to artificially produced magnetic fields even indicate an increase in plant growth.

The effect of water treated with electromagnetic fields on biologic systems is another interesting question. Presman (1970) reported changes in the physicochemical properties of water exposed to magnetic fields. According to his findings, such properties as surface tension, viscosity, and electric resistance can be strongly influenced by a field intensity of 0–800 kA/m. The dielectric constant of water or distilled water is also changed by a factor of 3 to 4 exposure to a magnetic field of 80–160 kA/m.

Effects of magnetically treated water (80–120 kA/m) on plants are mentioned by Dardymov et al. (1965). Sunflowers, corn, and soybeans exhibit a significantly greater plant growth and thicker stems than comparable control groups when the plants of the experimental group are watered with such magnetically treated water.

Studies by Varga (1973) contribute much to clarifying the mechanism of the observed biologic effects of nagmetic fields. He first repeated well-publicized experiments on the effect of magnetic fields on the growth of bacteria. The exposure of bacterial suspensions (*E. coli, Bac. subtilis*) to a static magnetic field and to a 50-Hz field led to an average increase in the growth rate of 27% and 33%, respectively, a shortening of the duplication time by about 18% and 22%, respectively, and an average increase in yield of about 20%–60% and 40%–80%, respectively. The effect of the magnetic field on the growth rate was optimal at about 7 mT. Investigations of the frequency-dependence of the growth effects of magnetic fields of about 10^{-5} T showed that the effect of the field increased with frequency in the range from 1 Hz to 1 kHz, doubling the growth rate over this range. A minimum occurred at 10 kHz (return almost to the value at 1 Hz), followed by a flat maximum between 100 kHz and 1 MHz.

Investigations of the optimal duration of exposure of the bacteria to the field showed no measurable difference over the range between 1 s and 1 hr.

Further experiments produced the following, very illuminating result. The growth of microorganisms can be accelerated simply by exposing the culture medium to a magnetic field. In other words, no microorganisms need be present during field exposure in order to achieve the described effects. It was also found that only about 3% of the total solution need be treated with the magnetic field to produce growth acceleration.

Directly related to these results are experiences reported by Weissenborn (1968) with strong magnetic fields. He reports on some 600 applications of a permanent magnet with a magnetic flux of $17–20 \times 10^{-5}$ Vsec (17,000–20,000 maxwells) directly at its 60-cm^2 surface. In his experiments, the permanent magnet was held near the heart of patients for a period of several seconds. After a few minutes a clear reaction could be seen in the form of changes in the ECG and pulse-pressure curves. He even reports some instances of unconsciousness occurring immediately after a brief exposure or several minutes later. In any case, the treatment reportedly effected an alteration of circulatory function toward more normal values.

Beischer (1969) did a study on the effects of extremely strong magnetic fields (8×10^6 A/m) on the ECG of monkeys. The change is clearly evident in the traces in Fig. 4-10.

A Japanese team of internists headed by Prof. Noboro Kimura (1967) reported on the therapeutic effect of a magnetic arm band on shoulder stiffness. For experimental purposes, the patients were issued two types of arm bands (Fig. 4-11). One was fitted with permanent magnets arranged so that the magnetic lines of force penetrated into the arm. In the other, the magnets were arranged on the outside so that the wearer was not magnetically influenced. All bands were identical in appearance, and none of the patients knew which type he wore. The patients were given the choice of wearing the band on the side with the greatest pain. Many patients wore them on both arms. The experiment lasted about 4 weeks. If shoulder stiffness did not improve after one week, the case was recorded as a failure. Statistically, the result was as follows: 41% of the patients with the "therapeutic" arm bands reported improvement, while 59% reported no improvement. In the control group with the "nontherapeutic" bands, the results were 7% success and 93% failure.

Because the improvements were observed only subjectively, it is premature to

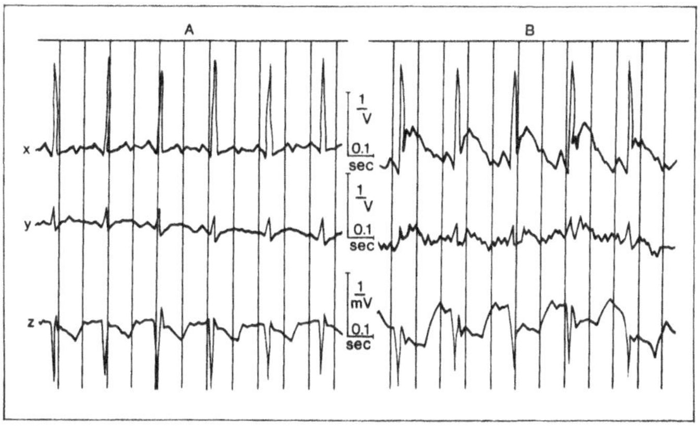

Figure 4-10. Excerpts from the vectorcardiogram (ECG) of a monkey. (*A*) Control outside the magnetic field; (*B*) animal in a static magnetic field of 8×10^6 A/m, or 10 T. After Beischer (1969).

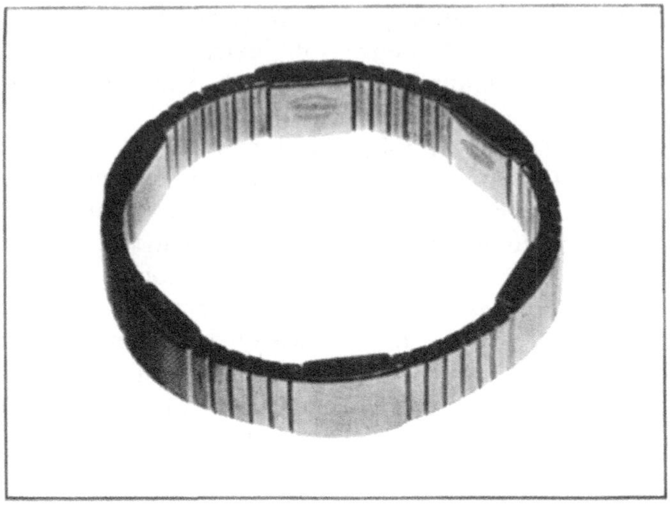

Figure 4-11. Japanese armband with permanent magnets, used in the treatment of shoulder stiffness. From Kimura (1967).

conclude that the use of the magnetic arm band cures disease. But since the band was effective in 41% of the cases, it would appear that it is as sensible a therapy for shoulder stiffness as any of the more orthodox symptomatic treatments.

Effects of Extremely Weak Magnetic Fields on Biologic Systems

Barnothy (1969) quotes a study by Conley (1969) done in connection with manned space-flights, because similar conditions would be encountered there. Table 4-2 presents an overview of the wide range of magnetic field intensities that occur in space.

Conley (1969) reports on experiments conducted with mice in magnetic fields of less than 0.065 A/m (corresponding to 80 γ in air). He found a lower acid phosphatase activity in cellular solutions from the animals kept in these weak magnetic fields during incubation than in the control animals. These in vivo experiments could not be successfully reproduced in vitro, however, Barnell and Brown (1961) also investigated the biologic effects produced by

Table 4-2. *A comparison of representative values of static magnetic fields. After Beischer (1969).*

	A/m	Tesla
Al-Ni magnet surface	$\sim 8.10^4$	$\simeq 0.1$
Solar flares	$\sim 8.10^3$	$\simeq 10^{-2}$
Quiet sun surface	~ 80	$\simeq 10^{-4}$
Earth's surface (at the pole)	~ 80	$\simeq 7.10^{-5}$
Interplanetary space during magnetic storms	~ 0.08	$\simeq 10^{-7}$
Moon, Mars, Venus	< 0.08	$< 10^{-8}$ (Moon)
Interplanetary space, normal	$\sim 8.10^{-4}$	$\simeq 5.10^{-9}$

relatively weak magnetic fields. They demonstrated the ability of snails to orient themselves in magnetic fields of 80 A/m. Table 4-3 presents a partial survey of the biologic effects which have been produced by very weak magnetic fields, along with the authors to whom the observations are attributed.

In a comprehensive study, Busby (1968) explored in detail the problems of the biologic activity of magnetic fields. Based on an extensive survey of the literature, the author dealt both with the effects of extremely weak fields on the order of about 100 γ (10^{-7} T) and with the effects of very strong fields, which may occur in connection with space flight.

Schneider (1974) also studied the orienting ability of animals in magnetic fields. He found that june beetles are able to avoid certain fields produced by permanent magnets.

Miericke (1972) devoted a study to the action of magnetic fields on humans. His work led him to the general conclusion that biologic effects—on both man and animals—can be demonstrated for almost all magnetic fields which differ in any respect from the natural geomagnetic field. Despite the fact that no serious illnesses or deaths have yet been observed in connection with magnetic fields, the author recommends that man not be exposed unnecessarily to magnetic fields, regardless of their intensity.

In judging the degree to which the human body can be influenced by external magnetic fields, it would be helpful to know the intrinsic magnetic field of the human body itself. Cohen (1975) investigated this problem with the aid of a highly sensitive measuring device with an intrinsic noise on the order of 1.6×10^{-10} G $\cdot$ Hz$^{-1/2}$. The principal sources of these endogenous magnetic fields are apparently the currents generated by muscles and nerves. Magnetic fields on the order of 5×10^{-8} G emanate from the human brain. The magnetic fields associated with cardiac activity are considerably stronger, with an intensity of about 5×10^{-6} G and a frequency of 0.1–40 Hz (Fig. 4-12).

Extremely Slow Field Fluctuations

In the transition zone between static and ELF fields lie the extremely slow field fluctuations in the ULF range. In investigations of the biologic activity of such fields, the experimental object is introduced into "rotating" magnetic fields, produced by rotating magnets.

Persinger (1969) studied the open-field behavior of rats exposed prenatally to a rotating magnetic field of low intensity and low frequency of rotation (field intensity 0.3–3 mT, corresponding to 3–30 G; frequency of rotating field 0.5 Hz). To monitor the movement of the 21- to 25-day-old animals, he measured the number of times they crossed squares painted in an open field. The animals exposed to the rotating magnetic field exhibited more movement than the control animals. However, the treated male rats crossed the squares much more frequently than the treated female rats. Young rats which were exposed prenatally to the rotating magnetic field but were nursed postnatally by a control mother did not differ in behavior from similarly treated animals which grew up with their own mother. In connection with these investigations, and to offer a hypothetical explanation for the observed biologic effect of these fields on rat embryos, Ludwig (1971) showed by calculation that the field intensities used in the experiment were sufficient to alter the ionic environment at synaptic membranes so as to increase membrane permeability. The shielding effect of steel structures against magnetic field fluctuations in this frequency range proved to be at least theoretically negligible. As for possible mechanisms of action, it is presumed that so-called ELF waves exert central effects on the organism, while VLF waves exert peripheral effects. But this does not take into account the ELF

Table 4-3. *Summary of selected reports on biologic studies of the effects of extremely low-level static magnetic fields. After Barnothy (1969).*

Object of study	Part investigated	Process observed	Special parameters measured
Algae (Euglena, Chlorella)	Cells	Growth	Growth rate
Bacteria (Staph. Albus)	Cells	Growth	Size and number of colonies
Winter wheat (seedlings)	Entire plant	Growth	Root orientation
Protozoa (paramecium)	Entire organism	Motor activity	Direction of locomotion and number of direction changes
Birds (sparrow)	Entire organism	Behavior	Amplitude and character of motor activity
Mouse	Entire organism	Aging, growth, reproduction and behavior	Life span, litter size, activity, behavior
Man	Entire organism	Behavior	Number of psychiatric cases
Man	Entire organism	Behavior and clinical data	Psychological tests, EEG, ECG, body temperature and weight, blood
Man	Entire organism	Behavior, nervous function, various functions	Psychological tests

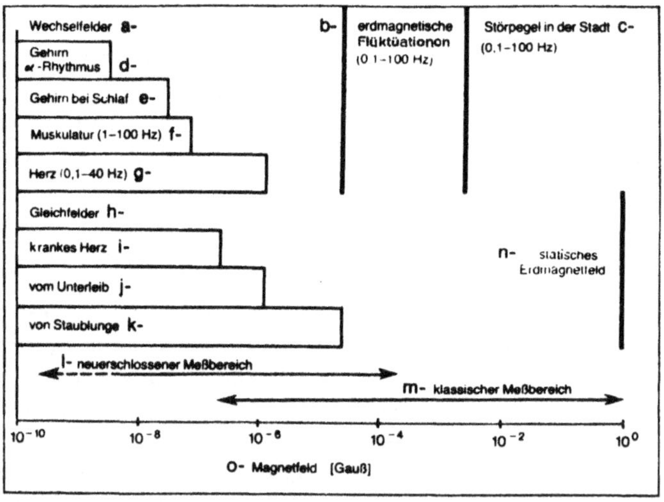

Figure 4-12. Measured intensities of the magnetic field emanating from various organs, compared with corresponding geophysical data. After Cohen (1975).
Key: a-Alternating fields; b-Geomagnetic fluctuations; c-Noise level in the city; d-Alpha rhythm of brain; e-Brain during sleep; f-Musculature (1–100 Hz); g-Heart (0.1–40 Hz); h-Static fields; i-Diseased heart; j-From the lower abdomen; k-From lungs in pneumoconiosis; l-Range now accessible to measurement; m-Classical measuring range; n-Static geomagnetic field; o-Magnetic field (G, or 10^{-4} T)

Magnetic field intensity	Duration of treatment	Effects observed	Author and year
0.08–8·10⁴ A/m	1 to 3 weeks	Growth accelerated in very weak fields, retarded in strong fields	Halpern (1966a, b)
4 A/m	72 h	Reduction in size and number of colonies	Becker (1963b)
Horizontal component of geomagnetic field 0.02 T (16 A/m)	Period of growth	Orientation parallel to geomagnetic or artificial field in all cases	Pittman (1964)
100 A/m	Several seconds	Statistically significant change of direction when magnetic field applied	Brown (1962)
about 50 to 135 A/m	2—9 h	Increase and change in motor activity	El'Darov and Kholodov (1964)
80 ± 40 mA/m	1 year	Shortened life span (6 mo.), diffuse tissue, cannibalism, supination	van Dyke and Halpern (1965)
Geomagnetic fluctuations	1 month	Positive correlation with intensity of geomagnetic field	Becker et al. (1961)
about 40 mA/m	10 days	Lowering of threshold in flicker fusion test, other tests showed no significant changes	Beischer and Miller (1964)
<40 mA/m	10 days	Lowered threshold in flicker fusion test, $p < 0.001$	Beischer et al. (1967)

components produced by VLF waves due to nonlinearity effects in biologic systems, even though the importance of such effects is considerable both from the theoretical standpoint and on the basis of experimental findings.

Persinger et al. (1972) worked with similar experimental conditions. In three experiments they exposed male, 115- to 150-day-old rats to either a 0.05–0.3 or 0.3–3 mT (0.5–3 or 3–30 G) rotating magnetic field with a frequency of 0.5 Hz for 5, 10 or 26 days. The rats exposed for 10 and 26 days showed a clearly higher water consumption than the control group. The same was observed for the 5-day rats, although these animals also showed a progressive decrease in relative thyroid weight. At the same time, a significant weight gain was observed during exposure up to 10 days, and a decrease in relative testicle weight during exposure up to 26 days. Differences in the number of circulating eosinophils and the relative weight of the adrenal glands were insignificant.

In another experiment with younger rats, which were only 80 days old at the start of the experiment, no differences were found after 21 days' exposure relative to the control group. The physiologic changes observed in the older animals are attributed to effects on the thyroid gland, including an alteration of its liquid and crystalline structure.

Persinger et al. (1978) also report the following, based on experiments with rats in low-intensity 0.5-Hz magnetic fields:

> Forty-eight adult female rats were caged in groups of 3 or 4 individuals

> under normal lighting conditions for 120–140 days. They were then exposed in pairs to one of 6 different variation intensities (0.001 to 10 G, i.e., 10^{-7} to 10^{-3} T) of a 0.5-Hz rotating magnetic field (RMF) or to one of 3 sham field positions for 10 days while under constant 45 lux illumination. Post-mortem examination and histological analyses demonstrated that RMF-exposed rats show no significant changes in thyroid follicle numbers, perifollicular mast cell numbers, wet adrenal and pituitary weights, body weight changes, or water consumption either between intensity groups or relative to sham-field controls. Significant interactions existed for thyroid and spleen weights between exposure intensities and pre-exposure caging conditions for the 10^{-5} T to 10^{-6} T groups; the effect was not linearly related to field intensity. Rats that had been maintained in the 4/cage situation showed significantly lighter body weights and more thyroid follicles; the significant alterations in thyroid, adrenal and spleen weights were related to body weight.

Persinger (1973) also studied the possibility of controlling cardiac activity by an external 0.5-Hz rotating magnetic field with an intensity of 0.001–0.002 T (10–20 G). For this purpose he performed two experiments to determine whether such fields could play a role as a control factor in the weakening (barbiturate-induced) of cardiac function in rats. Animals that received sodium pentobarbital at 30-min intervals and were exposed to the rotating field for 4 hr exhibited significantly slower cardiac contraction rates before extinction of the QRS complex than the control groups. No differences were found between the groups at the end of the experiment with regard to serum transaminase, serum iron or oxyhemoglobin levels. Persinger (1973) sees a connection between the ELF electrical signals of the heart and labile weather systems as well as associated solar eruptions, and states that natural fields in the ELF and VLF ranges (with pulse repetition frequencies in the ELF range) are similar to cardiac signals in the intensity of the E and H field. The question arises, whether natural ELF fields could influence electric and magnetic processes in the heart under certain circumstances. Such considerations are of interest primarily in connection with reports that relate myocardial infarction to the passage of weather fronts (prior to changes in air pressure and temperature). Susceptible persons would not be protected from such influences by the "controlled" climate conditions of a house. Extremely low frequency signals would precede ordinary meteorologic stimuli and penetrate readily into buildings.

The effect of VLF magnetic fields on human reaction time was investigated by Friedman et al. (1967). For this purpose they constructed a Helmholtz coil arrangement (two air-core coils which, due to their spacing, produce a region with a nearly homogeneous magnetic field) with which they could produce magnetic fields of the desired intensity and time variation. They found that the test subjects exposed to a 0.2-Hz field showed a statistically significant increase in reaction time relative to the two comparison groups (0.1-Hz and static fields). The same result was obtained in experiments with only one group of subjects who were subjected to the three experimental conditions in succession.

ELF Fields

In 1936 Schmid published a comprehensive summary of the literature on the biologic effects of low-frequency natural and artificial electromagnetic vibrations. We shall report on more recent findings in this area, drawing on such works as the survey compiled by Sheppard and Eisenbud (1977).

Effects on Man

The possibility that biologic effects are associated with ELF electric and magnetic fields has been recognized independently by numerous researchers in connection with various experiments and on the basis of special observations. For example, Hartmann (1957) reports on individual experiments known since 1939. In 1951 he began to work with Farenkopf and experimented with sawtooth electric field oscillations at frequencies of 1–20 Hz. He found very definite frequency-dependent biologic effects, which appeared to be type-specific. Sawtooth oscillations of 1–15 Hz apparently are biologically active. In man they can both trigger and alleviate disorders in the space of a second. They reportedly have an anti-inflammatory and spasmogenic effect in the lower-frequency range, and an anti-spasmodic and phlogogenic effect in the higher-frequency range. Patients should be treated with the field frequency most favorable for their individual reaction type. Presumably oscillations with specific frequencies can be associated with specific diseases, organs, parts of organs and parts of the body. Fields at 1.75 Hz are reportedly effective against acute inflammations, while fields at 10 Hz have a spasmogenic effect on the body.

The infrared body measurements made by Schwamm (1957) in a sawtooth oscillation field should be mentioned in this context. He was one of the first to use the infrared (IR) emissions of the human body to develop an IR-emission diagnostic technique, and found changes in the body's emission spectrum under various external conditions. Hartmann (1958) adopted the method of Schwamm in order to demonstrate the biologic activity of sawtooth oscillations by the spectrum of IR emissions from the human body. Figure 4-13 shows the time course of the sawtooth oscillations employed and the corresponding infrared spectrum. The greatest differences are seen between the all-positively and all-negatively polarized fields: the IR spectrum shows a clear shift from emission predominantly at 30 μm to peak emission at 6 μm.

Studies by König and Ankermüller (1960) originally dealt with the purely geophysical measurement of ELF atmospherics, as described in Chapter 2. For it was found on analysis of signal activity that various signals occurred in the range of 1–25 Hz which were apparently related to local weather phenomena. This was es-

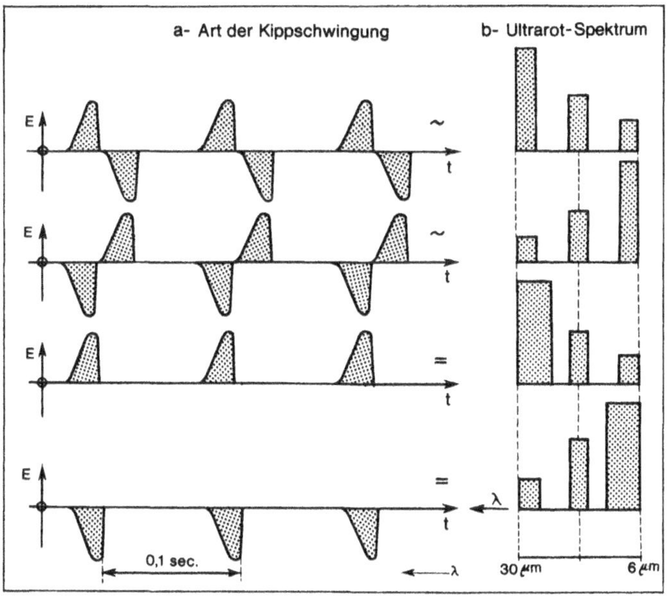

Figure 4-13. Effect of various ELF sawtooth-like waves on the infrared emission from the human body. After Schwamm (1957) and Hartmann (1958).
Key: a-Waveform; b-Infrared spectrum

pecially true of the so-called "type II" signals (see Figs. 1-10 and 2-12)—irregular processes in the frequency range of 3–6 Hz—as well as more or less sinusoidal processes of 0.5–2 Hz (type III). There were also days with extremely low signal activity, as was known from observations of VLF atmospherics of 4–50 kHz by Reiter (1953). Prompted by bioclimatic studies on the effects of VLF atmospherics on man and by our own observation that certain signal types frequently coincide with more or less typical local weather conditions, which are known to have general health effects, we performed various experiments.

The main question was, are the electromagnetic processes of 1–25 Hz that occur naturally in the atmosphere of purely geophysical significance, or are they also important biologically, especially for man? To answer this question, we attempted to find certain correlations by comparing signal recordings with the occurrence of events, and to reproduce the atmospheric signals artificially in the laboratory. The laboratory experiment allowed us to test the biologic activity of these signals so that results obtained by the comparative method could be substantiated if necessary.

Comparative investigations. A large number of reaction-time measurements were made among the visitors to an automobile exhibition in Munich in 1953, as reported by Reiter (1954). Reaction time was determined by measuring the time required for the subject to respond to a light signal by pressing a button. The extensive numerical material obtained in these measurements provided an ideal data base for comparing reaction times with signals recorded simultaneously in the ELF range (König and Ankermüller, 1960; König, 1962).

Reaction times and type II signals. Since observations suggested that type II signals would have the most marked effect, our investigations were begun in this area. Recordings of atmospheric signals during the course of the exhibition showed the presence of type II signals in 10 usable cases. The reaction times were determined for these 10 cases by the "n method" (critical hour n), and the mean reaction times were calculated for the half-hour intervals before and after the critical hour n ($n \pm t$, where $t = 0, 30, 60$ and 90 min). The time at which the recording of the type II signals began was taken as n.

The reaction-time curve during the critical period (Fig. 4-14), was obtained by taking the arithmetic mean of the corresponding reaction times during the 10 cases in which type II signals were recorded. The points making up the curve result from a total of 2000–3000 individual reaction-time values. The "type II signal intensity" curve shows the mean variation of signal intensity for the 10 cases of interest. The mean time of the onset of the type II signals was calculated at 2:30 P.M. For this point in time the "normal mean reaction time" curve shows the reaction time pattern and scatter typical of the exhibition as a whole. The points composing this curve result from about 45,000 individual reaction-time measurements.

The reaction times lie practically within the scatter up to the critical time n (Fig. 4-14). After the appearance of type II signals, the reaction time of the subjects increases considerably and lies well outside the scatter. It attains a maximum between $n + 30$ and $n + 60$ min, while the signal intensity steadily declines from its greatest value at n to $n + 90$ min.

Reaction times and type I signals. The undisturbed reception of type I signals was possible for 18 days during the month of September 1953. The mean signal intensities and mean subject reaction times were computed for this 18-day period. The resulting curves are superimposed in Fig. 4-15 to show the parallel between them. The reaction times entered in the graph pertain to the time between the half and whole hour, each point representing about 4500 individual reaction-time measurements. Type I signals apparently have an effect opposite to that of type II signals, for the reaction times shorten as the intensity of the type I signals increases (Fig. 4-15).

Despite statistical evaluation problems,

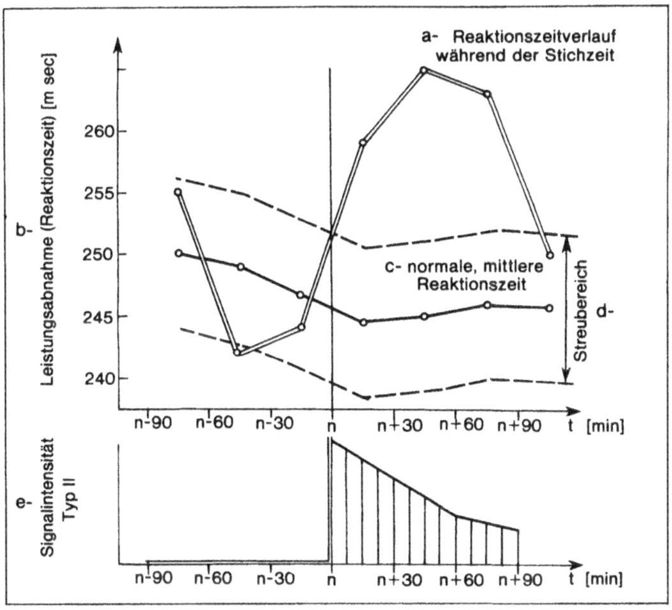

Figure 4-14. Relationship between the reaction time of test subjects and type II ELF signal activity (3–6 Hz). Reaction time measurements made during a traffic exhibition. Number of key cases: 10; mean time of key hour: 14:30. Number of measure values per point on curve: (*a*) about 2500 during the key hour, (*b*) about 40,000 for the normal, mean reaction time.
Key: a-Variation of reaction time during key period; b-Decreasing performance, reaction time (ms); c-Normal, mean reaction time; d-Scatter; e-Type II signal intensity

the significance of these findings was recognized immediately, for they clearly pointed to the biologic activity (possibly frequency-dependent) of atmospheric signals.

Experiments in Artificially Produced Electric Fields. The results of the previous investigations prompted us to conduct experiments with artificially produced fields similar in character to natural type I and type II signals. Electric voltages supplied

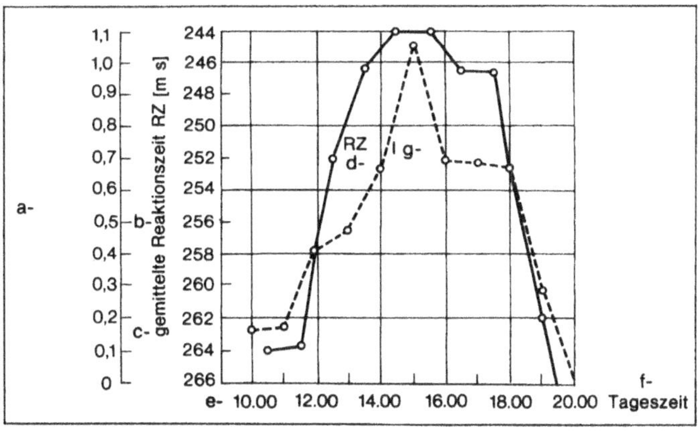

Figure 4-15. Relationship between the reaction time of test subjects and the occurrence of type I signals (8–9 Hz). Reaction time measurements made during a traffic exhibition. Number of days on which only this type of activity was recorded: 18; number of reaction time measurements per data point: about 4500.
Key: a-Relative intensity of type I signals; b-Increasing performance; c-Mean reaction time (ms); d-Reaction time; e-10:00, 12:00, (etc.); f-Time of day; g-Type I signals

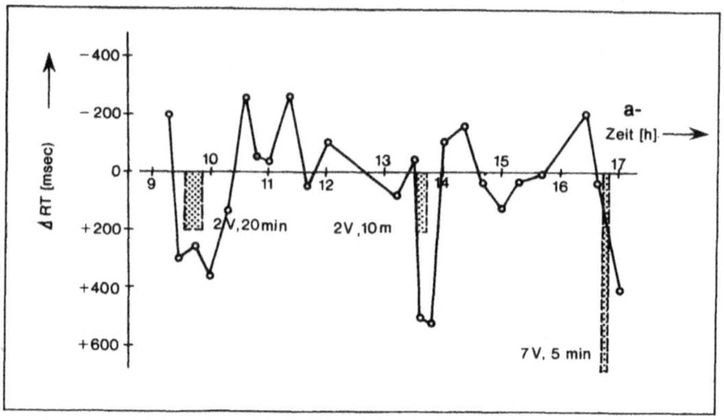

Figure 4-16. Typical effect of a vertical electric field (plate spacing 2.5 m) with a frequency of 3 Hz and a strong harmonic content corresponding to that of natural fields (type II) on the reaction time of an individual seated test subject in a single-blind experiment. The voltage measured at the plates and the duration of exposure to the field are indicated. The deviation of the reaction time (each point representing the mean value of 10 successive measurements) corresponds to a scatter of 235 ms. Key: a-Time (hr)

by a specially designed generator were connected to two wire screens, setting up an electric field between them. In one case, the voltage was sinusoidal with a frequency of 10 Hz; in another, it was distorted with a strong harmonic content, with a fundamental frequency of about 3 Hz. The screens were attached to the ceiling and floor of an experimental chamber and were separated by about 2.5 m. An arrangement was thus available for introducing individual experimental subjects into an artificial electric field which was known to correspond to natural conditions. The reaction times of test subjects were determined at 20-min intervals for 8 hr by the same method used at the Munich exhibition. During these tests the subjects could be exposed intermittently to the artificial field without knowing it.

The typical reaction of a single test subject to a type II field is shown in Fig. 4-16. The deviations of reaction time from the mean value are clearly seen in the graph. The field intensity values were on the order of 1 V/m, and the frequency was 3 Hz, which led as expected to an impairment of performance. The scatter, evaluated for the few data points, was equal to 235 ms.

The results for 8 cases of exposure to the type II field are summarized in Fig. 4-17 by the *n* method (critical time). The field was switched on for an average of 12 min at an intensity of about 1 V/m. An increase in reaction time, and thus an impairment of performance, is clearly seen after the field was switched on. The effect of exposure to an artificial type I electric field is shown in Fig. 4-18. Here the average field intensity was 2 V/m and the average duration of exposure 8 min for 4 cases. A decrease in reaction time (improvement of performance) occurred (Fig. 4-18). Of course the numerical material from these two experiments is insufficient to produce statistically convincing results, but it does confirm the results obtained at the Munich exhibition. Man is evidently influenced in a frequency-dependent manner by electric fields of the types described, regardless of whether they are natural or man-made.

Clinical Tests. To supplement the more subjective investigative methods previously described with results of more objective techniques, we performed electrodermatographic (EDG) tests using the same experimental set-up. Whereas the subjects were seated during the reaction-time tests, they reclined for the EDG measurements. Of 10 subjects exposed to electric field intensities over 5 V/m (3 Hz,

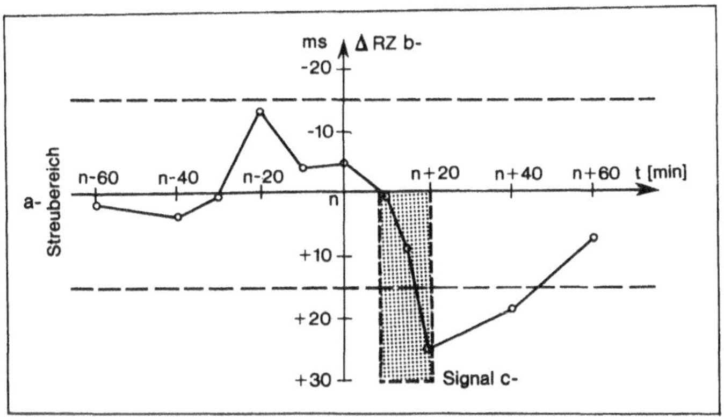

Figure 4-17. Effect of an artificially produced electric field corresponding to type II signals of natural origin (3 Hz, strong harmonic content) on the reaction time of test subjects. An impairment of performance was observed.
Key: a-Scatter; b-Reaction time; c-Electric field

distorted waveform), 5 showed a decrease in skin resistance after the artificial field was switched on (Fig. 4-19).

Subjective Reports of Test Subjects. Besides the results just described, some interesting subjective reports were obtained from subjects exposed to the artificially produced type II electric field. The subjects repeatedly complained of headaches a few minutes after short-term exposure. Repeated exposure led to exacerbation of these complaints or to signs of fatigue. Disappearance of the headache was often followed by fatigue. All complaints, including a "tight feeling in the chest" or "sweating of the palms," gradually disappeared within a few hours.

10-Hz Processes of General Significance. Also of interest are studies by Rohracher (1949, 1952) and Rohracher and Inaga (1969), which revealed mechanical vibrations of the human body and ground, both with a frequency of about 10 Hz (although the latter are somewhat uncertain). They strongly resemble type I signals in their time course. Wüst (1956) also dealt with the human body as a vibrating system at these frequencies. Thus it appears that the frequency of 10 Hz occurs in nature in many forms, and it was apparent even then that a type of natural constant was involved.

General Considerations. Originally the question was asked, Do atmospherics have

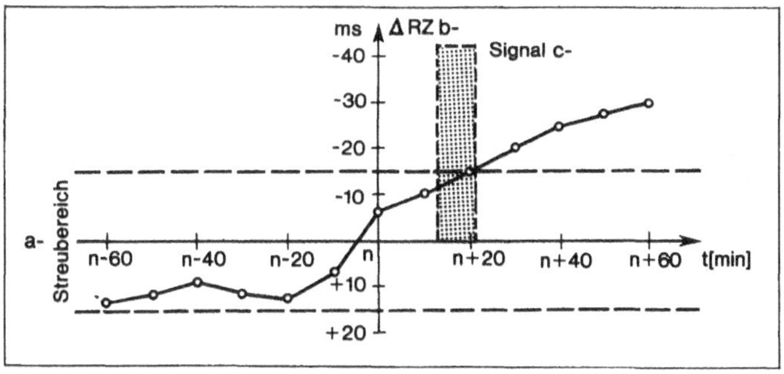

Figure 4-18. Effect of an artificially produced electric field corresponding to natural type-I atmospherics (8–10 Hz, sinusoidal) on the reaction time of test subjects. An improvement of performance was noted.
Key: a-Scatter; b-Reaction time; c-Electric field

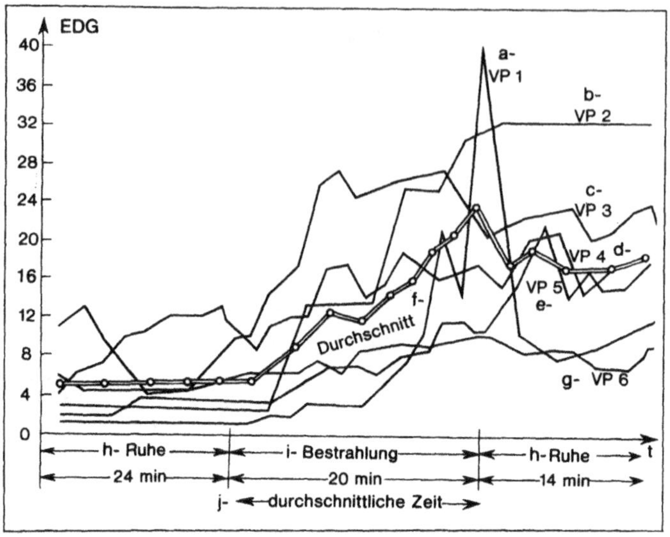

Figure 4-19. Effect of an artificially produced field corresponding to natural type II atmospherics (3 Hz, strong harmonic content) on the electrodermatogram of test subjects. Field intensity 1 V/m: 12 subjects, 1 positive response; field intensity 5 V/m: positive response in 5 of 10 subjects.
Key: a-S 1; b-S 2; c-S 3; d-S 4; e-S 5; f-Average; g-S 6; h-Rest; i-Field exposure; j-Average time

an effect on man? In the light of observations and experimental data which suggest an affirmative answer to this question, it would seem appropriate to add some speculative comments to our discussion.

One is struck, first of all, by the results of electroencephalographic studies. For example, the alpha rhythm of the brain is remarkably similar to type I signals in frequency and waveform, while certain disturbances manifested in electric brain activity (delta waves) closely resemble type II signals (see Fig. 1-10). In consideration of this fact on the one hand, and the assumption of bioclimatology that the "biotrophic factors" of weather phenomena affect the human autonomic nervous system on the other, it could be argued that atmospherics have an affect on man and should be regarded as one of the biotrophic factors.

The connection observed between daily fluctuations in atmospheric signals and the daily rhythm of the human organism also points to an effect of atmospherics on man. Type V signals are of special interest in this regard. These signals can be regularly recorded at sunrise. Along with type I signals, whose intensity increases about tenfold at that time, these signals may play a role in triggering ergotropy in the autonomic nervous system.

The objection can be raised that the EEG would primarily indicate activities producing the special waveform associated with motor function, sensory function and attention, with no significant involvement of the autonomic nervous system. Indeed, there may be no direct connection between atmospherics and this system, even though experiments by such investigators as Gavalas et al. (1970) strongly indicate it (see below). In any event, an alteration in the function of electric brain activity could have taken place during the course of evolution. Considerably more experimentation, research, and analysis are needed before these questions can be resolved.

While Hartmann (1957) obtained his results more on the basis of individual, unrelated cases, as reported earlier, our studies on connections between atmospherics and human function have succeeded on a more systematic (and natural) basis in demonstrating the biologic activity of such fields.

It was several years before other scientific institutions concerned themselves specifically with this problem. For example, Hamer (1965) also performed reaction-time measurements in human subjects in order to study the biologic effects of electric fields in the ELF range. He investigated the possible frequency-dependence of the effects as well as the signal intensities necessary to produce such effects. The capacitor-plate arrangement used by Hamer to generate electric fields is shown schematically in Fig. 4-20. An alternating electric voltage *U* applied to these plates set up a horizontally oriented electric field to which primarily the head region of the subjects was exposed. The frequency-dependent effects of the electric fields at a plate voltage of 2 V are also shown in Fig. 4-20; first by reaction-time differences between 3 Hz and 8 Hz, compared with the reaction time of the subjects with no field present; and second by reaction-time differences between 3 Hz and 12 Hz, also compared with zero-field conditions. The experiment confirmed what was already known: fields of relatively low frequencies prolong reaction time, while higher-frequency fields shorten it. The results of a follow-up experiment using other frequencies without a zero-field control show a reaction-time difference of about 6% between subjects exposed to fields at 2 Hz and 6 Hz, and at 3 Hz and 11 Hz.

The ideal of a further experiment was to reduce the frequency range defined by the upper and lower test frequency of the field until a statistically significant difference in the reaction times of the subjects could no longer be discerned.

The result of an analysis of the experimental data kindly provided by Hamer (1965) is shown in Fig. 4-21. Two series of experiments were conducted (upward-and downward-pointing brackets in the Fig. 4-21) with voltage values between the capacitor plates that differed in each case by a factor of 10 (0.2 V, 0.02 V, and 0.002 V). The end of each bracket in Fig. 4-21 marks the limit which the plate-voltage frequency could approach from above or below and still produce a statistically significant difference in reaction times. In other words, the brackets show the frequency band within which no statistically significant reaction-time difference could be discerned. In statistical terms, then, this range is associated with a particular uniform reaction time.

Two important conclusions can be drawn from this study. First, biologic effects are observed even in laboratory experiments with the weak field intensities characteristic of natural electric fields, especially type I

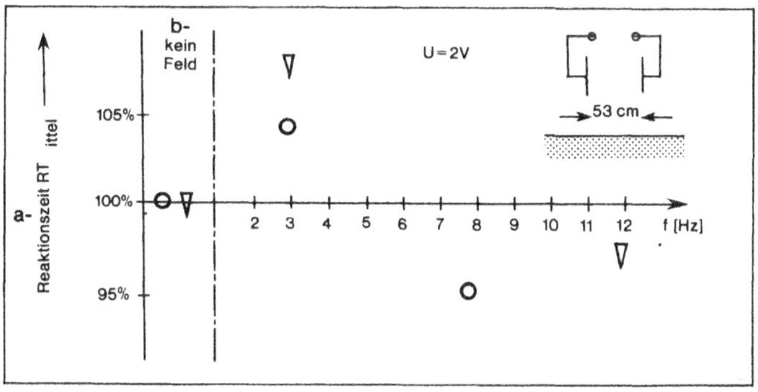

Figure 4-20. Effect of ELF fields on human reaction time as a function of frequency. Comparative measurements between fields at frequencies of 3 Hz and 8 Hz (circles) and 3 Hz and 12 Hz (triangles) show an increase in reaction time relative to the "no field" situation at the low field frequencies, and a decrease at the higher field frequencies. Numerical material from Hamer (1965).
Key: a-Reaction time RT_{mean}; b-No field

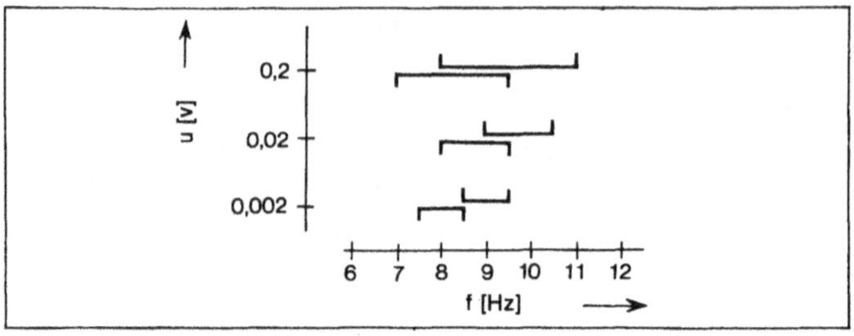

Figure 4-21. Reaction-time measurements in human subjects at various capacitor plate voltages *U* (see Fig. 4-20). Based on two series of experiments (up- and downturned brackets), the frequency range was determined in which a statistically significant difference in the reaction time of test subjects could no longer be discerned. The "selectivity" increases with diminishing field intensity. Numerical material from Hamer (1965).

signals (10-Hz resonance signals of the earth–ionosphere system). This contradicts one of the most important counterarguments regarding the significance of natural fields: that relatively high field intensities are employed in laboratory experiments, which are not characteristic of naturally occurring signals. Second, the apparent influence of field intensity on frequency sensitivity gives cause for further speculation. It appears, for example, that such fields can produce a predictable effect only if their intensity does not exceed a certain value. As the intensity increases, the "resolving power" of the human body with respect to the frequency-dependent action of such fields is diminished. This conclusion has been confirmed by other investigations and so is probably of importance for biologic systems in general.

One can only conjecture as to the consequences of such processes in susceptible biologic systems. The possibility cannot be ruled out, however, that certain previously unexplained complaints may have their origin in stray fields. We are speaking here of headache, dizziness, insomnia, etc. (familiar weather-related complaints), as well as complaints associated in some circumstances with man-made electromagnetic fields (lighting circuits, railway current, high-tension lines, etc.). Experience in this area has shown that many complaints of unknown etiology (often diagnosed as "neurodystonia") are environmentally determined, and even a slight change of location (changing one's sleeping place, for example) with the accompanying change in electromagnetic field conditions can often work wonders in such cases.

An important discovery was made by Hamer (1969) in connection with his study of the effect of sinusoidal electric fields on the ability of human subjects to estimate a time interval of 5 s. The double-blind experiment was conducted at a field intensity of 4 V/m, and the behavior of the subjects exposed to field frequencies of 8 Hz and 12 Hz was investigated and compared. A statistically significant difference was found in the ability of the subjects to estimate the passage of time. The 12-Hz electric field led to a shorter estimation of the 5-s interval than the 8-Hz field.

A conditioned defense reflex is reported by Petrov (1952) in subjects exposed to an electric field with a frequency of 200 Hz. The field was generated by an apparatus located above the head of the subject. The reflex observed was found to be unstable, however.

We should also mention in this context the observation by Knoepp et al. (1962) of an inhibitory effect of low-frequency electromagnetic fields on the growth of cultures of normal and diseased human cells. The cultures were exposed to a field with a frequency of 99–1000 Hz and intensity of 1.1–1.7 V/m for a period of 1–3 h. Although this treatment raised the temperature of the cultures by 2.3° C, their growth was nonetheless retarded, ranging from a very slight delay to a total cessation of culture growth and the death of the cells. The fact that these effects occurred only at certain frequencies specific for individual cell types is of particular interest.

Effect of Weak Electromagnetic Fields on the Circadian Rhythm in Man. At the Max Planck Institute for Behavioral Psychology, Wever (1968) did studies on the human circadian rhythm (from Latin: *circa* "about" + *dies* "a day," thus a period of about 24 h), with special emphasis on the question of the origin of the rhythm. This question is already answered in one sense: the circadian rhythm is of endogenous (internal) origin and is simply synchronized to the period of the earth's rotation by the periodically changing factors in our environment. It is known that in an environment kept artificially constant, the rhythm (under suitable conditions) continues with an undiminished amplitude but with a period different from 24 h.

The next question pertains to the physical factors in our environment which are capable of influencing the circadian rhythm. Of particular interest here are not just factors that are consciously perceived, but those for which the human body has no direct receptors. These include the electric and magnetic fields present in the atmosphere (excluding light, of course). Wever (1968) focused his investigation on the effects of 10-Hz fields.

In experiments on the circadian rhythm, it is essential to screen out as completely as possible all natural and man-made fields, so that the investigator will not have to work with a mixture of intentionally produced and other fields. To screen out natural electromagnetic radiation (especially low-frequency forms), two chambers were set up in an underground bunker specially designed for this investigation of human circadian rhythms. One of the chambers was lined throughout with several layers of a special iron shielding. This arrangement reduced the natural static field by at least 40 db (factor of 100), and by much more at higher frequencies. A vertical 10-Hz square-pulse electric field with an intensity of 2.5 V_{pp}/m was installed in this chamber for the experiments.

The subjects were informed about neither the shielding nor the artificial field. Since the subjects could not consciously perceive the switching on or off of the field, the conditions necessary for an objective study were met. The fact that the subjects had no sense of the duration of their circadian rhythm during the 3 to 4 weeks of the experiment and could notice no changes in the length of this subjective day contributed further to objectivity.

To investigate the effect of a permanent field, the 10-Hz field was switched on and off for periods of at least one week. The course of one such experiment is shown in Fig. 4-22. For all experiments, statistical analysis showed that the accelerating effect of the artificial 10-Hz electric field was highly significant ($p > 0.001$).

Another effect of the artificial 10-Hz field was observed in connection with experiments in which an "internal desynchronization" occurred. This refers to a state with abnormally prolonged periods of activity, with simultaneous periods of normal autonomic function. Cases were observed in which the rhythm was desynchronized immediately after the field was switched off; in one case, a previously manifest desynchronization was terminated immediately after the field was switched on, so that the rhythm was again stabilized. This effect of the artificial 10-Hz field also proved to be statistically significant. The observed effect of the artificially produced field on the period of the free-running circadian rhythm suggests that the same field might be capable of synchronizing the circadian rhythm if it were switched on and off periodically.

The results of these comprehensive investigations showed that if the difference between the two experimental chambers is indeed based only on the shielding of the first chamber, then the results of these experiments mean that the natural electromagnetic fields exert the same effect (as least qualitatively) on the human circadian

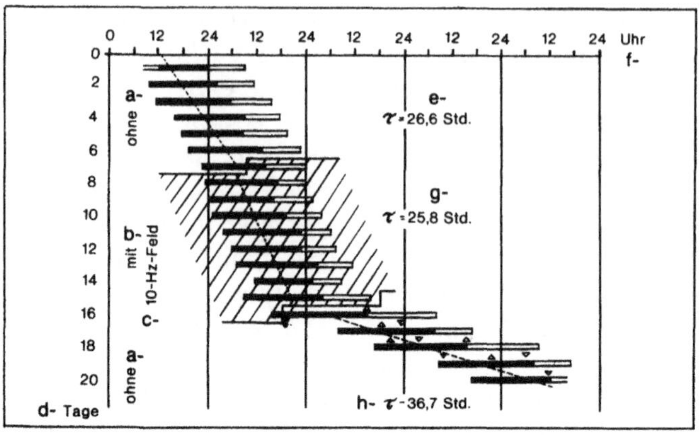

Figure 4-22. Circadian rhythm of an experimental subject without, then with, then without the influence of an artificial 10-Hz electric field (the period of field exposure is shaded). The activity rhythm is indicated by bars (filled: activity; open: rest). The body temperature rhythm is indicated by the position of its maxima (△) and minima (▽). Open triangles indicate same-time repetitions of maxima and minima already shown. The duration of the period (τ) is indicated for each phase of the experiment. After Wever (1968).
Key: a-Without; b-With; c-10-Hz field; d-Days; e-$\tau = 26.6$ hr; f-Time; g-$\tau = 25.8$ hr; h-$\tau = 36.7$ hr

rhythm as the artificial 10-Hz field. This effect can be characterized as follows:

1. Both fields accelerate the rhythm; the longer the period without a field, the greater the accelerating effect.
2. Both fields prevent internal desynchronization, which was observed only in the absence of both the natural and artificial fields.

In another discussion, Wever (1967) deals with the shielding of the experimental chambers. The qualitative agreement between the effect of electromagnetic shielding and that of switching off the artificial 10-Hz field raises the question of whether the shielding effect is based specifically on the elimination of the natural 10-Hz field. This question is not answered by the results obtained, first, because the shielding is not selective, but eliminates all static and dynamic fields of terrestrial origin; and second, because the intensity of the 10-Hz field employed was several orders of magnitude higher than that of the natural 10-Hz field.

Nevertheless, the importance of the artificial field experiments is obvious. They revealed for the first time a definite physical environmental factor which influences the circadian rhythm but does not act via specific sense organs and is not consciously perceived. They also demonstrated (once again) that the human organism can be influenced by relatively weak electromagnetic fields.

The remarkable finding that internal desynchronization in man can, when it appears spontaneously, be terminated by switching on a 10-Hz field prompted Wever (1973) to ask whether the effect of a 10-Hz field on the circadian rhythm is limited to man. Some preliminary experiments with animals indicate that this is not the case, if internal desynchronization is observed in the animals at all.

In experiments with the green finch, the free-running activity period could be shortened by switching the artificial 10-Hz field repeatedly on and off. When the field was turned off, the activity period was prolonged. This result coincides even quantitatively with the findings for man.

Lang (1973a) cites the interesting experimental results of Dowse and Palmer (1969) in this area, who desynchronized the circadian rhythm of mice by means of a Faraday shield, and resynchronized it with a 10-Hz field. Their experiments are thus similar to those of Wever (1967, 1968, 1973).

Special investigations with 10-Hz square-pulse fields. As experience has shown, a field with a square waveform with respect to time is of special significance, presumably due to its harmonic content.

An electrostatic field modulated by 10-Hz square pulses was used in studies by Altmann et al. (1976a) on students as well as employees in large office areas. Exposure to the artificial field conditions was relatively longterm: 162 students from four high-school classes were exposed alternately for 3-week periods to the field and to normal conditions (normal heating and ventilation). The students were in a seated position, and the field intensities at head level were 30 V_{pp}/m for the square-pulse field and about +50 V/m for the static field (negative pole grounded). Two classes served as controls. At the end of each phase of the experiment, the students were given the d-2 attention-loading test to determine attention-loading capacity. All the students had been given the Eysenck (1959) test for personality structure, as had the 53 office employees. The latter subjects were exposed for 6 months on the job to a static field (+ 150 V/m at head level, seated), modulated by a 10-V_{pp}/m square-pulse field; the controls were exposed to no artificial field. After that time the field was switched off for 3 weeks, while the controls continued to work under normal conditions (full air conditioning, 24°C, 50% relative humidity). At the end of each of these test periods and 3 weeks after the field was reactivated, the subjects were tested for attention-loading capacity, ability to concentrate, and subjective well-being. The students showed a continuous improvement in the attention-loading test over all 3 days of testing regardless of experimental conditions—a result of learning. However, this was accompanied by an additional improvement of attention-loading capacity in the students who had been exposed to the field conditions. This effect of field exposure was even more pronounced in the office employees. Attention-loading capacity decreased by 4.8% after the field was switched off, and improved by 11.7% after the field was switched on again. This finding was highly significant statistically ($p = 0.005$). The concentration test confirmed this result (Fig. 4-23). Subjective well-being also improved under field exposure: the subjects described themselves as significantly less sluggish, unhappy and absent-minded, based on responses to a list of 51 questions. The subjects tended ($p = 0.10$) to describe themselves as more energetic, happy, cheerful and pleasant, and less sluggish, incapable, anxious, fearful, sad, apathetic, careless and distracted. For some subjects the field conditions had a hyperexcitatory effect, manifested in heightened feelings of nervousness and irritability.

An examination of weather conditions during the course of the experiment showed the following correlation: the psychotropic action of the electric field conditions was always evident (measured by the shift in average performance), but was statistically significant only if a moderate to strong biotropism was present 6 days before and 6 days after the day of the test. Although no correlation could be found between psychic make-up and field effects, the correlation between biotropism and electric field effects can be interpreted in the manner of Schulz (1970): the organism subjected to prior stress from external weather conditions can be optimized in the manner indicated with regard to its functional capacity and well-being by means of electric fields, while an improvement under weather conditions which are already physiologically favorable cannot be demonstrated with statistical certainty.

The investigations of Altmann et al. (1976a) were repeated in a study by the author (unpublished). Electric fields of the type described above were installed in two high-school classrooms. Each was switched

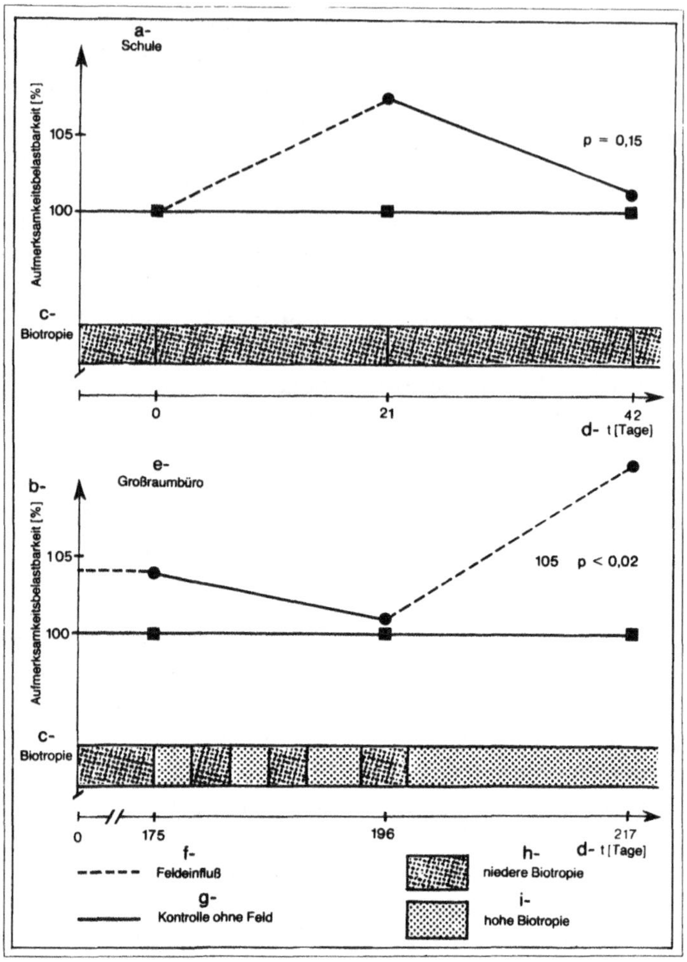

Figure 4-23. Changes in the attention-loading capacity of students (top) and office employees (bottom) under the influence of an electrostatic field modulated with a 10-Hz square-pulse electric field, as a function of exposure time. After Lang and Lehmair (1977b).
Key: a-School; b-Attention-loading capacity (%); c-Biotropism; d-t (days); e-Large office area; f-Field influence; g-Control without field; h-Low biotropism; i-High biotropism

on alternately for about 3 weeks, without the knowledge of the teacher or students. The experiment was continued for about 9 months, after which time it was found that the students classified as "labile" showed a particularly marked improvement in scholastic performance during the period of field exposure. On a grade scale of 1–6, an average improvement of 0.5 points in test performance ($p = 0.07$) was noted in this group during field exposure. This corresponds to a 13.5% grade improvement in the most favorable case.

In a similar series of experiments, Kröling (1976a) investigated the subjective well-being of 16 employees by obtaining their responses to a list of 20 questions each day for 10 weeks. Half the employees were exposed alternately for 5 weeks to the same type of artificial field described above (+150 V/m static field and 5-V_{pp}/m square-pulse field, at head level). Kröling (1976a) first interpreted the result of his evaluation of the most important questions to mean that the electric field effected no improvement in the well-being of the subjects. Later, a more differentiated evaluation technique showed an improvement of well-being between 1% and 18% during field exposure, depending on the method of scoring the answers.

Improved reaction times and heightened subjective well-being were also found by Fischer (1973, 1976a,b) and Fischer et al. (1977) in a group of patients exposed to such fields in 14-day intervals. No improvement in attention capacity could be observed in these mentally ill patients.

In a joint research project, Anselm et al. (1976) studied the effect of an artificially produced square-wave field of the type described above on the driving performance and reaction behavior of subjects in a driving simulator (Figs. 4-24 and 4-25). A group of 48 subjects were selected from about 300 persons as being representative of three age levels and two personality types with regard to mental lability or stability. A strictly standarized test program and fully automatic recording and encoding of experimental data ensured that the conditions for a double-blind experiment were satisfied. An electrode attached to the sunvisor of the simulator carried a direct potential of + 1000 V with respect to the chassis, with superimposed square-wave pulses (frequency 10 Hz, amplitude 20 V_{pp}). Field potential measurements at a distance of about 7/cm from the driver's head showed a direct potential of about 60 V and square-wave potential of about 4 V with respect to the vehicle chassis. During the simulated driving sessions, which took place both mornings and afternoons and on days of high and low biotropism, the field was automatically switched on for periods of 10 min, and the self-assessments, reaction times and driving errors of the subjects were recorded and evaluated. It was found that the reaction times remained virtually unchanged after the field was switched on, but that the number of driving errors (e.g., neglect of right-of-way, stop signs, pedestrian crosswalks, etc.) was reduced by an average of 8%–10% (Fig. 4-26). This tendency proved to be characteristic of all observations. In the mentally labile subjects there was an average decrease of 12% in

Figure 4-24. The driving simulator at the Alliance Center for Technology in Munich: View into open driver's compartment. In place of the windshield is a monitoring screen with simulated driving terrain which moves in accordance with the subject's driving actions.

Figure 4-25. Model landscape for the driving simulator in Fig. 4-24. The landscape is mounted on a conveyor belt, which moves it relative to a television camera mounted on a transverse guide rail (top left). The speed of the conveyor belt, and thus the apparent driving speed in the simulator, is governed by the position of the accelerator pedal. Turning the steering wheel moves the TV camera laterally across the roadway on its guide rail.

speed-limit violations and contacts with the edge of the road, but only a 2% decrease in the mentally stable subjects. When all driving errors were totaled, it was also found that the effect of the artificial electric field varied in degree between the mentally stable and mentally labile subjects. The overall improvement was 5% in the stable persons, and 8% in the labile persons.

The experiment with the driving simula-

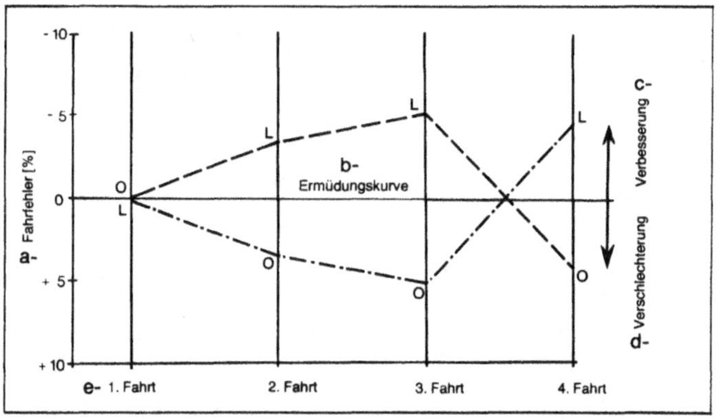

Figure 4-26. Effects of an electrostatic field modulated with a 10–Hz square-pulse electric field on the daily variation of the total number of errors made by subjects in a driving simulator. The deviation from the fatigue curve is shown with (L) and without (O) field influence. After Kirmaier and König (1976).
Key: a-Driving errors (%); b-Fatigue curve; c-Performance improved; d-Performance impaired; e-1st drive, 2nd drive (etc.)

tor was followed by a practical test, on which Kirmaier et al. (1978) reported:

> 240 male subjects ranging in age from 18 to 55 were available for the study. All applicants were tested for their suitability by means of the Eysenck personality inventory (Eggert, 1974) in a session preceding the experiment. Two groups of 50 persons were selected. These 100 subjects were each tested for one day (4 per day) according to a detailed schedule. They made 2 morning and 2 afternoon drives over a 46-km course near Munich, West Germany. The following tests were performed before and after the drives: reaction time, attention frequency, self-evaluation. Three sign recognition tests were given during each drive, and the subject had to negotiate a skill course at the end of the drive. An ECG was taken, and the latency of eye-focusing movements was determined. A two-part medical examination was also given (including pupillary reflex, visual acuity, blood pressure, pulse, neurological coordination, and an interview with a physician). Furthermore, the duration of the drive, the average and maximum speeds, and the economy of the drive were determined. A commercially available device was used to generate the necessary electrode potential for a positive electrostatic field with superimposed 10-Hz square-wave pulses. The field electrode (about 0.02 m^2 in area) was attached to the sun visor on the driver's side. The direct voltage applied was +60 V, the pulse voltage 4 V_{pp}. A letter-coded device was used to turn the field on and off in the sequence on–off–off–on, or off–on–on–off for the 4 drives. Neither the subjects nor the experimenter were informed as to the presence or absence of the field (double-blind experiment).

A total of about 700,000 individual bits of data (excluding ECG and vehicle-recorder data) were obtained and evaluated. The results are as follows:

1. Field-related improvements were most pronounced in the attention-loading test (d-2), i.e., they were highly significant in most cases ($p \leq 0.001$). This test showed an increase of 7.4%, 8.1%, 9.8% and 5.0%, respectively, in the number of recognized symbols at the end of drives 1 through 4 under the influence of the field. On a daily average, 22.7% fewer errors were made under exposure to the field. No field-dependent differences are evident in drives 2 and 3. If the number of errors is subtracted from the overall performance, the number of correctly marked d-2 blocks is obtained. Here, too, an improvement was found in drives 1 through 4 of 11.5%, 8.5%, 10.1% and 7.8%, respectively, under the influence of the field (Fig. 4-27).

2. The sign recognition test during the drives showed 3.7%–8.9% better results in drives 2 and 3 under the influence of the field (in all 3 sign recognition tests).

3. 15.9% fewer errors were made in the skill test of the second drive.

4. During the first drive driving times were 2.9% shorter.

5. 4.5% and 3.9% higher average speeds were found in drives 1 and 4, respectively, with no change in maximum speed.

6. 0.7% increase in flicker fusion frequency was found for the morning drives.

7. After the first drive reaction times were 2.7% shorter.

8. Field exposure produced no apparent change in the various physiologic parameters determined in the medical examination. The reports on subjective well-being obtained during the physician interview after the third drive (fatigue, headache, nervousness) were 19.4% more positive ($p \leq 0.024$) under the influence of the field. Of the subjects who had made drives 2 and 3 *without* a field, 26% felt worse afterwards than during the examination on the previous day, as opposed to 6% of the subjects who had made these drives with the field and were tired or nervous.

9. Self-assessments of well-being ("I feel good/bad") were 8.9% and 10.4% poorer at the end of drives 1 and 4, respectively.

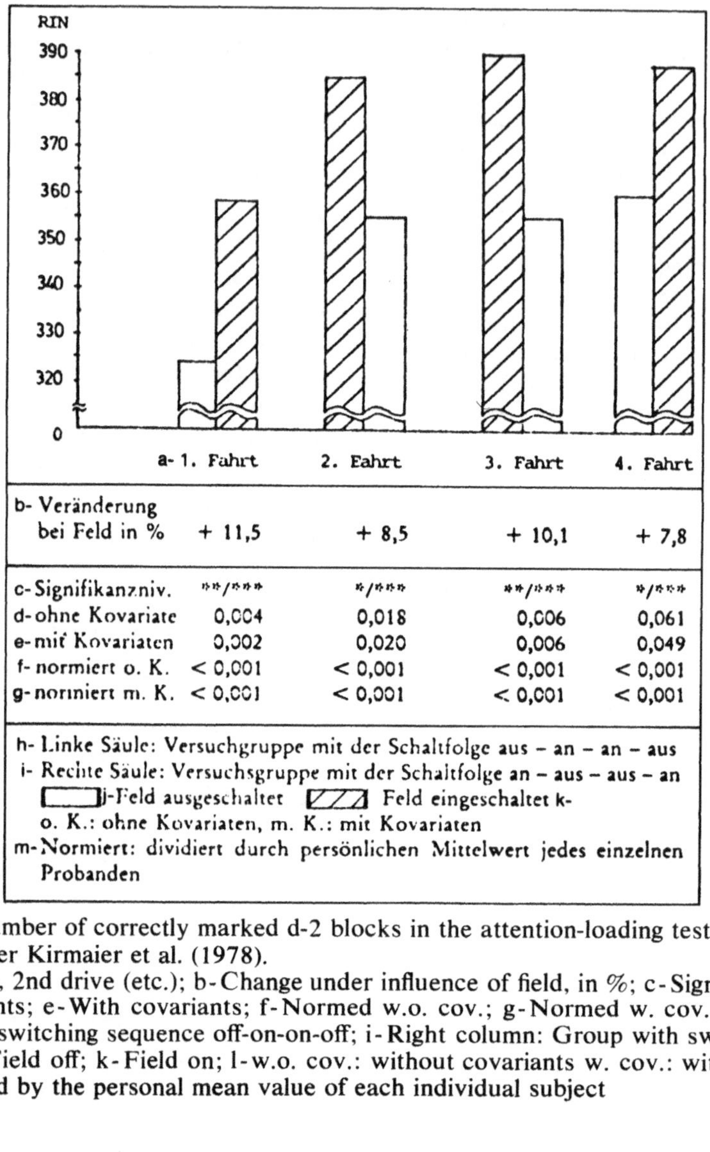

Figure 4-27. Number of correctly marked d-2 blocks in the attention-loading test at the end of the drive (RIN). After Kirmaier et al. (1978).
Key: a- 1st drive, 2nd drive (etc.); b- Change under influence of field, in %; c- Significance level; d- Without covariants; e- With covariants; f- Normed w.o. cov.; g- Normed w. cov.; h- Left column: Test group with switching sequence off-on-on-off; i- Right column: Group with switching sequence on-off-off-on; j- Field off; k- Field on; l- w.o. cov.: without covariants w. cov.: with covariants; m- Normed: Divided by the personal mean value of each individual subject

10. Self-assessments of inner tension ("I feel tense/calm") after the fourth drive were 16.5% poorer.

Animal and Plant Experiments

For obvious reasons, humans are not always available as subjects in scientific investigations. Consequently, plants and animals are often used to study the biologic effects of electromagnetic fields in the ELF range (and other frequencies). The results of such experiments are just as important from a scientific standpoint.

Of interest in this regard are experiments by Gavalas et al. (1970) with monkeys. The animals were trained by reward to press a button in regular 5-s intervals, while their EEGs were recorded by attached electrodes. After the animals had learned the task required, they were exposed to a weak electric field produced by applying a potential of 2.8 V_{pp} between two metal plates that were separated by 40 cm at head level. Thus, the animal's head was always completely within the electric field. The field frequency was either 7 Hz or 10 Hz. During exposure to the 7-Hz field, the monkeys showed in 5 of 6 experiments a significantly shorter interval between presses of the button. The average difference between the presence and absence of the field was 0.4 or more. The 10-Hz field had no apparent effect. At the same time, an analysis of the EEG data showed a relative performance peak at the frequency of the artificial field (7 or 10 Hz) in all experimental animals.

Reiter (1960) conducted exhaustive experiments with guinea pigs. While their conjunctival tissue pH was monitored, the animals were placed into a capacitor field which was either static or shielded by a rotating metal blade, and thus had the effect of an alternating field. The result of the experiment is shown in Fig. 4-28. One finds marked deviations of the tissue pH from the normal value associated with the alternating fields in the frequency range between 2 and 20 Hz. The reaction during a local thunderstorm is also shown for comparison.

The effect of low-frequency electromagnetic fields on the learning behavior of mice was studied by Amineev and Sitkin (1965). For this purpose the animals were made to run through a maze that was entered through a passage lying along the axis of a magnetic coil. In this way the animals could be exposed to a 100-Hz magnetic field with an intensity of about 24–37 kA/m. While this treatment did not influence the learning ability of the animals with respect to running through the maze in a specific direction, it did improve by 36%–40% their ability to relearn (to find the correct path) in comparison with the animals not previously exposed to the field. The authors regard this as evidence of the disinhibitory effect electromagnetic forces may have.

Comprehensive and exhaustive studies on the biologic activity of electric fields in the ELF and VLF range have been conducted at the Zoological Institute of the University of the Saarland, Germany. The experiments were done primarily on white mice. Lang (1971, 1972b), for example, reports on changes in the water and electrolyte balance of white mice exposed to a 10-Hz square-pulse field compared with mice kept under Faraday conditions. After 14 days under Faraday shielding, the mice displayed several anomalies of metabolism. Their oxygen uptake was reduced (1.79 ml O_2/g mouse/15 min versus 2.12 ml O_2/g mouse/15 min at 12° C), and with it, the oxygen pressure in the blood; the reduced energy supply led to a swelling of the cells (effect of anoxia). The water content of the tissue and the body in general was elevated (see also Fig. 4-29). At the same time, the elimination of water was reduced at a normal intake of food and water and normal sodium and potassium excretion. The increased water content of the interstitial space resulting from osmotic equalization between the intracellular space and interstitium apparently caused an isotonic increase in the blood plasma volume. The hematocrit and hemoglobin level was reduced per unit volume whole blood, although it remained unchanged per unit volume hemocytes. The sodium content of the whole blood was significantly increased, while the potassium content was reduced. Thus, shielding the animals from natural atmospheric electric fields caused not only a volume shift in their fluid systems, but also corresponding changes in their electrolyte balance. Apparently the organism adapted to the altered environmental conditions. The whole blood of the mice that were shielded from natural fields but exposed to an artificial 10-Hz pulsed field showed a somewhat reduced hematocrit relative to the normal animals. However, the measured sodium–potassium concentrations and the water content of the blood and blood plasma showed that exposing the mice to an artificial 10-Hz square-pulse field could largely offset the effects of shielding.

Since the effects of the Faraday shielding were not observed in animals exposed to

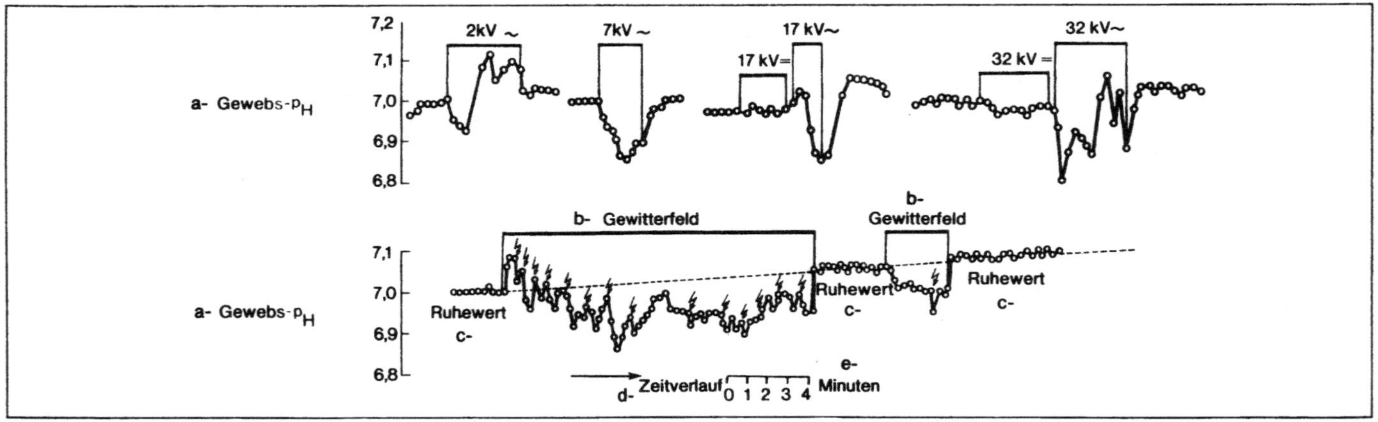

Figure 4-28. *Top:* The conjunctival pH of the anesthetized guinea pig in static (=) and alternating (~) electric fields over the frequency range from 2 to 20 Hz. *Bottom:* Behavior of conjunctival pH in the anesthetized guinea pig exposed to transitory electric field jumps during a local thunderstorm. Arrow = lightning strokes. Resting value = pH value during shielding. After Reiter (1960).

Key: a-Tissue pH; b-Thunderstorm field; c-Resting value; d-Time; e-Min

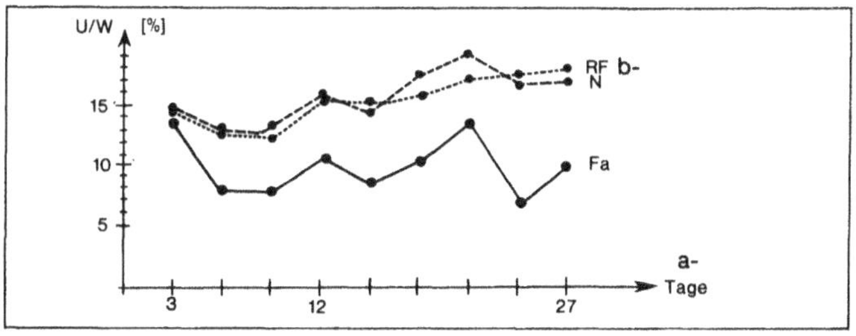

Figure 4-29. Urine output (U) of mice over a 3-day period (per 100 g body weight), shown as a percentage of water intake (W) under normal conditions (N), Faraday conditions (Fa) and artificial field conditions (RF). After Lang (1971).
Key: a-Days

the artificial 10-Hz field, they must have been completely altered or normalized by this field. Side-effects were also observed, however: protein consumption, measured by nitrogen excretion in the urine, was reduced, and anomalous though slight shifts in electrolyte concentration were detected in liver and kidney tissue.

Other investigaticns in this area by Lang (1972a) also dealt with the metabolic effects of Faraday shielding and an artificial 10-Hz electric field on white mice. Under the same experimental conditions as described above, it was found that sexually mature mice consumed less oxygen under Faraday conditions than under natural conditions. The water content of individual tissues and the body as a whole increased as a result of water retention. The extracellular space of the blood increased, while the volume of cellular material remained unchanged. Thus the blood hemoglobin level was reduced while the hemoglobin concentration in the erythrocytes remained unchanged. The sodium concentration in the blood was increased, the potassium concentration reduced. Lang offers the following explanation for these effects: The acid–base equilibrium in the cell system is normally maintained by regulatory systems. If perturbing influences act on the cell system, the organism first tries to compensate through corresponding changes in the body's water balance. According to Lang's theory, the Faraday effect places such a stress on the regulatory systems of the laboratory animals that the equilibrium state controlled by them can no longer be maintained.

In an attempt to relieve the metabolic effects in the mice under Faraday shielding, a 10-Hz square-pulse atmospheric electric field was artificially produced, and the metabolic functions of the animals were tested. If the working hypothesis was true—that natural 10-Hz fields constitute a critical component of environmental conditions—the artificial fields should produce largely natural conditions within a Faraday-shielded space. This logical conclusion was borne out by the experimental results. Oxygen consumption was higher in the mice in the square-pulse field than in those under Faraday conditions. The water content of the tissues as well as that of the whole body deviated only slightly from the normal value under natural conditions. The animals in the artificial field also showed largely normal values in the other metabolic indexes measured.

Studies in behavioral physiology were also performed on white mice by Altmann and Lang (1973) at the same institute. Their paper deals initially with the test of the biologic activity of electric thunderstorm fields by Schùa (1953) (Fig. 4-30). Schùa, who exposed nests of golden hamsters to electric fields with an intensity of 900 V/m and frequencies of 5 kHz–10 kHz, observed that the animals avoided the artificial fields and moved their nests out of the region of the artificially produced at-

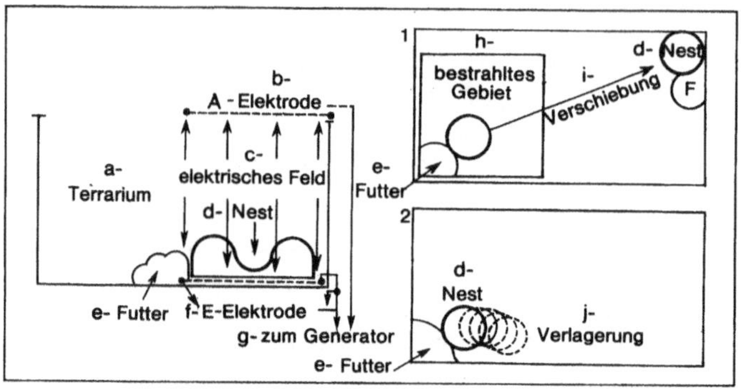

Figure 4-30. The nest-moving behavior of golden hamsters varies under the influence of an alternating electric field (1) and a static electric field (2). After Schùa (1953).
Key: a-Terrarium; b-A-electrode; c-Electric field; d-Nest; e-Food; f-E-electrode; g-to generator; h-Irradiated region; i-Relocation; j-Shift

mospheric electric field. He interpreted these findings as a physiologic reaction of the animals to the electric fields.

Similar studies were done by Zahner (1964), who pointed out various technical details in the experimental set-up which he believed weakened the results of Schùa (1953). Indeed, the reactions observed by Schùa did not reappear when the experiments were repeated. But the experimental results of Müller-Velten (1966) suggest than an olfactory orientation of the experimental animals may have played a role in the studies of both Schùa and Zahner.

Altmann and Lang (1973) therefore developed a preference-testing apparatus in which both the electrophysical conditions and the possibilities of orientation could be strictly controlled, in order to reinvestigate the reaction of small mammals to electric fields. Three identical plastic cages were interconnected by plastic tubes, and food, water and nest material (wood fiber) were available in esch cage. The food and water consumption and the location of the nests could be monitored and measured without disturbing the experimental animals. To preclude directional orientation by compass bearings, the entire arrangement was rotated by gearmotor 360° in 5 min and then back again. Means were available for creating normal conditions or shielded conditions in the three cages, as desired. In addition, an artificial electric field could be produced within a shielded cage. To prevent olfactory orientation, all experimental apparatus was thoroughly cleaned after each experiment.

In the first series of experiments, Faraday conditions were produced in the first cage, electrostatic field conditions in the second cage, and an artificial square-pulse field at 10 Hz and 3500 V/m in the third cage. The animals displayed a statistically significant preference for building their nests under Faraday conditions, but preferred the cage with the artificial 10-Hz square-pulse field as their exercise area.

In a second experiment the animals were given the choice between a cage with Faraday shielding, a cage with normal conditions, and a third cage with an artificial 10-Hz square-pulse field. The mice made their nests in the cage with normal conditions and often fed there as well, although the square-pulse-field cage was ultimately preferred as the feeding area. The Faraday cage was preferred from the outset for drinking, while activity measurements showed that the cage with the artificial field was preferred for play activity.

In a third series of experiments, the intensity of the artificial field was reduced to 180 V/m compared with the conditions in the second experiment. The animals now built their nests either in the cage with normal conditions or in that with the artificial field, with no apparent preference. The conditions in each cage were reversed several times during the experiment, but the animals did not react to this unless the compartments had not been thoroughly cleaned beforehand. Under the weaker field conditions, the animals apparently could no longer distinguish between the natural conditions and the artificial field conditions.

In summary, the authors conclude from their experiments that white mice are capable of orientation by means of suitable atmospheric electric fields. However, the sense of smell overshadowed orientation by atmospheric electric parameters in every case, unless special measures were taken to prevent it.

Jahn (1977) also reports on the significance of Faraday shielding:

> Experiments on the rearing of *Lymantria monacha* caterpillars caged under Faraday conditions showed that the shielding of electric fields had an adverse effect on the first instar larvae, but a favorable effect on older ones. This confirmed previous findings that electromagnetic forces tend to promote vitality in the initial developmental stages, but retard it in later stages.

Other experiments on the effects of electric fields on animals were conducted by Haine and König (1960). They observed the molting behavior of plant lice exposed to an artificial electric field at 3 Hz and 10 Hz. For this purpose the plant lice were exposed to the effects of the field for short times at irregular intervals. The number of moltings was determined on an hourly basis. Each electric field differed in intensity from the next by a factor of 10. A control group was kept in a Faraday cage or was exposed to natural field conditions. The result of the experiment is shown in Fig. 4-31; the field was switched on at time *n* according to the "critical hour" method described previously. An optimal, above-average number of moltings was found at an electrode voltage of 1 V (electrode spacing 20 cm). This maximum occurred more or less spontaneously with the switching on of the electric field. On the other hand, a reduction of the plate voltage, and thus the electric field, by a factor of 10 delayed the occurrence of the molting peak by about 1 hr compared with the experiment using a 1 V plate voltage. Evidently the field intensity was still sufficient to produce a maximum, but this did not occur as spontaneously as at the higher voltage. Excessive field intensities tended to inhibit molting in the plant lice. This was found for an electrode voltage of 10 V: no molting peak was observed in this case, and in fact molting tended to show a minimum. The result of this experiment may well be of general importance, because it shows once again that the intensity of such fields is not necessarily a measure of their biologic activity. It appears that there is a certain range (and possibly several ranges) of maximum activity, while excessively high field intensities may show no biologic effects, or effects that are different from those which experience would suggest.

Lott and McCain (1973) report on certain effects of static and pulsed electric fields on the brain-wave activity of rats. They divided the animals into two groups. In the first group, microelectrodes were implanted in the hypothalamus in order to record the electric activity there. In the second group, electrodes were applied to the scalp to record the EEG before, during and after the animals were exposed to either a static field (10 kV/m) or a pulsed field (20 V at 640 Hz/100 ms); this was followed by integration of the total brain-wave energies. Each experiment lasted at least 90 min, during which time all pertinent data, such as EEG, respiration, rectal temperature and ECG, were recorded.

It was found that a significant increase in brain activity occurred after the static field was switched on. This increase persisted throughout the period of exposure. After the field was switched off, the activity decreased in stages to the initial level. Surprisingly, the hypothalamic activity decreased during the exposure and the period that followed.

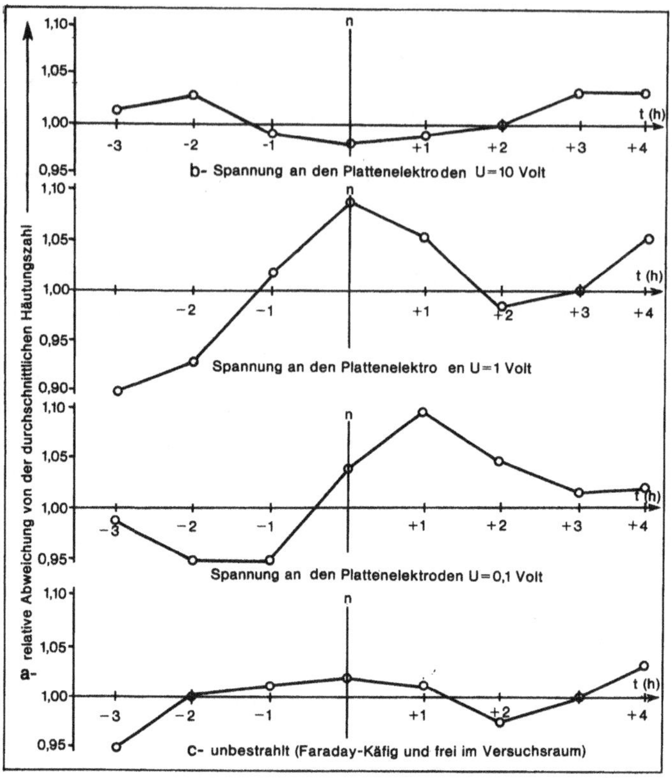

Figure 4-31. Effect of ELF electric fields on the molting of plant lice (*Mycus persicae Sulz.*). The field was produced by means of two electrodes with a spacing of d = 35 cm and voltage U. After Haine and König (1960).
Key: a-Relative deviation from average number of moltings; b-Voltage *U* at the plate electrodes = 10 V (etc.); c-Unirradiated (Faraday cage and normal laboratory conditions)

Exposure to the pulsed field (20 V at 640 Hz) also elicited an increase in electrical brain activity. Moreover, an immediate, statistically significant and persistent rise of activity was measured in the hypothalamic region during exposure. The activities returned to previous levels after the field was switched off. No significant alterations were found in respiration, rectal temperature or ECG during or after exposure to any type of electric field.

König and Krempl-Lamprecht (1959) report on the effects of low-frequency electric fields on the growth of plant organisms. They first experimented with lactobacilli. These bacteria were selected because the known biologic activity of low-frequency fields suggested that the souring of milk during abrupt weather changes and certain weather conditions may be associated with such natural fields.

Experiments with lactobacilli, rod-shaped bacilli of the species *Bacterium casei*, showed an increase in bacterial growth when the voltage was applied to the plate electrodes. The plates were metal with a spacing of 2.5 cm. The experimental solutions were introduced into the field formed between the plates, while control solutions were kept separate under equal light, temperature and other ambient conditions. Although the controls were for the most part outside the stray field of the experimental apparatus, they were apparently exposed to interference-field conditions present in the laboratory. This is evidenced by the fact that the control group displayed a growth rate which was higher than that of the specimen with the lowest field intensity. The treated groups were probably shielded from these interference fields by the electrode plates, so that, on the whole, the field intensity at the lower plate voltages exerted a smaller growth-promoting effect on the bacterial solution than that exerted on the control groups by the stray field present in the laboratory. The success of the experiments also depended to a certain extent on the shape of the glasses into which the solutions were introduced. Better results were obtained with glasses in which the solution had more surface area and less depth. The exact reason for this is unknown.

The same arrangement was used in electric field experiments on yeast cultures (*Saccharomyces cerevisiae*, or common brewer's yeast). The increase in the turbidity of the yeast solutions served as a relative measure of culture growth. The result of the experiment, shown graphically in Fig. 4-32, again demonstrates the effect of the electric fields. The cultures treated with 3.5-Hz fields with a strong harmonic content apparently displayed the greatest increase in growth. The 100-Hz sinusoidal fields lay between this first group and the untreated cultures. The points of equal turbidity (shown as circles in the graph) provide a certain absolute standard for comparison. They indicate that the turbidity (a measure of growth) which occurred after 10 days at a 100-V electrode voltage and frequency of 3.5 Hz with a strong harmonic content took 16 days to appear in the untreated and shielded cultures.

In another experiment, the effect of low-frequency electric fields on the germination of wheat was investigated. Wheat seeds were planted in Petri dishes using equal

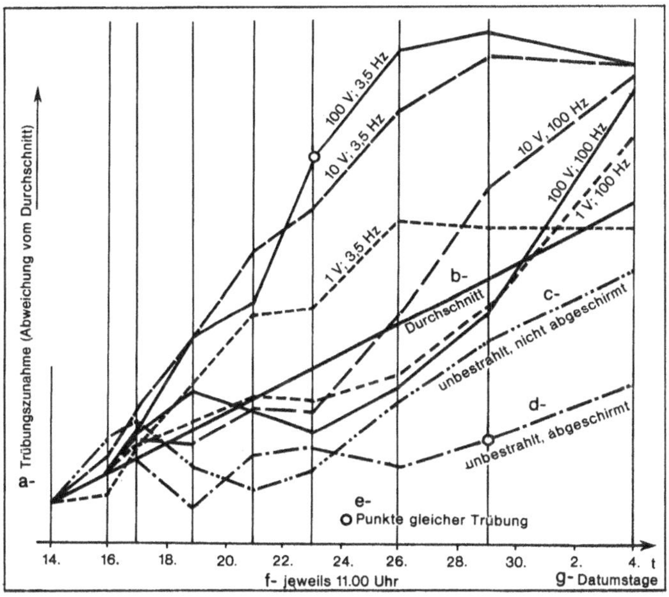

Figure 4-32. Treatment of yeast cultures (*Saccharomyces cerevisiae*) with various electric fields in the ELF range. The cultures were placed between two electrodes with a spacing of d = 7.5 cm. An increase in turbidity corresponds to an increase in growth. Points of equal turbidity (circles) indicate that the unirradiated, shielded controls attained the same state of growth 6 days later than the cultures treated with a 3.5-Hz, 100-V electrode voltage (strong harmonic content), which had reached that state in only 9 days. After König and Krempl-Lamprecht (1959).
Key: a-Turbidity increase (deviation from average); b-Average; c-Unirradiated, unshielded; d-Unirradiated, shielded; e-o = points of equal turbidity; f-at 11:00 on each day; g-Date

amounts of sterile sand and sterile water. Electrodes were attached to the Petri dishes so as to produce a vertical electric field within them. Then the growth of the wheat seedlings and the number of germinated seeds were compared in three different groups: electrode voltage at 3.5 Hz, strong harmonic content; 10 Hz and 25 Hz, sinusoidal. There was also a control. After a germination time of one week, it was found that the fields had practically no effect on the number of germinated seeds. However, the electrically treated germinated seeds had attained a 23% greater length on the average by the end of the experiment than the untreated seeds. The latter had grown to an average length of 67 mm after one week, compared with 81 mm for the seedlings grown in the sinusoidal field, and 84 mm for the seedlings in the harmonic field.

5

Project Seafarer

A large group of scientists in the United States has investigated the biologic effects of electromagnetic fields in the ELF range of frequencies in connection with a pending defense project. The resulting report ("Report of the Committee on Biosphere Effects," 1977) provides a detailed survey of the current state of the science in the opinion of the researchers and experts who prepared it.

The report was prepared by the Committee on Biosphere Effects of Extremely Low Frequency Radiation in response to a request from the United States Navy for a study of the possibility that plants, people, and other animals would be harmed by the electric and magnetic fields associated with operation of the transmitter of the Seafarer communication system proposed by the Navy.

Seafarer (formerly Sanguine) is a system designed to provide communication with submarines and other military facilities from a single transmitting location in the United States. The system would function in the ELF band of the electromagnetic spectrum at about 76 Hz, close to the 60-Hz frequency used in electric power service. At this frequency, the signal could be picked up below the surface of the water, allowing submarines to remain submerged, reducing substantially their vulnerability to detection and attack. Because this is a long-range system with high power, a single transmitter should enable maintenance of essential communication links to strategic ocean areas.

The Navy has specified an operating frequency of 76 ± 4 Hz and maximal field intensities of about 0.1 V/m and 3.7×10^{-5}T (0.37 G) directly over antenna cables in the grid. The fields would be less intense at greater distances from the cables. For the ground terminal areas, the Navy has indicated that the maximal fields would be 15 V/m and 1.1×10^{-5}T (0.11 G).

The report deals with the following problems in connection with the electric and magnetic fields which would be produced by the Seafarer system: shock hazards; genetic effects; fertility; growth and development effects; physiology and biochemistry; cell growth and division effects; serum triglycerides; circadian rhythmicity effects; effects on behavior; influence on fish, bees and other insects, magnetotactic bacteria, birds, plants, and soil organisms.

The Committee on Biosphere Effects of Extremely Low Frequency Radiation cites a large number of investigations in its report and discusses the results. The evidence confirms the rather well-known fact that while biologic effects from electromagnetic fields, especially in the ELF range, cannot be detected in every case, they are present to a significantly high degree. Reference is made to numerous studies in the literature which report specifically on the bioactivity of fields.

Genetics

Gann and LaFrance (1974) have examined the effect of 60-Hz fields on the growth and survival of cultures of subcutaneous cells of the rat. They found that exposed cells invariably died at high field intensities (600 kV/m), but there was no discernible effect of exposure on the cultures at lower field intensities (200 kV/m).

Coate and Negherbon (1970) studied the response of wild-type (Oregon-R) cultures of *Drosophila melanogaster* to exposure for 48 hr to 45- and 75-Hz electromagnetic frequencies at 2.0 G and 20 V/m. Again, the controls were reared under ambient conditions, i.e., 0.6 G and a 60-Hz field of 0.05-V/m intensity. Three treatment groups were involved and three corresponding controls of each. Five exposed males were placed in breeding bottles with 5 virgin Muller-5 females. Their progeny were examined, and then 1 heterozygous female and 5 Muller-5 males were placed together in each of 50 breeding bottles. (This technique varies, it will be noted, from the more common procedure of placing a single exposed male in a breeding bottle. The argument against the use of 5 is that it is not clear how many exposed chromosomes are being tested, because the mating behavior of the individual flies cannot be determined.) Both treated groups exhibited effects consistent with the induction of sex-linked lethal mutations; that is, 5.5% and 4.7% of the cultures in the 45- and 75-Hz fields lacked males (evidence of a sex-linked lethal mutation), whereas all control cultures contained males.

Tegenkamp (1969), also working with *Drosophila*, reported both specific mutations and deviations in the sex ratio in the offspring of flies exposed to fields of up to 520 oersteds (41,380 A/m) for 24 hr. Levengood (1966, 1967a) used a magnetic probe and claimed to find altered developmental times in *Drosophila* that emerged from "treated" pupae. This alteration, expressed only through exposed male pupae, was said to persist through many (up to 30) generations after exposure, but gradually diminished in importance with time.

Marino et al. (1976a) examined the effect of continuous exposure to low-frequency electric fields over 3 generations of Ha/ICR mice. Three groups were involved; a control, a horizontally exposed group (60 Hz at 10 kV/m), and a vertically exposed group (60 Hz at 15 kV/m). Random samples were drawn from each generation of animals to produce the next. The authors found no consistent effect on litter size, but reported an increase in mortality (in the vertical field only) and a decrease in weight gain (again, largely in the vertical field).

Fertility, Growth, and Development

Developing organisms may be more sensitive to environmental conditions than adults. The Committee reviewed reports concerned with possible alterations in animal fertility, growth, and development that might be caused by ELF fields. Table 5-1 summarizes the data from several experiments.

Exposures to 160 kV/m

In a study by Knickerbocker et al. (1967), 22 male mice were exposed to a 60-Hz field of 160 kV/m for 6.5 hr/day, except on weekends and holidays, for a total of 1,500

Table 5-1. Experiments on ELF effects on fertility, growth, and development.

Field intensity	Animal	Effect reported	Investigators
160 kV/m (air)	Mice	No	Knickerbocker et al. (1967)
15 kV/m (air)	Rats	Yes	Marino et al. (1976 b)
1–5 kV/m (air)	Rats	Yes	Dumanskiy et al. (1976)
1.4–3.6 kV/m (air)	Chicks	Yes	Krueger et al. (1972, 1975 b)
0.001–3.6 kV/m (air)	Chicks	No	Durfee et al. (1975)
0.1–100 V/m (air)	Rats	Yes	Noval et al. (1976)
2–100 V/m (air)	Rats	No	Mathewson et al. (in press)
100 V/m (air)	Rats	No	Dumanskiy et al. (1976)
100 V/m (air)	Mice	No	Krueger and Reed (1975)
10 and 20 V/m (clay)	Rats	No	Coate and Reno (1970)
12.5 and 25 V/m (water)	Tadpoles	Yes	McKinney (1973)
0.1 and 1 V/m (water)	Tadpoles	No	McKinney (1973)

hr during the course of 10.5 months. Each male was bred with 2 virgin females per month, and 8 of the litters produced each month were studied after being trimmed to 4 pups each. The weight of each mouse was recorded weekly from birth to 90 days of age. No alterations were observed in the general health, behavior, or reproductive ability of the exposed animals. Necropsies performed after exposure failed to show any pathologic changes.

The number of litters per month did not differ between exposed and control groups, except for a decrease in the number of litters from exposed males in the second month of breeding, which did not appear again when this phase of the study was repeated with a new stock of animals. The only other observed difference between the exposed group of mice and the parallel control group was a slightly smaller weight (4%) in the male (but not female) offspring of the exposed mice.

In a third study, Marino et al. (1976a) exposed 3 generations of mice to 60-Hz, 100-kV/m horizontal or 15-kV/m vertical polarized fields. Mice were allowed to mate and deliver and rear their offspring for 3 successive generations while continuously exposed to the electric fields. According to the authors, mice exposed to the vertical electric field had decreased body weight 35 days postpartum and increased mortality rates for three successive generations. Mice exposed to the horizontal electric field had decreased body weight for 2 successive generations.

Exposures to 0.1–5.0 kV/m

In a study by Dumanskiy et al (1976), male rats (strain and body weight not specified) were exposed to a 50-Hz electric field at 0, 0.1, 1, 2, and 5 kV/m for 4 months (22 hr/day). The rats were exposed in plastic cages (5–15 rats/cage) between parallel electrodes separated by 1 m. Measurements were made after 1, 2, 3 and 4 months of exposure. Clinical signs included alopecia and hyperemia of the mucus membranes of nose and eyes. There was a tendency toward increased weight gain. Necropsy revealed "dystrophic" and vascular changes in the brain, myocardium, liver, kidneys, adrenals, and thyroid. "Normalization" occurred 2 months after exposure. The authors reported that exposure to 1–5 kV/m affected the "functional state" of the nervous system, as indicated by reduction in blood cholinesterase content and reduced number of sulfhydryl groups in the blood, contrary to the authors' assertion, an altered blood cholinesterase content is not a reflection of modified nerve conduction (see also Davidson and Henry, 1969).

Dumanskiy et al. (1976) also concluded that exposure to electric fields (1–5 kV/m) increased adrenal activity and reduced thyroid function. The former conclusion was based on increased concentrations of hormone byproducts (17-ketosteroids) in the urine, and the latter on reduced metabolic activity of the thyroid (slow iodine-131 uptake and release) and lowered thyroxine blood content.

Exposures to 0.1–100 V/m

Noval et al. (1976) exposed young rats for 30–51 days to 45-Hz electric fields at 0.1–100 V/m to determine whether exposure to ELF fields has any effect on development. Exposed rats apparently had a 23%–30% decrease in weight gain, compared with controls. The effect did not appear to appear to depend on field strength. Exposed rats appeared to have less abdominal fat than controls. The activity of the neuronal enzyme choline acetyltransferase was lower in some, but not in other, portions of the brains of exposed animals. There was no change in choline acetyltransferase in controls. Liver tryptophan pyrrolase activity seemed to be increased in the exposed animals, but adrenal weight did not appear to be affected. The authors reported that their findings were consistent with those of Marino et al. (1976a). However, Noval et al. (1976) reported that exposure to a 45-Hz field caused an increase in plasma corticosterone in this experiment, whereas the serum 11-hydroxycorticoid content was generally decreased in the Marino study. The animals were sacrificed by "brief exposure to ether or halothane," both of which have been shown to cause an increase in corticosterone within 2 min. This may account for the high plasma corticosterone values even in the controls.

Development of Chickens

Krueger et al. (1972) studied the growth of chicks exposed to electric fields estimated at 3.4 kV/m (60 Hz) and 3.6 kV/m (45 Hz) or magnetic fields estimated at 1.2 G (60 Hz) and 1.4 G (45 Hz) from birth until 28 days of age. The electric fields were generated by applying 800 V between an aluminum cover plate and the metal cage, producing a nonuniform field distribution whose intensity was not readily measured. They reported that chicks exposed to electric fields at either frequency had a growth rate that was slightly, but not statistically significantly ($p > 0.05$), lower. Statistically significant results were found for the 45-Hz magnetic field exposure ($p > 0.01$). No difference in activity or behavior was detected.

Giarola and Krueger (1974) reported that chicks exposed to 45- or 60-Hz fields of 3.5 kV/m and at 1.3 G had 5% and 11% reductions in growth in the electric and magnetic fields, respectively. No differences in survival or behavior were noted. The cage design was such that a shock hazard existed when the chicks moved to the side of the cage to obtain food or water.

In a later report, Krueger et al. (1975b) exposed groups of four hens and one cock in metal cages (galvanized wire) for 12 weeks to 60-Hz fields of 1.6–3.6 kV/m at $1.4 . 10^{-4}$T (1.4 G). They found a reduction in egg-laying; weight gain, fertility, hatchability, and sex ratio were normal.

Tadpole Metamorphosis

McKinney (1973) exposed batches of tadpole eggs to electric-field intensities of 0.1–25 V/m at 45 Hz in laboratory tanks. The hatchability, time required for hatching, and mortality were determined. Metamorphosis was evaluated by the appearance of hind legs, loss of tail, and survival to adulthood. Tadpole and frog counts were made at 2-week intervals for 21 weeks; by that time, all tadpoles had either undergone metamorphosis or died. There was increased mortality and decreased metamorphosis in the 12.5- and 25-V/m groups compared with controls. Those exposed to 1 V/m or less were not affected.

Flatworms

Marsh (1968) reported that the normal regeneration pattern of flatworms was altered by 60-Hz electric fields of 310–420 V/m. Center sections (head and tail cut off) of two species (*Dugesia dorotocephala* and *Dugesia tigrina*) were placed in agar to regenerate, and some were exposed to 60-Hz electric fields by electrodes in conducting media. The decapitated, de-tailed piece normally regenerates a head at the "head" end and a tail at the "tail" end. In the presence of a 60-Hz electric field (310–420 V/m), this normal regenerative pattern was altered, so that "bipolars" were produced (heads at both ends). The current density to achieve this effect was 0.6–0.8 mA/cm^2—considerably greater than that associated with Seafarer. No abnormal regenerates were produced at lower field strengths (250–300 V/m) and corresponding lower current densities (≤ 0.5 mA/cm^2), which are also considerably greater than those associated with Seafarer.

Sudden Infant Death Syndrome

Eckert (1976) attempted to establish a relationship between sudden infant death syndrome (SIDS) and electric and magnetic fields. The first part of Eckert's paper dealt with investigations in Philadelphia, Pennsylvania, and quoted material from three unrelated references without presenting any data. The second, and major, part of the article reported the author's investigations in Hamburg, Germany. Eckert pointed out that close to each area of high SIDS there may be electric ground currents caused by railroads or well-conducting earth resulting from conditions such as high groundwater level. He indicated that infants living in higher apartments are less vulnerable: 73% of all cases occurred in cellar, ground-level, and second-floor apartments.

SIDS usually occurs in an infant who was thought to be in good health or whose illness appeared to be so mild that a fatal outcome was not anticipated (Adelson and Kinney (1956). There are many possible causes of sudden unexpected death in infancy. A review of over 100 publications related to SIDS has failed to reveal an acceptable cause-and-effect relationship for SIDS (Valdes-Dapena 1967).

The medical journal *Selecta* (1976) also addresses the problem of SIDS in a detailed survey article, with primary emphasis on anomalies of respiration; the acid, base and electrolyte balance; and cardiac activity during sleep. The analysis of cases of "near-miss infants" (those who could still be reanimated) shows that in most cases the children involved were only apparently healthy. If technical energy fields actually trigger a type of stress effect, then a possible connection between SIDS and these fields is entirely conceivable: the technical fields could exert a triggering effect by overburdening a child already subjected to a previous stress.

Triglycerides

The appearance of two reports in 1973 caused some concern about whether the electric or magnetic fields associated with Seafarer cause increases in human triglyceride concentration (Proceedings of the Ad Hoc Committee (1973), Beischer et al. 1973). One paper reported on men and women at the Wisconsin Test Facility, and another on men at an experimental facility at the Naval Aerospace Medical Laboratory in Florida.

Triglycerides constitute one of the three major lipid groups found in the plasma. Dietary fat is predominantly triglycerides. Approximately 1–2 g triglyceride per kilogram of body weight is consumed daily by the average person. These triglycerides undergo partial hydrolysis in the intestinal lumen and are taken up with other lipids into the mucosal cells, where they are reformed and assembled into particles called chylomicrons. The chylomicrons enter the systemic circulation via the lymphatic system and are removed from the blood by the action of lipoprotein lipases. Chylomicron triglycerides are described as exogenous in origin.

Triglyceride may also be of endogenous origin, in the form of very-low density lipoproteins (VLDL). These are synthesized mainly in the liver from such substances as free fatty acids and excess carbohydrate. VLDL are also removed from the circulation through lipoprotein lipase activity. This process is associated with the formation of low-density lipoproteins (LDL).

Three experiments were undertaken by Beischer et al. (1973). In the first, 2 subjects were confined to a chamber together for 3 days. A 1-G (45 Hz) field was applied during day 2 for about 10 hr. Preexposure triglyceride values were obtained 9 days before confinement; both subjects' triglyceride values were higher after the exposure. One subject was a type IV hyperlipidemic. There was no dietary control. There were no control (sham) subjects.

In the second experiment, 2 persons were confined and exposed; an additional subject served as a sham control. One experiment subject and the sham control were hyperlipidemic; the second experimental subject was apparently normal. Diet was restricted to a choice of meals prepared by a Navy hospital. Confinement was for 7 days, with a 22.5 hr 1-G (45 Hz) exposure halfway through. Preconfinement triglyceride values were minimal in number (one or two). One exposed subject showed a large "spike" in triglyceride values 24 hr after exposure amid otherwise normal values.

The third experiment (7 days long with a 1-G exposure for 22.5 hr halfway through) involved several sets of participants. In the first exposure set, 2 subjects' triglyceride values were ascending before the field was turned on and continued to rise until 48 hr after field application, placing them above the "normal" range. The next pair of exposed subjects had increased triglyceride 48 hr after exposure (with one above normal); both subjects had relatively stable values before field exposure. However, the 2 subjects had an argument during the fourth evening, and their increases could be attributed to the stress of the argument and confinement. Two subjects served both as controls and as experimental subjects, and their values show no consistent pattern with respect to whether the field was on; there was considerable and comparable variability in their triglyceride values. None of the control subjects had any abnormal values, but there was considerable variability.

Magnetotactic Bacteria

Magnetotaxis in bacteria (i.e., bacterial motility related to magnetic fields) was first reported by Blakemore in 1975. Bacteria responded to weak magnetic fields by orienting so as to align themselves with the field and then swam in the direction in which they had oriented. Such cells were first found in marine marsh muds and in surface layers of muds collected at a depth of 15 m in Buzzards Bay, Massachusetts. Similar forms were later found in the sediments of the Baltic Sea (R. S. Wolfe, personal communication). Magnetotactic bacteria also occur in freshwater pools and acid bogs on Cape Cod, Massachusetts, at population densities of 200–1000 cells/ml. Moreover, a surprising variety of morphologically different forms of these bacteria was observed in each environment. These observations suggest that magnetotactic bacteria are cosmopolitan, include a variety of species, and live in environments conspicuously different in salinity, pH, and other important characteristics.

By reversing the local geomagnetic field with Helmholtz coils, Blakemore (1975) demonstrated that the bacteria respond to both the horizontal and the vertical components of the earth's magnetic field. In more recent (unpublished) experiments, Blakemore and Kalmijn have studied the orientation capability of these bacteria as a function of the strength of applied uniform magnetic fields up to 10 times as strong as that of the horizontal component of the local geomagnetic field. They also found that magnetostatic bacteria behave in a manner expected of permanent magnetic dipoles. Cells immediately turned and moved in the opposite direction when a strong (0.03T = 300 G) remagnetizing pulse (1 μs duration) was applied. This response was assumed to reflect a change in the polarity of permanent magnetic domains associated with each cell. Thus, magnetotactic bacteria are, in reality, magnetic bacteria.

The instantaneous turning response of an entire cell by a change in the direction of a weak applied magnetic field (0.1 G or less) suggests that ferromagnetism is involved, and not diamagnetic or paramagnetic forces. Consistent with this, magnetic bacteria from both freshwater and marine environments were found to possess unusually large amounts of iron. This iron was contained in previously undescribed, structured particles in each cell. The particles have not been purified; consequently, their chemical composition and physical properties remain unknown.

Although the magnetotactic bacteria respond to constant magnetic fields as low as those associated with the proposed Seafarer antenna, the effects of alternating, 75-Hz, 0.13-G fields have not been studied. Thus, the possible effect on these bacteria of magnetic fields associated with the Seafarer antenna is impossible to assess from existing data. There might be interference with normal cell motility and cell distribution in the vicinity of the antenna or its terminals.

Blakemore (1975) speculated that magnetotaxis might direct these bacteria downward to anaerobic areas more favorable to their growth. In the absence of evidence of important ecologic changes associated with test environments, there may be no great cause for concern with respect to Seafarer. However, the need for answers to some questions, including the effects of altered and imposed small fields on the survival and behavior of magnetotactic bacteria, invites continued experimentation.

Bird Orientation and Navigation

To confirm sensitivity to the earth's magnetic field, Bookman (1977) designed a flight tunnel in which unrestrained pigeons could be trained to discriminate between a simulated field of 5×10^{-5}T (0.5 G) and a strongly reduced field of 2 ($\pm$ 1)$\times 10^{-6}$T (0.02 $\pm$ 0.01 G). The flight tunnel, about 3 m long and built inside a Faraday cage, terminated at paired feeding stations with concealed food bins. Birds were trained to travel the length of the tunnel and enter one feeding bin in the presence of the 0.5 G field and the other in the presence of the reduced (0.02 G) field. Mated pairs showed greater activity in the chamber than single birds, so they were used as test subjects; 3 pairs were tested. After 2 weeks to permit stimulus-reward association, a trial was initiated by release of the birds at the tunnel entry; they were observed with closed-circuit television. The first entry of the first bird into a feeding box was recorded as the datum. Incorrect response was punished by retaining the bird in the box for 30 s before permitting entry to the correct feeding box. The trial ended when both birds had entered and fed and were removed from the correct box. Birds were recorded as either walking ("no flutter") the length of the cage or "fluttering" (more than 3 s jumping, hovering, or in flight).

Records for all performances suggested significant discrimination between coil states.

Zugunruhe Experiments

An interesting set of experiments has been carried out by Southern with the ring-billed gull (*Larus delawarensis*), a gregarious species for which adults and young were available in large numbers at a Lake Huron site near Rogers City, Michigan. Banding studies (Southern, 1971) had shown generally southeasterly fall migratory movement of adult birds from that colony. Juveniles on their first flight showed a preference for east-southeast (mean angle, 112°; Southern, 1969, 1972) suggesting an innate approximate directional preference determined by the requirements of a future migration. Indeed, in work with younger and younger birds, to chicks 2–3 days old, a directional preference for southeast (walking or flying) continued to be demonstrable statistically. With a non-magnetic 8 ft (2.4 m) orientation cage at the Michigan site, but away from the colony, single chicks were released at the center of the cage; after 2 min, or at the perimeter, the direction of the bird relative to the release point was measured. Means for performance clusters were used to indicate preference direction. Southeasterly headings (mean angles) were selected to a statistically significant extent by chicks tested under both clear and overcast skies and during periods of moderate stability of the earth's magnetic field (K value 3 or lower; $< 4 \times 10^{-6}$G deviation; Southern, 1971, 1975). At greater magnetic disturbance (K value 4–7, 4×10^{-6} G to over 20×10^{-6} G), increased dispersion and randomness of headings were noted, the Rayleight test being used to demonstrate the absence of statistically significant directional heading.

Homing Experiments

In later experiments, Keeton (1971) attached a small bar magnet weighing about 3 g (magnetic field strengths: at poles, about 0.025 T [250 G]; at bird's head, about 5×10^{-5}T [0.5 G]) to the back at the base of the neck of each pigeon in the experimental groups. Controls had brass bars of similar size and weight. The "magnet" and "brass" birds had been raised and maintained in the same enclosures and were exercised together; the bars were attached just before testing. The birds were moved in closed vehicles from the loft to the release site, where they were tossed singly into the air in random directions. Compass bearing of the vanishing point, time to vanishing point (10×50 binoculars were used), and homing speed were recorded.

In a group of tests, experienced birds were released under sunny skies, some at familiar sites, others at unfamiliar sites. Within each group, the experimental birds carried magnets; the controls, brass bars. Both experimental birds and controls took appropriate homeward bearings, homing at similar speeds, although in one test the magnet birds took longer to reach the vanishing point. Thus, magnets failed to disrupt homing. Under conditions of heavily overcast skies, in 5 of 7 tests, the magnet birds scattered randomly (or nearly so), and the controls took the proper home bearing.

The evidence is now convincing that some birds are able to sense the presence of magnetic fields whose strength approximates the strength of the earth's field. The lower limits of detection have not been established; the initial finding that systematic deviation of mean orientation bearing in homing pigeons is an inverse function of the K value of the earth's magnetic activity suggests sensitivity to some small fraction of the natural earth field.

Plants

Rosenthal (1975) germinated and grew sunflower seeds in greenhouse and growth-chamber environments exposed to an electromagnetic field of 1 G and 10 V/m in air at 75 Hz. Measurements were made on many indexes of growth, such as percentage of germination and length and weight of roots and stems. The author concluded that there were small but statistically significant differences in seedling mortality, stem, and root length brought about by the ELF field.

Behavioral Assays of Possible Weak ELF Fields

Gavalas-Medici and Sagan (1977) reported that the behavior of monkeys exposed to weak (1–100 V/m) ELF fields (7 Hz–75 Hz) showed minor but systematic perturbations (shortening of interresponse times, IRTs). Furthermore, there was some indication of dose-dependency within a limited range of voltages (0–56 V/m p-p). The lowest threshold appeared for a field frequency of 7 Hz; other (higher) frequencies that were tested appeared to have substantially higher voltage thresholds. It was suggested that this reflected the biologic relevance of the 7-Hz field, since 7 Hz is in the range of hippocampal theta for alert monkeys performing this kind of task. An anomaly in the data was the absence of an effect for 60 Hz on the one occasion when that field was tested. A second point of ambiguity was the lack of a systematic effect at 100 V/m; this may be explained by a 24-hr "carry over" effect at that level.

Recent evidence for subtle but reliable behavioral perturbation as a function of ELF exposure comes from the work of Williams et al. (1977). They have observed that migratory birds near the test facility antenna showed a deviation in flight direction of 5°–25° when the N–S axis of the antenna was energized. The field was estimated to be 0.17 V/m rms at 10 m, perpendicular to the antenna. Indications were that flight direction was rapidly corrected after the birds passed the immediate area of the antenna. These results appear to be very compatible with the monkey ELF studies. In both cases, rather small but reliably measured behavioral changes could be discovered and in both cases the effects seem to be short-lived.

Many of the other studies that have been done on mammals have revealed only negative results (de Lorge, 1972, 1973b; Marr et al., 1973). However, a careful review of these studies in the light of some of the information available within the past few years suggests that some of these inconsistencies may be readily resolved if two major methodological issues are considered. The first of these is simply the differences in field parameters involved in the different studies, and the second centers on the precise nature of the behavioral assay used.

Optimizing Field Parameters For Behavioral Detection Of Weak Electric Fields

Field parameters may be readily subdivided into frequency (modulation frequency and carrier frequency), voltage or power, and duration of exposure. The data from the ELF monkey experiments suggest that behavior may be especially susceptible to EEG range frequencies. Results presented elsewhere suggest that there is a "tuning curve" of modulation frequencies for VHF fields used in elicitation of significant changes in calcium efflux from in vivo chick brains. Similar studies with ELF fields again implicate frequency-specificity in the interaction.

Thus, the background research to date suggests that frequency may be critical in assessing possible behavioral changes. A feasible working hypothesis is that the lowest threshold for behavioral change will be observed with frequency set at a level which is biologically relevant, e.g., within the EEG range of a given animal during performance of the specified task. Note that this cautions, for example, that dominant EEG frequencies observed in sleep may be without effect for an animal in training for a particular behavioral task.

Voltage levels are a second primary

parameter. In this case, the evidence is not clear-cut. In the monkey ELF behavioral studies, there was some evidence for dose-dependency with a larger IRT shift observed as voltage was increased. In the calcium efflux studies on neonatal chicks, on the other hand, there appeared to be a voltage window; no effect was observed at 1 V/m or 100 V/m but statistically significant changes were seen at 10 and 56 V/m p-p. In the shark studies of Kalmijn (1971), data were also presented that implicated a voltage window. These findings suggest that not only is voltage level critical but that one cannot assume that a negative result with relatively high ELF voltages implies no effect at lower levels.

Duration of exposure is an obvious variable that has often been overlooked in ELF studies. In view of the pervasiveness of weak ELF fields in human environments and the absence of catastrophic behavioral changes, it seems to be unreasonable to expect that brief exposure to weak ELF fields might perturb animal behavior. Yet exposure durations as short as 1 min have been used in studies cited as negative evidence (Marr et al., 1973). Our ELF studies have maximum exposure durations of 4 hr. Few studies have used days or weeks of exposure. The short exposure times used in some studies may reflect the fact that they have often been modeled on microwave research, in which behavioral effects appear to be more immediate.

6

Biologic Activity of Higher-frequency Fields and Chemophysical Reactions

Biometeorology and VLF Atmospherics

The contributions of Reiter (1953) to biometeorologic research must not be overlooked. In studies on the problem of weather dependency in man, he explored the basic problems of biometeorologic research using "biometeorologic indicators." Ever since the fundamental work of De Rudder (1952) and others, changes in air masses and the passages of fronts have been recognized as the meteorologic basis of bioclimatologic research. Reiter (1953) further claims that apparently quite specific atmospheric electric factors should be regarded as biometeorologic indicators and cites statistical studies and reaction-time experiments, notably by Düll (1935), in support of this. It has long been assumed that the higher incidence of deaths and suicides on days with chromospheric eruptions on the sun may have to do with the simultaneous increase in long-wave atmospheric disturbances. An extremely strong increase in radiation is also known in such cases. Reiter (1954) also reports on investigations with the following "biometeorologic indicators": (1) temporal disturbances in the static atmospheric electric field, (2) atmospheric long waves ("infralong waves") in the range from 10–50 kHz (now called VLF atmospherics), (3) atmospheric long waves of 4–12 kHz.

Based on correlative studies on the reaction time of healthy persons and VLF atmospherics, Reiter (1954) also gives a survey of the results of investigations (Fig. 6-1). The figure clearly shows correlations between VLF atmospherics and a variety of parameters based on numerous experiments and a large volume of data. As mentioned earlier, a large number of reaction-time measurements made at an automobile exhibition provided a suitable basis for statistical evaluations. The most important result of these is given in Fig. 6-2, which shows the times of the sudden onset of VLF atmospherics by the synchronization method. An increase in reaction time of about 20 ms is already seen within a period of 30 min.

Using data obtained by the recording of VLF atmospherics, Damaschke and Becker (1964) also did correlative studies on the respiration rate of termites as related to changes in the pulse repetition frequency of atmospherics. A clear connection was found between the O_2 consumption of the

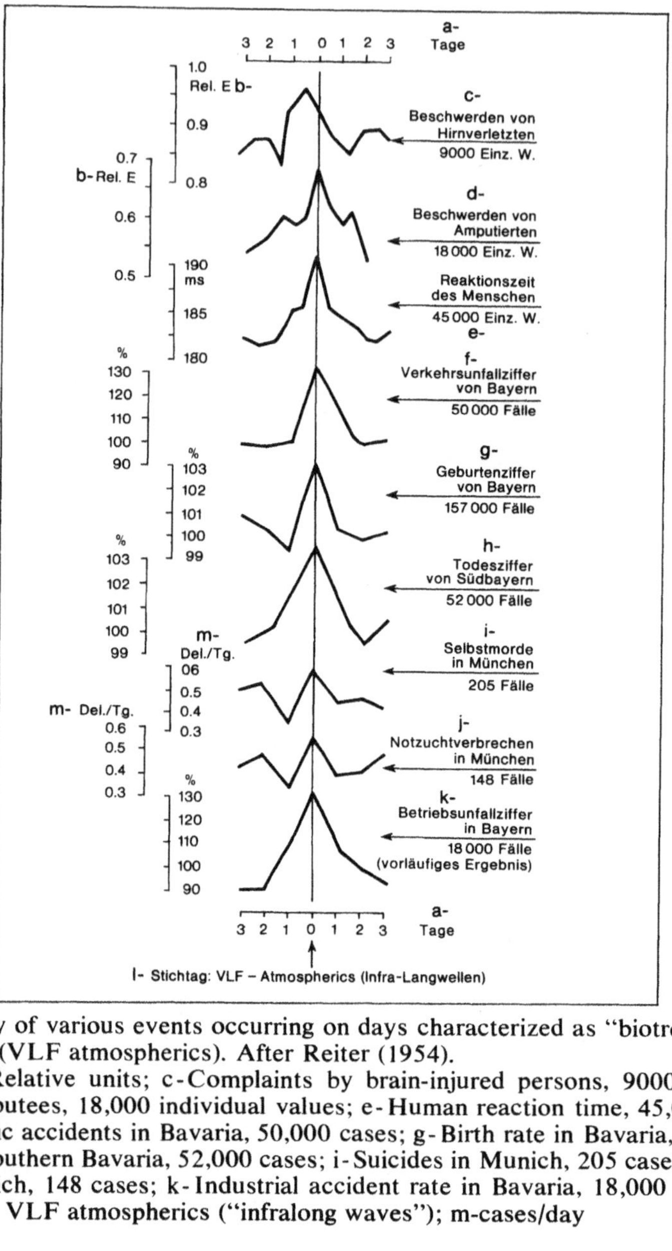

Figure 6-1. Survey of various events occurring on days characterized as "biotropic" by infralong-wave disturbances (VLF atmospherics). After Reiter (1954).
Key: a-Days; b-Relative units; c-Complaints by brain-injured persons, 9000 indiv. values; d-Complaints by amputees, 18,000 individual values; e-Human reaction time, 45,000 individual values; f-Rate of traffic accidents in Bavaria, 50,000 cases; g-Birth rate in Bavaria, 157,000 cases; h-Mortality rate in Southern Bavaria, 52,000 cases; i-Suicides in Munich, 205 cases; j-Crimes of sexual assault in Munich, 148 cases; k-Industrial accident rate in Bavaria, 18,000 cases (preliminary result); 1-Key day: VLF atmospherics ("infralong waves"); m-cases/day

termites and the pulse repetition frequency of the VLF atmospherics. Increases in respiration usually occurred when the atmospherics displayed minima independent of the daily cycle, while decreases in O_2 consumption were observed during maxima. Changes in respiration rate were also observed in artificial electrostatic fields 10 to 100 times more intense than the natural atmospheric electric field.

In a survey, Undt (1962) makes reference to studies by Wedler which showed an unusually high mortality rate in West Berlin from July 10–14, 1959, which could be clearly correlated with pulsed atmospheric radiation (VLF atmospherics).

Ranscht-Froemsdorff (1966b) and Ludwig and Mecke (1968) report on experiments with a golden hamster exposed to an artificial field corresponding to that created

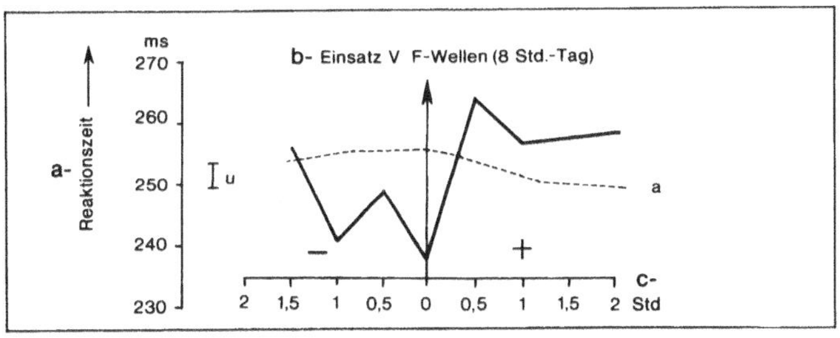

Figure 6-2. Increase in the reaction time of test subjects during increased VLF atmospherics activity (passage of 6 cold fronts and 1 trough). After Reiter (1954).
Key: a-Reaction time; b-Onset of VLF waves (8-hr day); c-Hr

by VLF atmospherics during winter weather. The running activity of the hamster decreased until the animal fell into a state resembling hibernation. When the artificial field was later modified to simulate the VLF atmospherics characteristic of summer, the hamster spontaneously awoke and worked off its artificially pent-up motor energy in a frenzy of activity that lasted 187 hr (Fig. 6-3). The result of this experiment is taken as proof that VLF atmospherics have an effect on higher nervous activity, at least in animals.

The artificial VLF atmospherics used here were produced in a room shielded from about 99.8% of natural long-wave radiation of this type, and had the following characteristics: electric field intensity 1 V/m; magnetic field intensity 2.65 mA/m; phase angle between the two field types 0°; carrier frequency 100 kHz; pulse repetition frequency 5 or 10 Hz, a range typical of biologic processes. It was found that the amplitude and carrier frequency of the artificial radiation were of minor importance, which is consistent with the theory of Ludwig (1968). Based on published data on the electric properties of nerve cords and synapses, Ludwig (1968) suggests a mechanism whereby pulsed atmospheric signals are picked up by the nerve conductors and processed further at special synapses. In the case of the hamster, the statistics of the repetition frequency (irregular pulse sequence) were of considerable importance, with the higher-frequency components apparently playing a special role. A small statistical distribution of the different

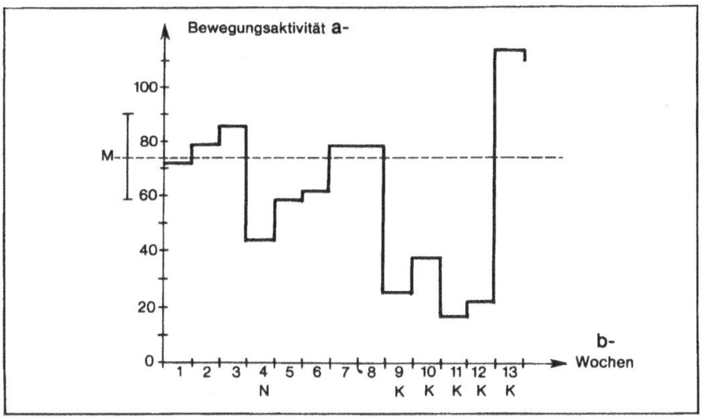

Figure 6-3. Motor activity of the golden hamster with and without a program of artificially produced VLF atmospherics. M: mean motor activity without program; after N: natural atmospherics switched on; K: artificial atmospherics. After Ludwig and Mecke (1968).
Key: a-Motor activity; b-Weeks

pulse repetition frequencies (i.e., a nearly constant repetition frequency over a prolonged period of time) resulted in a passivization of motor activity.

In contrast to the apparently sensitive response of mammals to atmospherics, no effect could be demonstrated in human subjects. Only when the subjects were subdivided according to a bipolar scheme could it be shown that the labile types responded reproducibly to artificial atmospherics. In labile vagotonics, for example, an increasing "giddiness" was observed after the artificial field was started. When the field was silently switched off, this state quickly disappeared (reproducibly to 1 s).

Such significant results prompted Ranscht-Froemsdorff and Weise (1969) to develop portable narrow-band receivers for measuring VLF atmospherics in the range around 10 kHz and 100 kHz. Measurements in terrain showed that the field intensity and number of 10- and 100-kHz pulses increased with elevation above sea level. VLF atmospherics were also found to vary with orographic conditions: rising ground screens out the shorter-wave atmospherics (100 kHz) more strongly than the longer-wave atmospherics (10 kHz)—an effect known primarily for the ultrashort-wave region (Fig. 6-4).

Certain results were puzzling. The field intensity of the VLF atmospherics was found to increase spontaneously over certain relatively circumscribed points on the ground. This was attributed to resonance zones caused by the presence of ground water or rock strata, thereby suggesting a possible new interpretation of divining-rod phenomena. However, the assumption of resonance zones could hardly account for these anomalies, because of the very small scale of the phenomena compared with the wavelength of the VLF atmospherics. It would appear that the explanation lies partly in the anomalies of ground or air conductivity which are known to exist in such areas.

Lotmar et al. (1969) showed significant correlations between VLF atmospherics and the respiratory quotient ($Q0_2$) of rabbit skin on 147 different days of experimentation. This led the authors to assume that pulsed atmospheric radiation (VLF atmospherics) is not only an indicator, but is also a biotropic weather factor that directly influences the degree of tissue respiration in vitro.

The liver tissue of white mice was similarly tested using artificially produced fair- and bad-weather pulse frequencies corresponding to natural VLF atmospherics ac-

	e-	f-	g-
b- 10 KHZ-Bereich	5-10 Imp/min	80-100 Imp/min	0.1-0.5 Imp/min
c- 100 KHz-Bereich	1-4	10-50	>0.01
d- Feldstärke	~12db	>20 db	0 db

Figure 6-4. Measurement of VLF atmospherics over water catchment areas shows a field intensity increase up to 20 db or more (factor of 10) within a small distance. After Ranscht-Froemsdorff and Weise (1969).
Key: a-Reception of sferics as a function of orography. Broad valley with stream, 300 m high, east-west course, weather phase 5–6, summer; b-10-kHz range; c-100-kHz range; d-Field intensity; e-5–10 pulses/min; f-80–100 pulses/min; g-0.1–0.5 pulses/min

tivity. The experiments were conducted in a chamber shielded from this type of radiation. Exposure to artificially produced bad-weather signals (3–100 pulses/s, intensity 0.1 V/m, fundamental frequency 10 kHz–100 kHz, simulating cyclonic weather conditions) resulted in a significant (42%) reduction of liver-tissue respiration. A magnetic-field program (Lotmar et al., 1969) and programs simulating anticyclonic conditions had no effect, however. On the whole, the results prove once again that the atmospherics which are often trivially termed "weather radiation" must be of significance biologically. It remains to be seen, however, whether the effects observed in vitro are analagous to cellular effects in vivo. In any case, the control values were obtained in appropriately shielded experimental chambers and have important implications for the problem of weather sensitivity and modern construction with metal and reinforced concrete.

Ranscht-Froemsdorff and Rinck (1972) report on the biologic effects of VLF atmospherics as a function of the occurrence of minimum or maximum pulse frequencies in connection with their study of the electroclimatic phenomenon of the foehn. They begin by stating that the "electric climate" created by atmospherics is known to have biologic effects when the intensity of the radiation increases. This suggests that similar perturbations may occur when radiation is deficient. Now, according to Ranscht-Froemsdorff and Weise (1969) and Reiter (1960), the pulse incidence tends to be extremely low under so-called zero weather conditions as well as under conditions of subsidence or Alpine foehn. As expected, the results of a study by Brezowsky and Ranscht-Froemsdorff (1966) showed that the incidence of myocardial infarctions is highest when the "electroclimatic radiation" is at a minimum.

In further experiments by Rinck (1972), young subjects were placed individually into special electroclimatic chambers, where they were exposed to various artificially produced atmospherics programs in three different temperature ranges. Blood samples taken during the experiment showed a significant change in blood coagulation under field exposure only at a moderate room temperature on the second and third days. However, switching off the electrical climate (foehn conditions) caused a significant change in coagulation values only at a relatively high room temperature on the second and third days. Blood coagulation was accelerated and delayed with the course of the simulated electroclimatic ideal cyclone, the temperature again playing an important role.

It should be noted, of course, that these results of blood coagulation studies in young, healthy subjects cannot be directly applied to the elderly or infirm. Nevertheless, it can be stated with certainty that as physical lability increases, the blood coagulstion disturbances caused by electroclimatic conditions may reach pathologic values within a matter of hours, as short-term experiments indicate.

Tromp (1970) also made observations of the biologic activity of VLF atmospherics over a period of years. He found striking correlations between atmospherics activity and blood sedimentation in human subjects.

Supplementing these studies are experiments performed by Senatra et al. (1978) to detect a possible effect on in vitro irradiated human blood samples. The erythrocyte sedimentation rate (ESR) and the bioimpedence of blood samples treated with a low-level 1.59-kHz electric field and untreated blood samples were measured. A statistically significant difference ($p < 0.001$) was found in the blood sedimentation rates between the controls and those exposed to VLF field conditions. The most striking result was that for 20% of the blood samples analyzed under the influence of the VLF field, the ESR tests were almost stopped by the latter, i.e., no blood sedimentation occurred. The bioimpedance values of "stopped" blood samples were one order of magnitude lower than those of unirradiated samples.

Jahn and Nessler (1971) report on the effect of ultralong radio waves (VLF range). Laboratory experiments based on observations in the field show that the mortality and fertility of the black arch moth can be influenced by such electric or electromagnetic fields. This is considered surprising by the authors, because the discovery that insects react to magnetic influences and are apparently capable of orientation by magnetic fields is a relatively recent one. It has even been shown that the life cycle of the black arch is controlled in part by these influences.

Besides these observations in the more classical VLF range, reports have also been published on investigations of somewhat higher ranges. Of particular interest in this regard are observations by Plekhanov and Vedyushkina (1966) on the production of a conditioned muscular reflex by means of electromagnetic fields at 735 kHz, which required field intensities less than 1 mV/m. These intensities are in a range which is characteristic of natural fields.

Budko (1964) investigated the effect of electromagnetic fields at frequencies from 9.5 to 9500 kHz on the blood sugar level in rabbits. In these experiments, only the animal's head or liver region was irradiated with a 1.5 kV/m (capacitor-produced) field for 20 min. The blood sugar values after such an irradiation are plotted in Fig. 6-5, which shows: (1) irradiation of the head caused a much greater change than irradiation of the liver region; (2) the sugar level rose at field frequencies from 10 to 100 kHz, while it fell at higher frequencies (> 1 MHz); (3) in the intervening frequency range, the field had practically no effect. In all cases the blood sugar level began to return to its normal value 20–30 min after the end of the experiment and reached it in 60–90 min. The cause of the opposite effects of irradiation at relatively high and low frequencies is presumed to be the different frequency-dependent depths of penetration of the radiation into the liver and brain tissues. The blood sugar level could also be influenced by the use of decimeter and centimeter waves.

Attempts have also been made to show biologic effects of electromagnetic fields in isolated tissues and cells. Budko (1964), for example, employed frequency-dependent irradiation to test the glucose content of the liver in the range from 500 kHz to 21.5 MHz at a field intensity of 1.5 kV/m. He showed that the glucose level was somewhat higher than the control at the lower irradiation frequencies, initially fell to 80% of that value with increasing frequency, but then rose continuously with increasing frequency to 100% of the control at 1 MHz and even to 120% at 21.5 MHz.

The effect of field intensity was also tested at four fixed frequencies: 9.5 MHz,

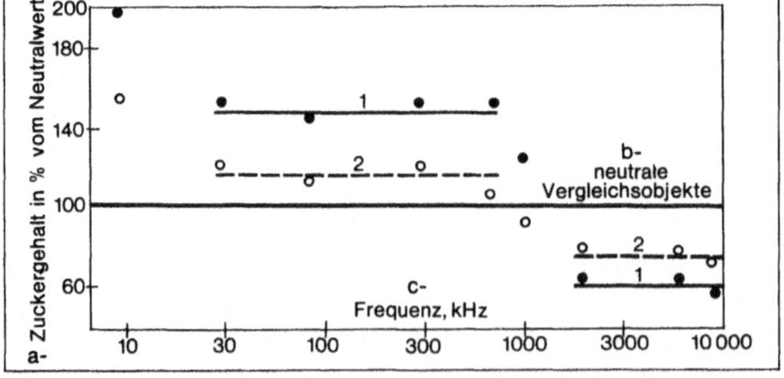

Figure 6-5. Change in the blood sugar level of rabbits irradiated for 20 min with electromagnetic fields in the frequency range from 9.5 kHz to 9.5 MHz; (1) irradiation of head region; (2) irradiation of liver region. After Budko (1964).
Key: a-Sugar level in % of neutral value; b-Neutral controls; c-Frequency (kHz)

730 kHz, 85 kHz and 29 kHz. An intensity-dependent effect was found with regard to the glucose level. Values from 1000 to 4000 V/m served as comparative field intensities. At the highest frequency, a glucose maximum could be shown at about 1.5 kV/m, which exceeded the other values by 6%–7%. At all other test frequencies, a glucose minimum (10%–15% below other values) was observed at the same field intensity (1500 V/m).

High-frequency and Microwaves

Since high-frequency energy and water can apparently be of importance in connection with location-dependent biologic effects, we shall begin our discussion with experiments by Peschka (1974) which deal with dynamic effects in test specimens consisting of high-frequency oscillating circuits or circuit elements, with water as the dielectric.

Essentially, the specimens represent oscillating-circuit elements (i.e., inductances and capacitances for high frequency) or high-frequency circuit components such as quarter-wavelength waveguides. The oscillating circuits or circuit components were set up in a water medium and, together with a container of insulating material (generally Plexiglas), comprised the entire experimental appangement. The object was to demonstrate that the specimens show dynamic effects during the inductive or capacitive coupling of high-frequency energy to them. Each specimen was attached to a torsion balance, so that any dynamic effects would produce a deflection. As a precaution, extensive trial runs were carried out with two torsion balances prior to the actual experiment. No coincidence was found either with temperature fluctuations or movements of the building. Electrostatic effects were also ruled out.

In the actual experiment, de-aerated water (primarily tapwater, but also desalinated water) at room temperature was introduced into the specimen under a water-jet pump vacuum. The water remained in the sealed specimen throughout the duration of the experiment, generally 4–6 weeks. After checks were made by means of additional trial runs, the actual experiment began: activation of the specimen by capacitance-coupled high-frequency energy (generally for about 1 min). In some cases the activation was repeated several times in intervals of a few minutes. The power input was about 20 microwatts to 100 milliwatts. Over 200 such experiments were performed over a 3-year period using two torsion balances, one of which carried only ballast. No coincidence was observed in any case.

It was found that activation led to maximum deflections of the torsion balance with torque values up to 10^{-7} N × m, but that most deflections were in the range of about 25 × 10^{-9} N × m. These deflections, which persisted for 1–2 hr at minimum power imput even after the high-frequency energy was switched off, can be attributed to no previously known effects: neither electrostatic, magnetic, nor hydrodynamic influences can satisfactorily account for them. Moreover, it is unclear whether the deflections occur solely on the basis of a force or solely on the basis of a turning moment. The energy frequencies suitable for use (which lie in the range of electronic frequencies characteristic of the experimental apparatus) must be kept constant to about 10–100 Hz for 1 min. Evidently there are several critical frequency ranges which appear to be discrete and can produce a response in the specimen. These are 30–40 MHz, 120–130 MHz and 200–350 MHz. A pure sine-wave input produced no dynamic effects; apparently harmonic components are necessary.

A record of the experimental results revealed previously unexplained phenomena in activated specimens. Specimens activated once or several times still show effects even after several weeks, notably a strong response to high-frequency disturbances (caused by the switching on of fluorescent lamps, resistor overloads, spark discharges at great distances, high-frequency oscillators located outside the laboratory, and lightning flashes). The associated phenomena, which result primarily in deflections lasting from 1–2 s, are called short-term effects. Long-term effects can also be observed in activated specimens, however. In such cases the torsion balance undergoes periodic deflections about 12 s in duration, upon which are superimposed other short-term effects which may originate from external sources of high-frequency disturbances. In many cases (60%–70% probability) the mere presence of persons in the room containing the activated specimens is sufficient to cause deflections of the torsion balance, which may persist for up to 2 hr.

Moreover, the feeding of high-frequency energy to activated specimens is accompanied by mood changes in some experimental subjects which are clearly perceptible but difficult to describe (feelings of irritation, slight disturbances of concentration) and which are not at all comparable to the effects observed in subjects exposed to high-power, high-frequency generators.

The phenomena observed are undoubtedly associated with electromagnetic energies that cause changes in matter by various possible mechanisms of action. A lower limit for the necessary high-frequency power would appear to be present when the high-frequency power input reaches the portion of the thermal noise output falling in the frequency interval of the half-value width. At half-value widths of several hertz, these powers are on the order of 10^{-20} W. In view of these very low powers, it is unlikely that further light will be shed on the problem until extremely narrow-band receivers are available which respond to extremely weak signals below the thermal noise level.

The importance of these findings, which according to Peschka (1974) may open the door to entirely new and unexplored regions of human knowledge, is increased still further if one considers their biologic implications. The production as well as the absorption of electromagnetic energy of such low intensity is possible in the molecular structure of the cell and may well have played a key role in the evolution of life. According to papers published in the U.S.S.R. on this subject (Presman, 1970), this would enable cells to exchange information by the use of electromagnetic waves. Popp et al. (1979) deal with this problem in detail. They report that biological systems can emit electromagnetic waves and also show sensitive responses throughout the whole spectral range, from the extreme long-wave "slow fluctuations" up to the short wave UV area. Resonance activity serves as a basis for communication in several ways: biological systems communicating with each other and their environments; communication with the organism itself; communications on an inter- and intracellular level. We thus see the beginnings of a science which lies in the borderline area between physics, psychology and biology, and which might well play a central role in the future.

Other reports and discussions on the biologic significance of high-frequency energy have also been published. Blanco and Romero-Sierra (1973) examined the possible role of microwave radiation as a means of communication in biologic systems. According to these authors, biologic systems are highly complex generators and receivers of various forms of energy and produce electromagnetic fields spanning a large portion of the electromagnetic spectrum as well as the static electric and magnetic field. These systems employ a variety of methods to communicate with one another and glean information from their environment. Besides the ordinary methods of communication, such as sight, hearing and smell, other unusual methods would be employed, such as the use of microwave radiation, to transmit bioinformation. Their reasoning was based on the premise that all bodies in our physical universe with temperatures above absolute zero are characterized by an electromagnetic radiation.

Table 6-1. *Penetration depth in cm (reduction of intensity to 1/e) of electromagnetic waves in various types of tissue. After Presman (1970).*

Reflection coefficient K for the following transitional layers:	Frequency (MHz)							
	100	200	400	1000	3000	10,000	24,00	35,000
Air-skin	0.758	0.684	0.623	0.570	0.550	0.530	0.470	—
Skin-fat	0.340	0.227	—	0.321	0.190	0.230	0.220	—
Fat-muscle	0.355	0.352	0.3004	0.261	—	—	—	—

For theoretical reasons, an appreciable portion of this energy must be emitted in the microwave range. This radiation from the human body was the subject of further investigations, therefore. It was found that radiation from the body in this frequency range increases strongly as a function of environmental conditions. Further preliminary results justifiably raise the question of whether such radiation could not supply an answer to the questions surrounding bioinformation transmission between biologic systems in normal and pathologic states.

Presman (1970) conducted an intensive, detailed study of the biologic effects of high-frequency electromagnetic fields in particular. One phase of his study dealt with the reflection of such energy. According to Presman, the biologically active power P_e can be described in terms of the power flux density P_o at the surface of an object as $P_e = P_o \times (1 - K)$.

Some values of the reflection coefficient K are given in Table 6-1. Presman (1970) also investigated the percentage absorption of electromagnetic energy in various tissues of the body. The results of his calculations (Fig. 6-6) show that high-frequency energy beyond 4000 MHz is almost completely absorbed by the skin.

The depth of penetration of electromagnetic waves into various types of tissue refers to the depth at which the intensity decreases to $1/e = 1/2.7$. In the skin, for example, the depth of penetration is 3.8 cm at 100 MHz, but only 0.07 cm at 24,000 MHz. In fatty tissue, by comparison, the penetration is 20 cm at 100 MHz and 0.3 cm at 24,000 MHz. In the brain, 3.6 cm at 100 MHz and 0.04 cm at 35,000 MHz are typical values.

Also of interest in this regard: standing waves may form between individual tissue layers as a result of reflections due to dif-

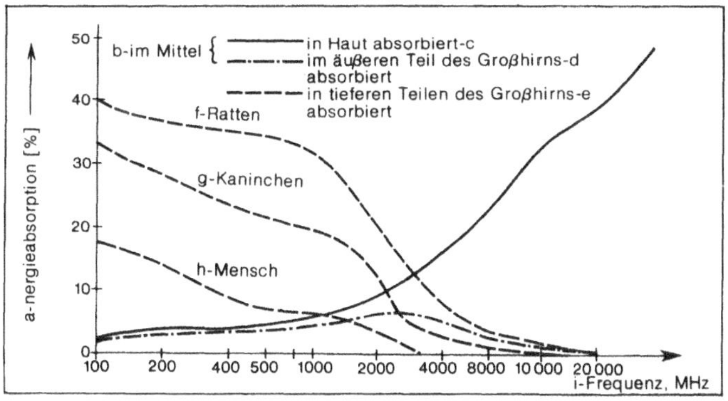

Figure 6-6. Frequency-dependence of the absorption of high-frequency energy by brain structures during irradiation of the head with electromagnetic energy in man, in rabbits and in rats. After Presman (1970)

Key: a-Energy absorption (%); b-Mean values for energy; c-Absorbed by the skin; d-Absorbed by the outer portion of the cerebrum; e-Absorbed by the deeper portions of the cerebrum; f-Rats; g-Rabbits; h-Man; i-Frequency (MHz) ferences in parameters. This phenomenon is explored in greater detail by Fleming et al. (1961), who demonstrates the possible formation of standing waves in the tissue layers of man and large animals under the influence of electromagnetic radiation above 3 GHz.

Where very high-frequency electromagnetic fields are involved, there is always the problem of distinguishing whether the observed biologic effects are caused directly by the action of the field, or by thermal effects. In the present case, of course, nonthermal effects are of principal interest. This presupposes energy values which do not exceed a certain maximum. According to Presman (1970), highly diverse reactions are observed in animals, which in every case can be produced by nonthermally active radiation as well. Interestingly, it could be shown in experiments with dogs that the reactions observed at very high field intensities are opposite to those observed in the presence of correspondingly weak fields.

In another experiment, reported by Pressman (1970) selected groups of rats were irradiated with 1-decimeter, 10-cm, 2-cm and 1-mm waves at 1 mW/cm^2 for 1 hr/day and the reaction of the animals to an audible signal was observed. In all cases the sensitivity of the rats to the audible stimulus was diminished. The degree of this effect decreased with the wavelength.

Presman (1970) gives the threshold values for thermal effects in the tissue of various living organisms (Table 6-2).

Considerable discrepancies have been found between the theoretically estimated and experimentally determined magnitude of the electromagnetic forces necessary to produce responses in biologic systems. This is pointed out by Persinger (1973), among others. For example, electrostatic field intensities of 10^{-5} V/m are sufficient experimentally to elicit electroreceptor responses in fish. This contrasts with values of about 10^6 V/m that considered necessary in theory. Static magnetic field values of 50 A/m have proved sufficient to increase motor activity in birds experimentally, as opposed to values of 10^5–10^7 A/m determined theoretically. Only in the high-

Table 6-2. *Intensity threshold of electromagnetic energies for thermal effects in the tissue of living organisms at various frequencies. After Presman (1970).*

Frequency range	Organism	Type of thermal effect	Intensity threshold
500 kHz	Rat Rabbit	Rise of rectal temperature	8 kV/m 160 A/m
14.88 MHz 69.7 MHz	Rat Rabbit	Rise of rectal temperature	2.5 kV/m 200 V/m
Decimeter	Rat Rabbit	Rise of rectal temperature	40 mW/cm^2 (390 V/m)
10-centimeter	Man	Sensation of warmth	10 mW/cm^2
		Heating of irradiated region	10 mW/cm^2 (190 V/m)
	Rat	Rise of rectal temperature	10 mW/cm^2
3-centimeter	Man	Sensation of warmth	1 mW/cm^2 (61 V/m)
	Rat	Rise in rectal temperature	5—10 mW/cm^2 (135—190 V/m)
	Rat	Heating of the irradiated region	1.5 mW/cm^2 (75 V/m)
Millimeter	Rat	Rise in rectal temperature	7 mW/cm^2 (170 V/m)

Table 6-3. Lethal effect of electromagnetic radiation of various frequency ranges in animals. After Presman (1970).

Electromagnetic parameters	Animal	Field intensity	Duration of exposure, min.	Temperature increase, °C	% lethal outcomes
Constant magnetic field	Fly	9600kA/m	60	—	100
50—500 Hz		650,000 V/m	60—120	—	70—90
50 Hz	Mouse	650,000 V/m	270	—	50
500 Hz		650,000 V/m	90	—	50
14.88 MHz	Rat	9000 V/m	10	—	100
	Rat	5000 V/m	100	—	80
	Rat	4000 V/m	100	—	25
69.70 MHz	Rat	5000 V/m	5	—	100
	Rat	2000 V/m	100	—	83
200 MHz	Dog	330 mW/cm^2	15	5	50
	Dog	220 mW/cm^2	21	4.1	25
	Guinea pig	590 mW/cm^2	20	5.9	100
	Guinea pig	410 mW/cm^2	20	4.2	67
	Guinea pig	330 mW/cm^2	20	4.1	100
	Rabbit	165 mW/cm^2	30	6—7	100
2000—3000 MHz pulsed	Dog	165 mW/cm^2	270	4—6	100
	Rabbit	300 mW/cm^2	25	6—7.5	100
	Rabbit	100 mW/cm^2	103	4—5	100
	Rat	300 mW/cm^2	15	8—10	100
	Rat	100 mW/cm^2	25	6—7	100
	Rat	40 mW/cm^2	90	—	100

frequency range are the theoretically predicted field intensity values on the same order of magnitude as the values derived experimentally. In the microwave range of frequencies, large discrepancies again arise. The intensity of 10 mW/cm^2 considered necessary in theory to produce thermal effects contrasts with the 2×10^{-2} mW/cm^2 necessary to produce a change in the rabbit EEG experimentally.

The question of the biologic activity of electromagnetic fields is not easy to answer. The main reason for this is that in many cases "biologically active" is equivalent to "injurious." Electromagnetic fields are considered injurious in themselves if their use leads to the injury, if not the destruction, of biologic systems, and thus the death of living organisms. Corresponding experiments were performed with animals. The results are compiles in Table 6-3, based on values given by Presman (1970). The lethal outcome of the experiments was independent of thermal effects up to about 70 MHz. Only at frequencies from 200 to 24,000 MHz was death obviously associated with the generation of heat within the body. Power densities of several hundred mW/cm^2 were sufficient to produce this effect.

Presman (1970) also observed that electromagnetic fields can perturb the regulation of blood composition. In dogs, for example, a single exposure to a relatively strong field (2800 MHz, 100 mW/cm^2 and 200 MHz, 165 mW/cm^2) for 6–8 hr led to a 25%–55% increase in the leukocyte count 24 hr after the start of exposure. At the same time, the lymphocyte and eosinophil count fell immediately after the start of exposure, but rose to higher than normal values after 24 hr. The hemoglobin content underwent similar relative changes. Other experiments with dogs using a similar exposure (100–300 mW/cm^2) for 25 min led to a two-phase alteration of the leukocyte count: a reduction immediately after the start of exposure followed by an increase after 2–4 hr.

According to Gorodetskaya (1964), the irradiation of rats with shorter waves (10,000 MHz) at 400 mW/cm^2 for 5 min effected a reduction in the leukocyte and erythrocyte count which persisted for 5 days after the irradiation ceased. Exposing rats to still shorter waves (24,000 MHz) at 20 mW/cm^2 for 7.5 hr led to an increase in the erythrocytes and the hemoglobin level, but to a reduction in the leukocyte count immediately after the start of the irradiation. However, all blood indices began to increase slowly aafter 16 hr and again reached normal levels within 14 days.

The following results were obtained with regard to the thermal effects of high-frequency radiation:

1. Rectal temperature exhibited 3 phases during irradiation: (a) a rise of temperature by about 2°C during the first 25 min; (b) maintenance of this elevated temperature for about 40 min; (c) sustained irradiation apparently leads to an irreversible breakdown of thermoregulation, and the continued rise of body temperature results in death after another 20 min.
2. Irradiation with increasing intensity produces a greater thermal effect at 2800 MHz than at 200 MHz.
3. Anesthesia heightens the sensitivity of the animals to the irradiation.
4. If only the head region is irradiated for a certain period of time, the same rise in body temperature occurs as when the body is subjected to a 25% higher irradiation intensity for an equal period of time.
5. Long-term irradiation (3–6 hr) with field intensities of 100–165 mW/cm^2 induces leukocytosis.

Based on these experimental results, Presman (1970) states that, at least under high-frequency irradiation, the death of the animals cannot be attributed simply to an overheating of the body. It appears that pronounced disturbances also take place in the body's regulatory processes which depend not only upon the quantity of heat generated by the radiation, but also upon the frequency of the radiation, the region of the body irradiated, and the physiologic state of the animal.

Finally, Presman (1970) gives a survey of the effects of microwave electromagnetic radiation on malignant tumors in connection with gamma radiation. An astonishing cure rate was achieved in experiments with mice, although experience indicates that this therapy is still unsuitable for use in humans.

Electromagnetic energy was found to have an effect on the blood pressure of rats even at a frequency of 500 kHz. Over a 10-month period, Nikonowa (1964) exposed a group of animals for 2 hr/day to an electromagnetic field whose electrical component was predominant (1.8 kV/m). In a second group, the magnetic component was predominant (50 A/m). In the first group the blood pressure began to fall after 5 months' exposure and reached a level about 5% below that of the control animals. In the second group the blood pressure fell only after 7 months' exposure, reaching a 10% lower value. This pattern of change in a single phase suggests that the nature and degree of the effect of electromagnetic radiation on the cardiac muscle and thus on circulatory function depends primarily upon whether the peripheral or central nervous system is acted upon by the field.

More precise information on this question was provided by pulse rate measurements in rabbits, when various regions of the body were exposed to a field (Presman, 1970). In an initial series of experiments, the animals were exposed to pulsed fields of 10-cm waves (1 μs in duration, 700 pulses/s) with an intensity of 10–12 and

3–5 mW/cm^2 (average power). This treatment caused the pulse rate to fall (negative chronotropic effect). By contrast, exposure of only the dorsal side of the body or of both sides of the head had the effect of accelerating the pulse rate (positive chronotropic effect). Two special observations were made: (1) the negative chronotropic effect appappeared immediately after the start of irradiation; (2) pulsed irradiation had a greater effect than continuous irradiation, even though the field intensity was somewhat higher in the second case. An important conclusion can be drawn from these results: the negative chronotropic effect is probably caused by an action on the peripheral nervous system, and the positive chronotropic effect by an action on the central nervous system.

Such considerations also point to the importance of Presman's survey of the effect of electromagnetic fields on the brain and on the sensitivity of the central nervous system to other stimulants. Various EEG studies were performed primarily in animals; the principal result was that electromagnetic fields influence the electrical activity of the brain as evidenced in the EEG over a very broad range of frequencies and intensities. Specifically, Presman observed:

1. Increased synchronization: an increase in the number of high-amplitude slow waves. This change could be detected within a fraction of a second after exposure was begun.
2. Prolonged desynchronization: a decrease in the amplitude of the main rhythm of the biopotentials and an increase in the number of higher-frequency spectral components of the currents. This change appeared after some delay.
3. Short-term desynchronization, which appeared soon after field application or the termination of exposure.
4. An after-effect that was similar to increased desynchronization but was not evident until some time after the cessation of exposure.
5. The occurrence of spasmodic, epileptic discharges (relatively high frequency and amplitude).

These EEG changes associated with exposure of the head to fields ranging in frequency from a few hundred kilohertz to 2000–3000 MHz are similar to the effects observed in connection with a constant magnetic field.

Various investigations have shown that high-frequency fields also influence the development of chicken embryos. Van Everdingen (1940) reports on the exposure of eggs with 5-day-old embryos to high-frequency fields at 1875 MHz. The metabolism of these embryos was subsequently reduced by a factor of 1.5, which resulted in their death. Exposure during a later stage of embryonal development had less effect on metabolism and caused no deaths. In 11-day-old embryos, no effects were observed. High-frequency fields also influenced the pulse rate, which declined from 90–110/min to 10–20/min. In addition, the amplitude of the ECG peaks was somewhat greater than normal.

Based on further experiments with pigeon embryos using fields at 2450 and 24,000 MHz as well as weak magnetic fields (320–560 A/m), and with frog embryos using magnetic fields at 560 kA/m, Presman (1970) concludes that magnetic and microwave fields exert a general perturbing effect on normal embryonal development. They adversely affect metabolic processes and interfere with cell multiplication and differentiation.

Presman also used one-cell organisms to investigate the effects of electromagnetic fields. At frequencies of 5–7 MHz, the experiments showed that one-cell organisms apparently orient themselves along electric field lines. The prolonged exposure of one-cell organisms to alternating currents at various frequencies and voltages led in stages to three different types of movement a- Verhältnis der Titrierungen von behandelten zu nichtbehandelten Proben

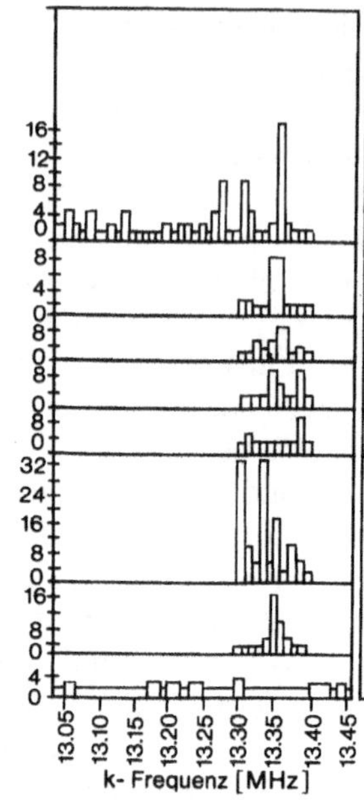

Impulsfrequenz Hz b-	Impulslänge, µsec c-	Feldstärke, V/cm d-	mittlere Intensität, mW/cm² e-	Behandlungsdauer, min f-	totale absorbierte Energie, J g-	Lösungsvolumen, cm³ h-	totale Dauer der Behandlung, sec i-
525	60	130	2.11	20	5680	0.9	38
230	10	40.0	0.013	30	63	0.8	4.5
100	10	40.0	0.005	30	24	0.8	1.8
200	10	20.0	0.0025	30	11	0.8	3.6
300	10	13.3	0.0016	30	7.8	0.8	5.4
kontinuierliche Bestrahlung j-		13.3	0.54	10	875	0.8	600
		6.7	0.135	30	655	0.8	1800
		4.7	0.07	30	330	0.8	1800

Figure 6-7. Changes in the gamma globulin activity of human blood caused by high-frequency fields at specific frequencies. After Bach et al. (1961a,b).
Key: a-Ratio of titrations of treated to untreated specimens; b-Pulse frequency, Hz; c-Pulse length, µsec; d-Field intensity, V/cm; e-Mean intensity, mW/cm²; f-Duration of treatment, min; g-Total absorbed energy, J; h-Volume of solution, cm³; i-Total duration of treatment; j-Continuous irradiation; k-Frequency (MHz)

and finally to death. A broad frequency band (20 Hz–3000 MHz) was employed in these experiments. The high-frequency fields were pulsed.

Of particular interest are experimental results which point to a resonance-like behavior of biologic systems. Thus, Bach et al. (1961a,b,c) reports on a change in the gamma globulin activity in human blood (after titration of irradiated and unirradiated solutions) brought about by high-frequency fields (Fig. 6-7). These fields are especially active within a very narrow frequency range, and the effects observed were evident even at relatively low intensities. In this case, the action of the field as such is apparently dominant, and the energy absorbed in the experimental solution is insignificant.

Genzel (1978) also dealt with the question of the nonthermal effects of millimeter-wave radiation with a power density of about 10 µW/cm² on biologic systems. In growth experiments with yeast cultures in the 40–60 GHz range, small frequency differences of 41.667 GHz, for example, resulted in significant growth differences of − 20% to + 12% relative to the controls. Calculations based on a thought model for a possible explanation of such phenomena, taking proteins 5 nm in size as an example, open at least theoretically the probability of oscillating processes in such molecular combinations, excited by the applied microwave radiation. This offers an approach to interpreting the surprising experimental results from a biologic standpoint.

Based on experimental results in the high-frequency range, Presman concludes that the direct action of such high-frequency electromagnetic forces on the central nervous system—either via the peripheral elements of the nervous system or directly via the brain structure—must be accepted as fact. This is a consequence of experiments with electromagnetic fields of various frequencies and inferences drawn from the depths of penetration known for these frequencies. At the same depth of penetration, different nerve centers are reached in small animals than in large animals. This also explains the differentiated reactions of animals of different sizes to waves of the same frequency.

A second characteristic feature is the connection between the effects of fields on the central nervous system, and the field intensity necessary to produce them. In some cases a low field intensity led to much more pronounced effects than a high intensity, while in other experiments an opposite effect was observed when the intensity of the field was correspondingly reduced.

A third and very essential point is the cumulative effect of low-intensity radiation, which occurs on multiple exposure. Apparently the phases of opposite reactions are involved: an increase in the excitability of the central nervous system after the first exposure and a decrease in excitability after subsequent exposures.

Field effects have also been convincingly demonstrated in studies on the effects of high- to ultrahigh-frequency electromagnetic fields on cardiac function. A large number of clinical studies on persons exposed to such fields are available, and these have shown similar changes in cardiac function, such as a reduction of blood pressure, pulse rate and intraventricular conductivity. However, these changes are pronounced only if centimeter waves play a role during the irradiation, that is, if electromagnetic energy is absorbed in the surface of the tissue of the human body. According to Presman (1970), this is evidence that the changes are caused by the direct action of electromagnetic forces on superficial receptors. This assumption is consistent with our knowledge of the physiology of the neural regulation of cardiac function. It is known that this type of change is typical of vagotonic changes in autonomic neuroregulatory processes, assumed to occur from the action of various factors on the regions of the superficial receptors. Experiments with animals confirm this hypothesis.

Exposure to microwave radiation modifies the behavior and physiology of animals and humans. This is a foregone conclusion that has been proven beyond any doubt.

Radiofrequency and microwave radiation at sufficiently high intensity and exposure duration can produce harmful biologic effects. To protect the general population and the persons occupationally exposed to microwave radiation, two types of standards are promulgated: personnel protection guidelines (exposure standards) and device performance regulations (emission standards). An exposure standard refers to the maximum permissible exposure level to the whole body or body parts, and generally incorporates a safety factor of at least 10. An emission standard refers to equipment and specifies the maximum emission (radiation leakage) from a device at a specified distance. The maximum permissible emission (or leakage) is such that the exposure resulting from it is usually several orders of magnitude below the personnel exposure standard. The standards for device emission and personnel exposure are designed to protect the general public and the worker, but do not apply to the utilization of microwave energy for medical purposes (e.g., microwave diathermy).

Stuchly (1978) discussed in detail the problems of "Microwave bioeffects and radiation safety" in a special volume of the *Transactions of the International Microwave Power Institute*. Stuchly and coworkers present papers on "Radio waves and microwaves—basic definitions and concepts," "Microwave biophysics," "Biologic and pathophysiologic effects of exposure to microwaves," "Microwave and radiofrequency protection standards," "Control of microwave exposure in Canada," and "Quantifying hazardous microwave fields." The following is reported on emission standards:

> Emission or performance standards restrict the emission of radiation from devices that are used in large numbers, often in uncontrolled environments. A typical example of emission standards are the microwave oven regulations in Canada and the United States.
>
> The U. S. regulations apply to microwave ovens used in homes, restaurants, food vending and similar establishments which operate at frequencies between 890 MHz and 6 GHz in the allocated ISM bands. The power density of the microwave radiation emitted by a microwave oven at a distance of 5 cm from the oven must not exceed 1 mW/cm^2 prior to acquisition by a purchaser, and not more than 5 mW/cm^2 thereafter. For the tests 275 ml of water is placed in the oven. Additionally, the regulations require several other safety features, such as interlocks, to protect against possible exposure due to malfunction of the oven.
>
> Human exposure levels that may result from a microwave oven having leakage of 5 mW/cm^2 can be determined reasonably accurately. The power density in the vicinity of a leaking microwave oven decreases rapidly, approximately as the inverse square of the distance from the oven. At a distance of 30 cm from an oven that operates at 2450 MHz, the power density is about 0.1 mW/cm^2 while at 1 m it decreases to 0.01 mW/cm^2
>
> *Exposure Standards*
>
> Despite the recognition of a need for unified, internationally accepted exposure standards, and substantial efforts in achieving an international agreement, there still exists a great discrepancy between the standards in Western countries and those in the USSR and East European countries. Standards issued by the U. S. Army and Air Force, United Kingdom, France, Germany, Canadian Standards Association and American National Standards Association all give essentially the same recommendations as the ANSI C95.1 standard.
>
> The latest version of the ANSI C95.1 of November 1974 applies to frequencies from 10 MHz to 100 GHz and states that, for normal environmental conditions, and for incident electromagnetic energy of frequencies 10 MHz to 100 GHz, the cw radiation protection guide is 10 mW/cm^2 and the equivalent free-space electric and magnetic field strengths are approximately 200 V/m rms and 0.5A/m rms, respectively. For modulated fields, the power density and the squares of the field strengths are averaged over an 0.1 hour period; that is none of the following should be exceeded, when averaged over any 0.1 hour period:
>
> | | |
> |---|---|
> | Mean squared electric field strength | 40,000 V^2/m^2 |
> | Mean squared magnetic field strength | 0.25 A^2/m^2 |
> | Power density | 10 mW/cm^2 |
> | Energy density | 1 mWh/cm^2 |
>
> The maximum permissible levels in this standard are to serve as the guide numbers appropriate for moderate environments. Under conditions of moderate to severe heat stress the guide numbers should be reduced.
>
> In 1971 the American Conference of Governmental Industrial Hygienists recommended the following limit values for occupational exposure to the frequencies from 100 MHz to 100 GHz:
>
> 1. For power density levels not greater than 10 mW/cm^2, the total exposure time shall be limited to the 8-hour work day (continuous exposure).
> 2. For power density levels from 10 mW/cm^2 to 25 mW/cm^2, total exposure time shall be limited to no more than 10 minutes in any 1 hour period during an 8 hour work day.
> 3. For power density levels in excess of 25 mW/cm^2, exposure is not permitted.

For comparison, Stuchly (1978) also

sites the lower limits of permissable power density in certain other countries: Sweden, 1 mW/cm^2; U.S.S.R., 10 μW/cm^2; Czechoslovakia, 10 μW/cm^2; Canada, 1 mW/cm^2.

Modulated High-frequency and Microwave Fields

Combining the knowledge of biologic effects of the low-frequency (ELF and VLF range) electromagnetic fields and those of high-frequency or microwave fields, one is able to use high-frequency fields modulated by low frequencies. By this method it is possible locally to apply energies at the low biologic frequencies by the use of a selected high-frequency carrier as described by Hartmann et al. (1967).

This possibility was apparently also recognized by Bawin et al. (1973), and so their investigations must be considered of particular importance. To explore the biologic effects of weak (1 mW/cm^2 or less) electric fields, they combined a high-frequency carrier (147 MHz) with signal information at frequencies of 0.1–100 Hz. The effects of such amplitude-modulated fields were investigated in cats. These animals were conditioned to produce a specific brain impulse, which was recorded by EEG, at 30-s intervals in response to a light flash. The action of the field was monitored by comparison with a control group. It was found in these experiments that the irradiated animals showed a marked difference from the control group in terms of functional capacity, accuracy of the reinforced patterns, and resistance to their extinction (at least 50 days versus 10 days in the controls). A study of the specific effect of the modulation frequency in a further experiment clearly showed that fields act as "reinforcers" (increase the number of manifestations of spontaneous rhythms) only if the modulation frequency is close to the experimentally related, biologically dominant frequencies of the natural EEG. The paper concludes with a discussion of various possible effects of the external experimental fields on the central nervous system, and it is hypothesized that the amplitude-modulated VHF fields may influence the excitability of neuronal membranes.

Reiter (1960) reports on the effect of metric-wave radiation (diathermy device) modulated with 50 Hz on the pH values of conjunctival tissue. The effect is clearly seen in Fig. 12-4 (see Chapter 12 for more details).

Gordon (1964) found an effect of long-term exposure to pulsed centimeter-wave energy on blood pressure in experiments with rats. Two hundred twenty animals, divided into four experimental groups and one control group, were subjected to irradiation with decimeter-range, 10-cm, 3-cm and 1-mm waves. The intensity for all groups was 10 mW/cm^2, the duration of exposure 1 hr/day. The treatment was continued over a period of 6–8 months. Blood pressure exhibited a two-phase response to decimeter- and centimeter-wave irradiation: a rise in the first weeks of irradiation, followed by a continuous fall from the twentieth through twenty-fourth week. In the case of the 3-cm and 1-mm waves, however, only one phase was observed: a fall of blood pressure, which began during the first week of irradiation. When the effects produced by pulsed and continuous 10-cm waves were compared, no differences were found. However, in the case of continuous exposure the change appeared much earlier (during the eighth week) and consisted of only a single phase. After irradiation was terminated, the blood pressure returned in stages to its normal value within 8–10 weeks in all cases.

There was no difference between the millimeter- and centimeter-wave radiation with regard to the intensity employed. Irradiation with 3-cm waves produced no marked effects either at 1 mW/cm^2, or when the intensity was increased from 10 to 40–100 mW/cm^2. Decimeter and centimeter waves, on the other hand, produced reactions even at 1 mW/cm^2. One peculiarity was the diminished effect of 10-cm waves when their intensity was increased by a factor of 10.

The effects of pulsed microwave radiation were also studied by Tanner (1966) in birds. He proceeded from the assumption that microwave radiation not only produced thermal effects, but also influences the activity of the animal nervous system by way of induced electric currents. Animal tissue absorbs microwave radiation diffusely, and since such tissue contains intermediate membranous layers which possess semiconductor properties, a polarization effect can occur. If the microwave radiation is pulsed, then the resultant electric currents are modulated by the corresponding pulse duty factor. There is a possibility that the microwave radiation will exert an influince on the nervous system that is dependent on the penetration depth of the radiation.

To investigate this problem in more detail, 14-week-old chickens were exposed to microwave radiation at an intensity of 10–30 mW/cm^2 and a frequency of 16,000 MHz, with a pulse modulation of 8000 pulses/s. With the aid of a horn antenna, it was possible to irradiate the animals with microwaves from above, from the side and from below. It was found that the radiation was effective only if directed from above. Regardless of whether the head or the body was shielded during exposure, the chickens always reacted to the irradiation within 1 s by extending a wing or leg.

Flocks of birds constitute a well-known hazard to air traffic in the vicinity of airport runways. In an effort to solve this problem, Tanner et al (1967) conducted experiments to determine whether birds could be driven from such critical areas by the use of microwave radiation, and how the radiation could be optimally applied to evoke an effective flight response. For test purposes, chickens were irradiated with a pulsed 9.3-GHz field (416 pulses/s at a pulse width of 2.3 ms). The peak output of the horn antenna was 94 kW, with an average output of 90 W. Calibration of the field intensity in a test cage situated 15 cm above the floor gave an average intensity value of 46 mW/cm^2.

The following reactions were observed in both young and full-grown chickens. Lateral irradiation led to no wing or leg movement on the exposed side, but produced an extensional effect on the contralateral side. The birds raised their heads and oriented them in the direction of the strongest point in the field. If the animals were already excited as a result of previous handling, they tended to run part way along the edge of the focused beam, keeping one side of the body tangent to it. The diverse reactions of the animals depended on which part of the body surface had come into contact with the beam. Only when irradiated from below did the animals display little or no response. Otherwise a stimulative effect always gave impetus to a change of position, which could cause an alteration of flight or even a type of collapse in birds.

Kamenskii (1967) reports exhaustive studies on a neuromuscular preparation exposed to 3000-MHz fields. He employed both a continuous field with an intensity of 10–1000 mW/cm^2 and a pulsed field with a pulse duration of 1 μs and a pulse repetition frequency of 100–700 Hz with a mean intensity of about 10 mW/cm^2. He also used DC pulses with a duration of 0.1–1 μs and a frequency of 2 pulses/s. He obtained the following results:

1. The continuous high-frequency field heated the nerve by 2°C within 30 min, which led only to an increase in the velocity of stimulus conduction in the nerve. This effect increased at higher field intensities. Further research showed that, contrary to expectations, the effect is not thermal in nature.

2. Experiments with pulsed high-frequency fields, which also heated the nerve by 2°C within 30 min, led to an increase in excitability (lowering of the threshold of excitability).

Presman (1970) investigated the genetic effects of modulated electromagnetic fields. He showed that garlic chromosomes are affected by 5-min exposure to fields with a frequency of 5–40 MHz, pulse length of 15–50 μs, pulse frequency of 500–1000/s and intensity of 250–6000 V/m. He also found in a larger series of experiments that such electromagnetic fields affect bacteria at a frequency of 31 MHz and the spores of certain penicillin strains at frequencies of 2–11 MHz.

When mature fruit flies (*Drosophila*) were treated with fields at approximately 25 MHz, the first generation contained 8 times more female offspring than male, while exposure to 30-MHz fields led to a change in the second generation. In this case there were twice as many males as females.

All these effects occurred primarily when the field parameters lay within certain ranges: frequency, 1–250 MHz; pulse length, 1–10 μs; pulse frequency, 30–10,000 pulses/s; and field intensity (of the pulses) between several hundred V/m and several tens of kV/m. The parameters were selected such that no significant heating would occur in the specimens. The frequency and intensity were applied in accordance with the desired effect. The fields were produced in air, and the specimens had no direct contact with the field-generating electrodes.

A very comprehensive survey of the state of research in the U.S. with regard to the biologic effects of nonionizing radiation is given by Taylor (1975). He reports on effects involving the nervous system, special sensors, biochemical and biophysical processes, genetic parameters, developmental processes and the behavior of systems. A special chapter is also devoted to the problem of dosage.

Measurements of the centimeter waves that are emitted from the interior of the body were made by Barrett and Myers (1975). According to their study, electromagnetic waves in the centimeter and decimeter range not only penetrate into organic tissue (as in the application of diathermy), but also follow the opposite path. The waves are extremely weak (on the order of 3×10^{-12} W/cm² at 3 GHz), but can be measured by special techniques developed for radio astronomy. A correlation exists between radiation intensity and body temperature. A comparison of measured temperatures showed that microwave thermography is capable of revealing temperature structures and changes beneath the body surface.

Questions on the biologic activity of microwaves were addressed at a meeting of the U.S. National Committee of the URSI (1975). Cain and Rissmann reported on the effects of microwaves on mammals at 3 GHz; West and Regelson on biologic effects of pulsed electromagnetic high-frequency radiation (27 MHz); Lin on effects of microwave radiation on mammalian cells in vitro; Varma and Trabculay on dominant lethal tests and DNA studies on mutations induced by non-ionizing radiation (1.7 GHz); Lotz and Michaelson on adrenocortical reactions in rats exposed to microwaves (2.45 GHz); Oscar on effects of electromagnetic radiation on the brain-blood system in rats (1.3 GHz); and Kritikos and Takashina on nonthermal effects of electromagnetic fields (frequency 2.45 GHz, pulse frequency 1–30 MHz) on the central nervous system. Effects of ELF fields on biologic systems were also discussed: Bawin and Adey reported on the influence of weak low-frequency electric fields on calcium loss in the isolated chicken and cat brain (frequency range 6–30 Hz); Greenebaum et al. on long-term effects of weak electromagnetic fields 45–75 Hz on the mud salamander; Bliss and Heppner on effects of zero-field conditions on the daily activity rhythm of the house sparrow; Mathewson et al. on effects of a 45-Hz electromagnetic field on rats; Durfee and Plante on the treatment of chickens with electric and magnetic ELF fields at frequencies of 45–75 Hz; Medici on the effects of weak elecxtric ELF fields on the monitored behavior of monkeys (frequency range 7–45 Hz); and Sugyama and Mizuno on the effects of alternating electric fields on the visual threshold in man (frequency 20–60 Hz).

We shall conclude this section by pointing out an interesting observation. A chance discovery made by American scientists during the course of an experimental study on the danger of high-frequency radiation to man indicates the importance that further research on the biologic effects of electromagnetic fields may have in the future (*Medizin-Elektronik,* 1970). One experimental subject reported that he was able to sense, unaided, the presence of 3.6- and 10-cm electromagnetic waves. This prompted the investigators to perform experiments with 1-mW pulses (modulation frequency 100–1000 Hz) with several subjects; copper plates were used as shields to "switch" the radiation on and off. According to reports, the subjects not only were aware of a humming or hissing sound, they could also sense the exact moment at which the copper shield was used or the frequency of the pulse modulation was altered.

Light and Cosmic Radiation

Amazing results that suggest that living cells may be capable of transmitting information by photons—specifically by light in the ultraviolet (UV) range—were reported by Kasnatschej et al. (1973), Ruth and Popp (1976), and Popp and Ruth (1977).

Living cells were cultured in two quartz glass vessels containing suitable nutrient media. The walls of the vessels were placed in contact. Then one of the cell cultures was infected with a virus. Almost simultaneously the cells in the neighboring colony also showed signs of viral infection. The same thing happened when the cells received lethal doses of UV radiation or were poisoned with sublimate (mercuric chloride). In each case the neighboring cells were afflicted with precisely the same symptoms, even though the quartz glass should have protected them from the effects of the poisoned cells. Only when ordinary glass was used in place of quartz glass did the neighboring cells remain unaffected. The authors surmise that photons in the UV range, which can penetrate quartz glass but not ordinary glass, are capable of transmitting metabolism-regulating information from cell to cell.

In conclusion, they make the following statements:

1. The ultraweak photon emission from biologic systems really exists and shows in the spectral range from visible radiation to the UV wave range intensities in the order of $10-10^4$ photons/(s.cm^2).
2. As experiments on yeast cells suggest, proliferating cell cultures exhibit a more intense photon emission than those untreated cultures for which an increase of the cell number could not be proved.
3. At first, the photon emission from dying cell cultures increases to about a hundred times the intensity of untreated cells. Later on, it drops according to a quasiexponential decay with increasing decay constant. The total number of photons emitted is in the order of the cell number of the poisoned cell population. The decay time amounts to at least some 10 min and may not depend on the frequency of the mode under examination.
4. The spectral distribution shows resonance character, that is: it has certain peaks which do not necessarily shift after treating the cells with toxic agents, although the peak intensities show a considerable increase to about the factor 10^2-10^3.
5. The photon emission flucturates over longer periods of time.

A very interesting study was done by Bonka (1975) on the radiation dose received by persons in the Federal Republic of Germany. As explained earlier, radioactive nuclides in the earth's crust which emit alpha, beta and gamma rays are the source of this radiation, as well as primary cosmic radiation which penetrates the atmosphere.

In studying the radiation dose to man, it is useful to divide natural radiation according to its origin into cosmic, terrestrial and internal (endogenous) components. Cosmic radiation includes that caused directly by charged particles from space or by secondary charged particles (such as electrons) which arise from reactions of primary particles with the air. The terrestrial component includes that portion caused by gamma radiation emitted from the ground. Finally, the internal component includes radiation caused by the decay of natural radionuclides (tritium, radiocarbon, potassium-40, etc.) incorporated into the body through food intake and respiration.

At the latitude of the Federal Republic of Germany, the cosmic radiation dose depends almost entirely upon elevation above sea level. It is equal to 31 mrem/a at sea level, 36.5 mrem/a at 500 m (Munich), 54 mrem/a at 1400 m, and 164 mrem/a at 3000 m. The mean terrestrial radiation dose outside of cities fluctuates between 30 and 45 mrem/a (e.g., in Upper Bavaria, Lower Bavaria, Württemberg and the Frankfurt/Cologne area), with maximum values of 105–150 mrem/a (Bavarian Forest, Black Forest); the mean value for the population as a whole is approximately 55 mrem/a. The mean natural radiation dose to the population ranges from 95 to 125 mrem/a, with extreme values of 65 and 300 mrem/a.

Searle (1966) reported on the importance of particle radiation with regard to the susceptibility of genes. In a series of experiments with 100,000 mice, he found that mutations were 20 times more frequent after neutron bombardment than after prolonged gamma irradiation, and 5 times more frequent than after X-ray irradiation. The animals were exposed to the neutrons for a period of 3 months. Dominant mutations occurred in 1 out of 2000 mice. These consisted mainly of changes in coat coloration, curled fur, a ringed tail, deformed ears and extremities, and unusual eye colors. If carriers of mutations were crossbred, the offspring were invariable stillborn. In animals exposed to short-term but high-dose neutron irradiation, considerably fewer genetic defects were noted.

Chemical and Physical Reactions

Piccardi has studied the effects of extraterrestrial radiation on chemical reactions since 1935. He observed that a copper screen placed over a reaction vessel can alter the chemical reactions occurring within it, presumably because certain rays which can influence the course of the reaction are absorbed. As proof, Piccardi (1970) investigated the precipitation of a polymerization product from a solution of monomeric acrylonitrile in water. The solution was poured into several flasks, and half the flasks were covered with a copper screen 0.1 mm in thickness. It was found in 80 of 108 experiments that the weight of the precipitated polymer was less in the screened vessels than in the unscreened vessels.

In a further series of experiments involving 1000 separate tests, Piccardi (1970) observed the precipitation of bismuthyl chloride. In 1000 control tests a Gaussian distribution curve was obtained when it was measured how often the sedimentation of bismuthyl chloride occurred more rapidly in 10 red-tagged reaction vessels than in 10 blue-tagged control vessels. When the red-tagged vessels were covered with a copper screen, however, sedimentation occurred more rapidly in about 70% of these vessels than in the blue-tagged vessels. According to Piccardi, these experiments show that chemical reactions are influenced by external (and perhaps extraterrestrial) environmental factors. These investigations prompted other researchers to investigate the extent to which a simple chemical reaction (corresponding to Piccardi's test) can be influenced by artificially produced electromagnetic radiation of various wavelengths in the laboratory. Eichmeir and Büger (1969), for example, studied the effect of infrared (IR) radiation, visible light, UV, X-ray and gamma radiation on the coagulation rate of BiOCl corresponding to Piccardi's test reaction and conducted additional experiments with copper and lead shielding. They found the following: The coagulation of BiOCl is accelerated by IR radiation and by copper and lead shielding, but delayed by UV, X-ray and gamma radiation. Visible light had no apparent effect. These observations can presumably be explained by the influence of electromagnetic radiation on the formation of a hydrate shell on the surface of the BiO^+ ions formed by the reaction. Since it is this shell which controls the coagulative mechanism, the coagulation rate is correspondingly affected.

Prompted by the precipitation reactions reported by Piccardi (1970) in various colloidal systems, and in order to explore further the effects of electromagnetic radiation of natural origin, Fischer et al. (1968) performed selected experiments with artificially produced fields. It was hoped in this way that new knowledge could be obtained on the reaction processes which underlie the phenomena described. For this purpose, the sedimentation of silicon dioxide in potassium lye was observed under the influence of various electric and magnetic fields (as well as bismuth chloride). Maximal effects were found in the range around 10 kHz and between 10 and 100 MHz, followed by a minimum at 500 MHz. This was followed in turn by an increasing effect up to frequencies of about 50 GHz (wavelength ~ 6 mm).

In further experiments, only the water used for the precipitation reaction was exposed beforehand to an electromagnetic field, and the resultant effect on the precipitation reaction was observed as a function of frequency. A distinct minimum was observed near 10 kHz, followed by a pronounced maximum extending up to 100 kHz. Then came a broad minimum from 1 to 10 MHz. Finally, the effect showed an increasing tendency from 10 to 500 MHz, which probably stems from the electrophoretic mobility of silicon dioxide (siliceous earth).

A test of specific electric resistance was also conducted. When water was treated with electromagnetic fields at frequencies of 10 Hz–10 kHz, it showed a minimal electrical resistance in the range of 100–5000 Hz. This was followed by a pronounced maximum at 10 kHz, then another flat minimum at about 1 MHz. Another, relatively sharp conductivity maximum was observed when the water was treated with fields at about 200 MHz. These findings are consistent with the results obtained by Presman (1970), Dardymov et al. (1965) and Varga (1973) in similar experiments.

In summary, it can be said that electromagnetic and electric alternating fields in the intensity range of $1-10^4$ V/m or around 3.75 mT (37.5 G) that act on colloidal systems or their suspension media apparently lead to an increase in the sedimentation rate, electrophoretic mobility, specific electric resistance, viscosity and shock-freezing temperature, while magnetic fields and static electric fields tend to slow the sedimentation rate. Plaksin (1966) published information on the effect of higher-frequency fields on the physical properties of water. Treating the water with fields at 100 kHz–8 MHz for 30 min led to an increase in its optical density in the range of 380–691 nm (10^{-9} m). Valfre et al. (1964) further reported a weight decrease in guinea pigs and mice that received only water that had been activated with a 10-kHz electromagnetic field in comparison with animals that received normal water. A similar effect was seen in the second generation of the animals, even if it was given normal water. Moreover, the effect produced by the treated water was particularly marked during periods of solar activity.

7

Biologic Activity of Electromagnetic Energies: General Considerations

Action of Fields

Presman (1970) makes the following statements on the basic actions of (high-frequency) electromagnetic fields as determined experimentally. The experiments lead in most cases to disturbances in the regulation of physiologic processes. Such disturbances are particularly marked during embryonal development and during growth, and thus during periods when resistance is nonexistent or not yet fully developed. An analysis of the precise nature of these disturbances indicates that the actions of electromagnetic fields are manifested primarily in the electromagnetic processes associated with the regulation of physiologic functions.

In most experiments, unfortunately, the selection of electromagnetic field parameters is influenced mainly by the type of generator available to the experimenter, rather than by biologic requirements. Nevertheless, certain conclusions can be drawn.

Every living organism effectively protects itself against external natural and artificial electromagnetic disturbances, apparently employing two different defensive systems, one passive and the other active. The former is a very fast-reacting peripheral system that responds to a certain parametric range of electromagnetic disturbances (the main parameter being field intensity). This passive system could be responsible for the two or more "optima" observed for field effects, and may be related to the resulting biologic disturbances and the intensity of the biologically active electromagnetic fields. If this passive system is overburdened, the active system is called upon. The fast-reacting peripheral system signals the arrival of disturbances to the slow-reacting, central system (which regulates physiologic functions). The latter then protects itself by lowering the body's sensitivity to such disturbances. Another, higher stage of the defense reaction is attained when the central system modifies physiologic processes so as to produce a counter-reaction to the undesired, externally evoked reaction of the organism. This thought model is apparently confirmed by experimental data on the response of animal organisms to the action of electromagnetic fields on the central or peripheral nervous system or of whole-body irradiation on both systems.

To explain the mechanism of this action, it is justified to assume that the cerebral cortex and diencephalic (between-brain) structures, especially the hypothalamus, respond very sensitively to electromagnetic fields (Presman, 1970). The action of electromagnetic fields on the nervous system is apparently based either on a stimulation of nerve cells or on a change in the parameters of their functional state (excitability, amplitude of biopotentials, propagation velocity of impulses, etc.).

But what are the physico-chemical processes that underlie the action of electromagnetic fields on nerve cells? Because relatively little is known as yet about the physico-chemical nature of nerve excitation, available theoretical and experimental material is still insufficient to provide an exact answer to this question. Nevertheless, several points have been established based on the experimentally demonstrated sensitivity of nerve cells to electromagnetic fields over a broad range of frequencies:

1. Electromagnetic fields can be detected in the membrane of nerve cells. As a result, a direct component of the rectified current which is produced by external fields may lead to a stimulation of the cells or to a change in their excitability at various frequencies.

2. Electromagnetic fields can influence the mobility of ions involved in the process of nerve excitation. Vibrations of the ions at the frequency of the external field will affect their ability to penetrate neuronal membranes, with a corresponding change in neuron excitability.

3. There is evidence to indicate that changes in the physico-chemical properties of the aqueous medium of the nerve cell play a role.

4. This change in the aqueous medium could alter the permeability of neuronal membranes.

5. Electromagnetic fields may also influence the "spontaneous activity" of receptors. Little is known about this effect as yet, but it is apparently related to thermal disturbances of ionic processes in the membrane.

6. The additive effect of electromagnetic forces on certain properties of the blood indicates that information or a certain noise signal may be stored there (as well as in other biologic systems).

The mechanism of action of electromagnetic fields on the regulation of physiologic processes in the organism as a whole can also be viewed from a different standpoint, however. External electromagnetic fields may determine the internal electromagnetic regulation which is active in various frequency ranges. According to this hypothesis, electromagnetic fields are capable of influencing or disturbing the function of a certain branch of this regulatory system—by increasing its activity, for example, in cases where the frequency of the external field is close to the natural frequency of this special biologic structure, or by damping or reducing the activity of the system when it is initiated by a signal with an uncharacteristic frequency that is foreign to the system.

Based on his experimental results on neurohumoral regulatory processes, Presman (1970) makees the following general points regarding the effects of electromagnetic forces on the body as a whole:

1. Changes in the organism resulting from exposure to electromagnetic fields are nonspecific.

2. The cause of the changes based on electromagnetic forces apparently lies in their effect on various parts of the nervous system.

3. When an electromagnetic field acts on a specific part of the nervous system, the nature of the resulting change is practically independent of frequency.

4. The nature and magnitude of the changes effected by the exposure of peripheral parts of the nervous system to electromagnetic fields is practically independent of field intensity. However, if internal parts of the body are acted upon by the fields, the resulting effects apparently depend chiefly on the intensity of the fields. It is characteristic of the central nervous system to respond more strongly to relatively low field intensities than to higher ones. In

some cases this was true to the extent that a response was evoked only at a certain low field intensity, and was absent when the intensity was increased.

5. When electromagnetic forces or fields act on both the central and peripheral portions of the nervous system, there are usually certain optimal intensities (normally two) at which the response to the field is strongest. The strength of the response varies in a simple manner with the duration of exposure.

6. Responses to field exposure under equal conditions have two phases, which depend on the intensity and the duration of exposure. At low intensities (or a short exposure time) the changes in the organism are opposite to those occurring at high intensities (or a long exposure time).

7. The effect of multiple exposure to fields is cumulative. Exposure to strong fields normally leads to an adaptation during the course of subsequent exposure. Correspondingly weak fields, by contrast, produce gradually increasing changes in the organism, which frequently exhibit two typical phases. The entire process depends ultimately upon the number of exposures, however.

Significance of Fields

Summarizing the findings on electromagnetic fields and the regulation of the vital activity of organisms, Presman (1970) states that there are sufficient experimental observations and theoretical arguments which directly or indirectly point to the existence of three types of information-transfer links in nature: (1) the transfer of information from the environment to the organism; (2) the transfer of information within organisms; (3) the transfer of information between organisms.

Apparently these links can have developed only during the course of evolution, giving biologic systems the ability to receive, transmit and transform information by means of electromagnetic fields. This hypothesis is in agreement with the speculations of König and Ankermüller (1960) with regard to the 10-Hz Schumann resonance oscillations observed in nature as well as in certain biologic rhythms in the same frequency range (EEG, ECG, muscular vibrations, etc.).

This hypothesis is supported by numerous experimental results that show statistically valid correlations between solar activity and certain biologic processes. For example, a connection has been shown between solar activity and the incidence of meningitis in New York, or the recurrence of fever in the European U.S.S.R. There is also a probable link between magnetic storms and mortality from neural and circulatory diseases in Copenhagen and Frankfurt. The connection between the "biologic clock" and natural electromagnetic fields is as firmly established as the orientation of living organisms by the daily cycle of the earth's magnetic or electric field, the motor activity of snakes in electrostatic fields or the geomagnetic field alone, the link between reaction time and the daily variation of natural ELF signals, the orientation of root growth in seedlings in the natural magnetic field, and the statistically significant dependence of chemical reactions on astronomical data and constellations. All these factors point to a coupling between external electromagnetic fields and internal "receptors" in biologic systems that are apparently capable of responding to them.

Much is known already about the role of other physical parameters (sound, ultrasound, pressure, light, heat, etc.) as agencies of information transfer between organisms, but here, too, radiations play a role. Planaria, for example, are capable of orientation even in weak gamma radiation fields (only 6 times more intense than the natural radiation level) and are capable of perceiving the radiation source. Experiments with ants demonstrate the possibility of intercommunication based on ionizing radiation.

In any case, the assumption that electromagnetic fields at a variety of frequencies are utilized for signal transmission by living organisms on a much broader basis than had been previously supposed is justified. This is evidenced by several facts:

1. Organisms are extremely sensitive to most electromagnetic fields, which in particular can serve as a conditioned stimulus for producing reflexes.
2. Persons exposed to electromagnetic fields experience various sensations. Moreover, some animals possess special receptors for such fields.
3. Electromagnetic fields in various frequency ranges have been observed in the vicinity of isolated organs and cells as well as near complete organisms.

One phenomenon supporting this claim is the phosphenes (discussed further in Chapter 12) which appear under the influence of electromagnetic fields.

Acoustic sensations are also produced by electromagnetic energies. Thus, persons exposed to a pulse-modulated electromagnetic field have reported hearing various sounds. Similarly, it is known that certain fish respond to various types of electric fields.

It has been shown that electromagnetic fields form at times in proximity to organs and cells. For example, an electric field will form about a nerve that is stimulated while in a conducting medium. Direct-contact measurements showed low- and high-frequency emissions ranging from 10 to 150 kHz during muscular contractions. The greatest effect was noted in the small muscles. The results of noncontact field measurements made with an antenna located about 1 cm from the specimen are shown in Fig. 7-1.

Very convincing measurements were carried out by Gulyaev et al. (1967). Using a specially developed probe amplifier, they were able to measure electromagnetic fields in the vicinity of active nerves, muscles and the heart in frogs, as well as near the human heart and muscles. They measured the

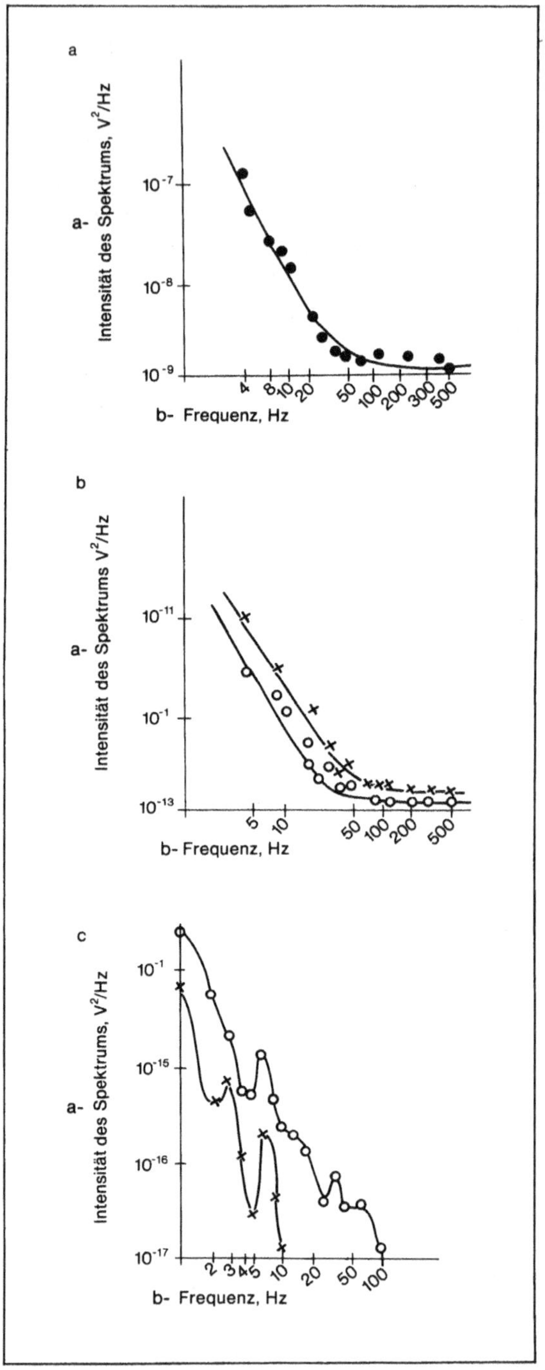

Figure 7-1a–c. Measurement of the spectral characteristics of electric field energies of living organisms by antenna at a distance of 1 cm; ***a*** tensed forearm muscle; ***b*** dragonfly earlobe; ***c*** frog brain (circles: dark-adapted eyes; crosses: illuminated eyes). After Malakhov et al. (1965).
Key: a-Intensity of spectrum (V^2/Hz); b-Frequency (Hz)

"electroauragram" of an isolated frog nerve at a distance of 25 cm (where the voltage, presumably with respect to reference potential, was about 1 mV), that of an isolated frog muscle and heart at a distance of 14 cm, and that of the human heart and muscles at a distance of 10 cm.

There is also clear evidence to indicate that transmission of information between organisms takes place by means of electromagnetic forces. For example, the group maneuvers of flocks of birds or schools of fish plainly indicate that a certain signal must be given by a lead animal, which, in view of the circumstances, can only be of an electromagnetic nature.

The synchronous maneuverings often observed in various bird species, particularly the starling, led Heppner and Haffner (1973) to suspect that electromagnetic fields were used as a means of communication within flocks—for studies had shown that birds sometimes execute flight maneuvers within 5 ms when flying in flocks. Since flocks of starlings may be hundreds of meters across, the question of how the birds flying at the outer edge of the flock are able to receive the appropriate information so quickly occurs. Signal transmission by sound would be much too slow; sight of the lead animal is obscured by the other members of the flock; and a type of chain reaction passing from one bird to the next is ruled out due to the quickness of the measured reaction times. Evidence of the ability of birds to navigate by the terrestrial magnetic field led to the hypothesis that the flight maneuvers in flocks may be coordinated by means of electromagnetic signals. Such a signal could reach all the birds in a flock instantaneously, regardless of their position relative to the signal source, and would also contain sufficient information for flight maneuvers.

Besides this rapid form of signal transmission, there is another mode of bioinformation transfer which is relatively slow. It is exemplified by the ability of animals to find their way home over very great distances. Typically, such navigational powers are dependent upon the emotional state of the animal.

Harmfulness of Environmental Factors

In discussing the biologic effects of an environmental factor on man, it is often unclear whether health-impairing effects are meant, or simply any biologic effects, which may even be positive. The transition from adverse effects to general biologic effects with no far-reaching consequences is a smooth one. The transition zone is determined by two main factors. One is the nature of the external influence or environmental factor. Where the biologic effects of electromagnetic fields are concerned, the field intensity is most critical; but the frequency of the field, its modulation and other parameters also appear to play a significant role.

The second main factor is the sensitivity of the biologic system in general, and of man in particular. This sensitivity varies from object to object, from system to system, and from person to person. But the initial state of a system or person, as well as other environmental parameters (the weather, for example) are continually changing, continually exhibiting both long- and short-term effects. Also to be considered is the location factor, or the location-dependence of a response, which is known to play a major role. In view of the many variables involved, it is no wonder that a specific experiment with a certain person may yield an entirely different result when repeated the following day. This naturally makes it difficult to conduct such experiments—a fact which is known, or should be known, to every experienced doctor, psychologist and biologist, but which sometimes causes problems for the unversed physicist or technician due to his essentially different way of thinking.

The critical importance of the "transition

zone" between harmfulness and harmlessness with regard to certain environmental factors is well illustrated by the problem of "noise pollution."

Effenberger and Jatho (1973), for example, report that the number of persons in the Federal Republic of Germany suffering noise-related hearing impairment is reported at 500,000–800,000. Surveys point particularly to street noises that are perceived as being unpleasant.

It is now known that noise tends to induce ergotropy in the autonomic nervous system. The peripheral blood vessels contract, the blood pressure and pulse rate increase, the skin temperature falls, metabolism is accelerated, the adrenal cortex secretes more hormones, and electric muscle activity increases spontaneously. In most cases these autonomic responses (some of which are transitory) can be evoked by acoustic signals of very short duration, such as an automobile horn.

One of the most important findings of recent EEG studies in sleeping persons is the observation that the depth of sleep is significantly decreased by noise. If disturbing noises are repeated at intervals during the night, deep sleep can be entirely prevented, in fact. Persons so affected are chronically tired, listless and susceptible to disease. Here is a case of an environmental factor which can clearly have adverse health effects, even though the cause, noise, is not consciously perceived (the disturbing noises being much too weak to awaken the subjects). Despite their harmfulness, however, it is unlikely that the conscious perception of such noises by healthy persons is associated with such a pathogenic effect.

Conclusions

The problem of the biologic activity of electric and magnetic fields has been explored by numerous researchers for many decades. The basic difficulty with such research is that it is highly complex and given the current state of knowledge in the area, it appears that this research can be effectively continued only on an interdisciplinary basis. Moreover, increased public interest in such scientific problems has produced an increasing demand for information by the mass media.

The following remarks can be made on the current state of knowledge in this field of research:

1. Previous experiments clearly indicate that electric, magnetic and electromagnetic fields exert significant biologic effects, even at low field intensities.

2. Prompted by previous results, research efforts should be continued and intensified on a broad basis, making use of the most modern technical facilities and interdisciplinary cooperation. The increasing interest in obtaining scientific evidence and the desire to learn more about man's natural and technical environment are a further impetus.

3. The highest possible standards should be applied in the publication of such scientific results, in order to satisfy the customary strict rules of classical science.

It should be added, however, that the courage to publish still-unproved findings and results can sometimes serve to advance research. Charles Darwin once remarked that erroneous facts, which often persist for many years, are a great hindrance to scientific progress. Erroneous theories, on the other hand, cause less trouble due to the great pleasure derived from their refutation. Furthermore, the new paths that are often opened up in this process will aid in the search for the facts.

The problem of the biologic effects of electromagnetic fields can be approached from various standpoints. On the one hand the natural electromagnetic fields existing in our environment can be regarded as given factors. But there also exist internal fields within organims, which are utilized in the coordination of physiologic processes and especially in the transfer of information between individual organisms. The electromagnetic forces present in the environment can be subdivided even further into those of natural origin and those which are man-made.

Presman (1970) is surely correct when he postulates that the natural electromagnetic climate must have played a role in the evolutionary development of living organisms. Of special significance in this regard are natural electromagnetic processes in the ELF range, which show a remarkable similarity to signals of "biologic frequencies" in terms of their frequency and amplitude.

Presman points out that in some cases living organisms respond to electromagnetic forces only at certain optimal intensities, while in other cases the effects increase when the intensity of these forces is reduced; or, exposure to low and high intensities may lead to opposite effects. Cumulative biologic effects were also produced by repeated exposure to electromagnetic fields whose intensity lay below the known biologically active threshold for a single exposure. It was also found that at the same mean energy absorption by organic tissues, the nature of an organims's response depends largely on the specific modulation of the field. The same applies to the direction of the electric and magnetic field vector (with respect to the body axis of the organism, for example). Such questions and problems are especially pertinent with regard to high-frequency irradiation.

It was also found that periodic changes in the natural electromagnetic climate could exert a regulative effect on vital functions, such as the rhythm of the major physiologic processes, the ability of animals to become oriented in space, etc. Normally the internal systems of living organisms which transmit information by means of electromagnetic forces are largely shielded from natural electromagnetic phenomena and the attendant disturbances. In pathologic states, however, spontaneous changes in external electromagnetic conditions (solar flares, lightning discharges, etc.) can severely disrupt the regulation of physiologic processes.

It is becoming increasingly evident the extent to which nature has utilized electromagnetic fields during the course of evolution to enable organisms to perceive changes and processes in their environment—for of all the geophysical factors, electromagnetic forces represent the most reliable agency of information transfer. They allow information to be transmitted through any medium where living organisms dwell, under all meteorologic conditions, during any time of the day or night, in rivers, seas, within the earth's crust, and finally in the tissues of organisms themselves. It is necessary only that the appropriate frequency range and field type be employed.

Previous research results justify the assumption that the diverse manifestations of the biologic activity of electromagnetic forces in living organisms have specific properties that were shaped during the course of evolutionary development. The experimentally observed high sensitivity of organisms to electromagnetic forces—from protozoans to man—can be explained only on the basis of this hypothesis. This is also true for the reactions of most biologic systems and the sensitivity of living systems to changes in natural environmental conditions brought about by electromagnetic fields.

One further observation should be noted. The regulatory effect of natural electromagnetic fields on living organisms, and the various disturbances of regulation and of communication links in the organism resulting from mismatched artificial electromagnetic fields, were observed in connection with intensities that preclude any kind of energy effects in the tissue. It is now clear that the nature of an organism's response to such fields depends not only on the quantity of electromagnetic energy absorbed by the tissues, but primarily on the modulation and time-behavior of these fields and on which parts of the organism are affected by them. Drawing on experience, we know, for example, that the strength of a certain response is not proportional to the intensity of the acting field forces, but actually decreases significantly, in some cases, when the intensity is increased.

There is, of course, no dearth of scientific reports on failures in studies of the biologic activity of electromagnetic fields. De Lorge (1973a, b, 1975) and de Lorge and Grisset (1977), for example, report on experiments with two monkeys that were exposed to the effects of extremely low frequency magnetic and electric fields of low intensity. These authors employed a magnetic field at 10^{-3} T (10 G) concurrently with a corresponding electric field at a frequency of 60 Hz or 10 Hz. Neither animal displayed the slightest anomaly in either blood analysis or general state of health during the various experiments performed.

The "Report of the Committee on Biosphere Effects" (1977) made in connection with Project Seafarer also deals with the biologic inactivity of electromagnetic fields. The position is taken in this report that electromagnetic fields are innocuous, and corresponding evidence is presented, in an effort to vindicate the planned Seafarer project. However, an objective, scientific appraisal of the facts, shows:

1. There is an overwhelming body of evidence, gathered by a large and diverse body of international scientists in many fields, that electromagnetic fields are biologically active. All these researchers cannot be wrong.
2. Unsuccessful experiments do not contradict these results, but merely show that the fields are not biologically active in all cases.
3. In the light of current knowledge, such failures can only be seen as indications that such research work, which is extremely delicate and complicated, can be properly interpreted only by a team of researchers possessing years of experience in this complex area. The danger of overlooking essential points is great, indeed, and many such experiments are conducted primarily from a medical or biologic standpoint with little regard for critical physical and technical factors.
4. Conversely, there is an equal danger that biologic investigations will be approached from a purely physico-technical standpoint. But biologic systems cannot be measured in the same way as physico-technical quantities, which is why purely technical thinking with regard to the quantity measured, energy conditions, reproducibility, the effect of external circumstances on the quantity measured, etc., will almost always lead to erroneous results.
5. Aside from this, the difficulties involved in such research have not yet been completely clarified scientifically. This applies in particular to biologic mechanisms of action. Experimental failures whose causes stem from biologic function cannot be ruled out; but they do provide incentive for investigating these causes more thoroughly. In any case, these failures in no way cast doubt on the great many reports on successful experiments that have been published.

Summary

It was demonstrated in our survey that, on the basis of indisputable experimental evidence, a variety of biologic effects exist in connection with electromagnetic fields over the entire frequency spectrum. These effects are observed at all levels in the plant and animal kingdom, in fact, in all biologic systems from macromolecules to complete organisms.

Field intensity represents a special problem in this regard. As the field intensity increases, biologic effects become more pronounced only within certain ranges. Very intense fields often evoke no response, or an entirely unexpected one. In many cases there apparently exist one or more optimal ranges in which electromag-

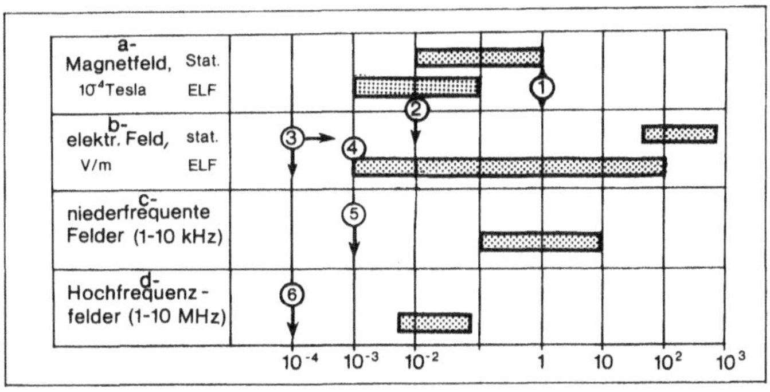

Figure 7-2. Intensities of natural electromagnetic fields (shaded areas) and lowest response threshold of biologic systems as a function of various frequency ranges: (1) increase in the motor activity of birds; (2) turn of the divining rod; (3) conditioned reflexes in fish with electric organs; (4) reaction time effects in man; (5) conditioned reflexes in fish without electric organs; (6) conditioned muscular reflexes in man. Baseline: Magnetic field in G (10^{-4} T), electric field in V/m. Data from Presman (1970), with additions.
Key: a-Magnetic field; b-Electric field; c-Low-frequency fields; d-High-frequency fields

netic field effects are maximal (Bachmann and Reichmanis, 1973; Murr, 1966a; Möse et al., 1973; Schulz, 1970; Levengood, 1967b; Varga, 1973; Haine and König, 1960; Gordon, 1964).

Agreement between observed sensitivity values and the field intensities occurring in nature is not absolutely necessary, for many systems may have altered during the course of evolution. With this in mind, consider the graph in Fig. 7-2. It shows the intensity range of natural electromagnetic fields in various frequency ranges and the lowest threshold of intensity at which a response has been experimentally evoked in biologic systems at various frequencies. As can be seen, biologic systems show sensitivities that are on the same order of magnitude as the intensity values of natural fields, and are even better in most cases.

These data on the biologic effects of electromagnetic fields and forces thus indirectly confirm the hypothesis on the important role played by electromagnetic fields during the course of evolution and their important influence on the vital activity of organisms. Both direct and indirect evidence supports the hypothesis of Presman (1970) and Popp et al. (1979) that electromagnetic forces play an important, hitherto-unrecognized role in the transmission of information to and between living organisms.

8

*Small Air Ions As Biologically Active Agents**

The concept of air ion effects on living forms is based on an unusual pattern. It has a "long past and a short history." The observation that "certain winds occasion disorder and sickness" (Petersen, 1946) dates back to 400 B.C. and Hippocrates's Greece. Failing to identify the specific mediators of weather sickness frustrated medical practice at the time, but does account for the long past of air ions. Their history was to wait another 2000 years. For it wasn't until that property of air identified with biologic activity was formally specified and recorded in some fashion, that the phenomenon would be accorded any true history. It began in the mid-1700s when, on opposite sides of the Atlantic, Benjamin Franklin (1752), in America, and d'Alibard (1752), in France, independently demonstrated the existence of atmospheric electricity. Very shortly thereafter, Becarria (1775), Gardini (1782), and Bertholon (1783), again independently, were moved to view electric phenomena as one of nature's modes of stimulating plant growth. L'Abbe Bertholon also theorized that it affected the course of various diseases in man. These ideas continued to surface from time to time as natural philosophers from the ancients on studied the world around them, but it wasn't until the publication of Lord Kelvin's paper of 1859 that electric phenomena associated with atmospheric conditions became truly amenable to scientific investigation. "History" was imminent. Thomson (1898) in England, and Elster and Geitel (1899) in Germany, discovered air ions, and their conduction of electricity in air became apparent. Sufficiently reliable apparatus for their quantification became available early in this century, making the recorded history of air ions scarcely seven decades old. The successful generation of small air ions free from oxides of nitrogen and ozone was not achieved until the 1950s (Krueger, 1969).

Once equipment was developed for generation and quantitation of air ions, laboratory studies of their properties were undertaken. Biologists initiated research on their effects on living forms and physicians started to use them in treating disease. In general, physical research progressed with dispatch and entailed few critical episodes. On the other hand, great controversy attended the biologic and clinical investigations. While the volume of work ac-

*This chapter is co-authored by A. P. Krueger and S. Sigel.

complished by the biologists and physicians is a tribute to their industry, the results were not clear-cut. Much of this ambiguity arose from defects in experimental design, e.g., experiments performed with corona discharges as ion sources neglected consideration of the ozone and oxides of nitrogen inevitably produced along with the ions; ion concentrations, temperature and relative humidity were not monitored; experimental subjects were not grounded; as a rule the air was not purified and combination of ions with pollutants led to widely fluctuating ion densities. Physicians, assessing the value of air ions as a therapeutic modality, faced the additional problems of placebo effect, the usual difficulty of case selection and application of a double-blind technique. Because of these formidable roadblocks, a considerable body of data in the literature is open to question. Nevertheless, there remains a rich vein of evidence, both old and new, capable of withstanding critical scrutiny.

In addition to the technical elements of uncertainty, the evaluation of air ions has been hampered by the construction of conceptual barriers. The study of air ions as biologically active agents has not been undertaken in a neutral atmosphere. The very term "ion" has historically evoked incredulity. Arrhenius, who proposed an ion concept to his doctoral committee to account for the migration of aqueous molecules in an electric field, was granted his degree with the greatest reluctance and the lowest possible grade. The idea was condemned as bizarre. In spite of a slow start, he went on to become the father of modern physical chemistry and the concept of charged particles in solution is no longer derided. In fact, it is the basis upon which much of modern physical science is founded. These days, there is no disagreement about the disparate physical nature of air ions and nonionized gaseous molecules. However, that such particles in ambient air are biologically active elicits the same kind of resistance encountered by Arrhenius almost a century ago. We have already indicated that prudence with regard to early experimental procedures is called for. However, the criticism that charged ions are present in too dilute an ambient air concentration to be biologic agents (Andersen, 1972) contravenes the acknowledged sensitivity of certain biologic sensors: naturally refined systems often initiate physiologic responses which far exceed the intensity of the environmental stimulus. For example:

1. The threshold value for eliciting a response in a male silkworm (*Bombyx mori*) with the sex attractant pheromone of the female is 2600 molecules: the concentration required is less than 200 molecules/cm^3 of air (Bossert and Wilson, 1963). Ion densities in clean country or mountain air fluctuate around 2000/cm^3 or 10 times that concentration.

2. Amoebae exhibit shifts in shape and mobility when exposed to fields of approximately 10–15 V/cm at 10 Hz (Friend et al., 1975), an electromagnetic frequency for which biologic sensitivity was not previously known. Ion and atmospheric disturbances can generate potential gradients on the order of 50,000 V/m and these phenomena have been implicated in stimulating plant growth. Ions have already been identified as conductors of atmospheric electricity, but of even greater interest is that the field effect can be separated from the ionic influences on plant growth (Krueger et al., published in 15B, 1978).

3. Prenatal exposure of rats to an extremely low frequency (0.5 Hz) rotating magnetic field with intensities ranging from 1 to 30 G results in diminished ambulation, increased conditioned suppression and increased weights of thyroid glands and testicles (Persinger, 1969; Ossenkopp, 1972; Ossenkopp et al., 1972; Persinger and Pear, 1972). Though the mechanism of action has not been established, the findings once again demonstrate biologic effects mediated by previously unassumed modalities.

4. The human eye can detect a flash of light when a single active quantum reaches the retina (Pirenne, 1958).

5. Anthopleurine, the alarm pheromone

of the sea anemone, has a median effective concentration of 3.5×10^{-10} mole/l of sea water or about 1 molecule of active agent to 1.6×10^{11} molecules of water (Howe and Sheikh, 1975). This speaks for an analogy between the "oceans of air" that surround us and potentially effective ion concentrations where an air ion is dispersed among $1-3 \times 10^{16}$ nonionized gaseous molecules.

These and other physiologic responses are indicators of the heretofore unexplored limen of biologic sensor systems. In light of these examples, it is difficult to dismiss the capability of small air ions to affect living forms solely on the basis of their presence in low concentrations. There remain, however, more subtle barriers to their acceptance as biologic agents. Air ions are enigmatic entities for other reasons. They do not fit into categories of the classical disciplines, and for this failure they run askant of currently acknowledged scientific paradigms. History has given us some fascinating examples of systems in interface: electrophysical, electrochemical, psychochemical events; bizarre phenomena whose characteristics led in sequence to phases of extensive experimentation, heated controversy and finally to recognition of their spurious or veridical nature. Roentgen's X-rays instigated a dramatic paradigmatic shift and ushered in the era of small particle physics, an *electrophysical* field whose limits are yet to be specified. Blondlot's N-rays similarly defied existing phenomenologic explanation, but the failure to define criteria for the establishment of a new discipline doomed Blondlot and N-rays to historical obscurity (Blondlot, 1904, 1905). Eccles's (1953, 1957) demonstration of the transmitter properties of ACh firmly established the *electrochemical* nature of the nervous system, paving the way for the age of psychopharmacology and the study of *psychochemical* events, in the middle 1950s through 1970s. Though there are still unresolved conflicts (just how does a drug, a material entity, effect changes in ideation, by definition, a nonphysical event), its value as a model is adhered to on at least some implicit or empirical level. Air ions are unique in that they have a history of transecting the boundary lines of several disciplines. They are naturally occurring *electrophysically* charged particles, capable of altering *electrochemical* processes, including serotonin, a "powerful neurohormone" implicated in the modulation of emotional, *psychochemical* states. In so doing, air ions tend to violate several territorial imperatives, that is, implicit rules or paradigms (Kuhn, 1962). Not only does this invasion (of territory) create confusion, it may, as any ethological observer may affirm, elicit aggression from those whose boundaries have been encroached upon. Thus air ions upset the applecart and a vigorous response to the challenge is appropriate if scientific disciplines are to resist the vagaries of fad. The onus of proof falls on the challenger and one must respond to the question, are the putative biologic effects of air ions in the same category of phenomena as Blondlot's illusory discoveries? Our purpose here is to determine whether sufficient evidence exists to respond to the question in the affirmative or the negative, though it is the "peculiar difficulty of every synthesis that its end is already implicit in its beginnings" (Teilhard de Chardin, 1959).

Origins and Properties of Small Air Ions

Small air ions are commonly formed by several processes in nature. For example, radon and thoron produced in the soil are released and carried into the atmosphere by eddy diffusion. At the molecular level, emissions from such radioactive sources displace an electron from a molecule of one of the ordinary atmospheric gases and leave the nucleus with a positive charge. The freed electron is promptly captured by an adjacent molecule, which then becomes negatively charged. In a cubic centimeter of air, where there are about 2×10^{19} molecules present, ions collide with neutral molecules at a rate of about 10^9 impacts/s. This interaction readily leads to charge transfer with the result that positive charges eventually reside on molecular species of the lowest ionization potential, while electrons are held by species of greatest stability. Over dry land, in each cubic centimeter of air about 10 pairs are formed in 1 s. Within 2×10^{-7}s after this molecular ion formation, 1–8 neutral molecules, including water, the common gases and trace gases cluster about and the whole molecular aggregate is now properly called a small air ion. These ionic "rafts" are not uniform in composition or size and are subject to further changes during aging; recently there has been emphasis on the fluctuating incorporation of trace gases and on the potential of small air ions to serve as catalysts in the conversion of trace gases into chemical compounds possessing varying degrees of toxicity. However, since the size of the molecular cluster does not vary excessively and ions do carry a unit charge of 1.6×10^{-19} C, they are considered to exhibit a mobility of .01–1 $cm^2\ s^{-1}\ V^{-1}$. Thus ions are categorized by charge and mobility.

There are other sources of energy in nature that are also capable of producing small air ions, e.g., the shearing of water droplets in waterfalls or the pounding of the surf (the Lenard effect; Lenard, 1892), cosmic rays, corona discharges, etc. In areas remote from foci of pollution, small air ion densities will approximate 1200 positive ions/cm^3 and 1000 negative ions/cm^3. The concentrations may be considerably greater in some locations, e.g., on White Mountain in California at the 14,000 foot (5,200 m) level careful measurements have yielded averages of 2500 positive and 2200 negative ions/cm^3. In contrast, the total of small air ions in an industrial area has been found to be less than 80 ions/cm^3. The ion deficit in commercial zones is attributable to two properties of small air ions. They readily unite with (1) condensation nuclei to form large (Langevin) ions, and (2)

Table 8-1. *Comparison of ions with other particles encountered in the air. From Vasiliev (1966).*

	Size in cm
Molecules	10^{-8}
Light, or normal, air ions	10^{-7}
Average, or intermediate, air ions	10^{-6}
Heavy, or large, air ions	10^{-5}
Fog particle	10^{-4}–10^{-3}
Raindrops	10^{-2}–10^{-1}
Limits of visibility to the eye	5×10^{-3}
Limits of visibility to the microscope	2×10^{-5}

many classes of particulates, aerosols and pollutants. In both instances the biologic activity of small air ions is considered to be lost. Additional circumstances may also reduce their biologic activity, such as when their ionic nature is lost or deflected. For instance, when positive and negative ions collide, their charges are neutralized and they cease to be ions. Or, ions of like charge (unipolar ions), by repelling on another, tend to force the charge flow to enclosing surfaces where again their ionic nature is dissipated. Finally, since ions are small and carry a charge, they are readily deflected by electric fields. All of these characteristics make it difficult to maintain high ion concentrations during a biologic experiment; in practical terms this means a maximum concentration of 5×10^5 to 1×10^6 small air ions/cm^3.

Air Ion Measurement

It is useful to have available a means of measuring the numbers of ions emitted by a generator or the concentration of ions impinging upon a given surface per unit of time. A simple device with this capability was developed by Beckett (Fig. 8-1; Beckett and Krueger, 1961). The disc, which is surrounded by a grounded wire shield to protect against distortion of the electrostatic field, is attached to a well-insulated and shielded conductor for connection to a fem-

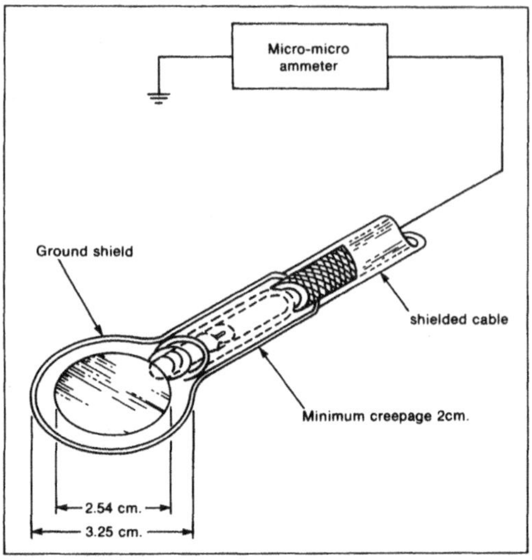

Figure 8-1. Schematic of air ion target probe as designed by John C. Beckett.

toammeter. Ions reaching the target are measured in terms of numbers of ions $cm^{-2}s^{-1}$ according to the equation:

$$N = \frac{I}{qA}$$

where: N = number of ions $cm^{-2}s^{-1}$
I = ion current in amperes
q = charge of one ion: 1.6×10^{-19} C
A = area of probe in cm^2.

Air ion density, i.e., numbers of ions/cm^3, can be measured with collectors of various types, all based on the assumption that each ion has one unit charge; Loeb has demonstrated that this is highly probable for small air ions. A simplified type of collector designed by Beckett (Fig. 8-2; Beckett and Krueger, 1961). consists of 4 parallel aluminum collecting plates 7 cm wide, 33 cm long, spaced 1 cm apart and grounded. Located between and on the outside of these collecting plates are five polarizing plates of the same size. The collecting plates are set on Teflon insulators so placed that they are well protected from surface charging and contamination. The plate collector assembly is enclosed inside a rectangular metal duct which serves as a shield against stray electric fields and provides a channel for the air containing the ions. A motor-powered blower mounted at one end of the duct moves the air through the duct at a velocity of 75 cm/s. Ions traversing the areas between the charged polarizing plates and the grounded collecting plates move to the latter, where the current thus developed is measured with a femtoammeter. Since ion mobility is inversely proportional to size, three different potentials (22.5, 67 and 200 V) provide a choice corresponding to minimum mobilities. Only small air ions will be collected at the lowest voltage, small and medium-sized air ions at the second voltage and small, medium and large ions at the highest voltage. The equation for volumetric measurement is identical with that for the ion probe, except for the inclusion of air velocity in the denominator. The overall accuracy of the equipment is about 15%.

Air Ion Generation

We have noted that a variety of processes in nature operate to produce small positive

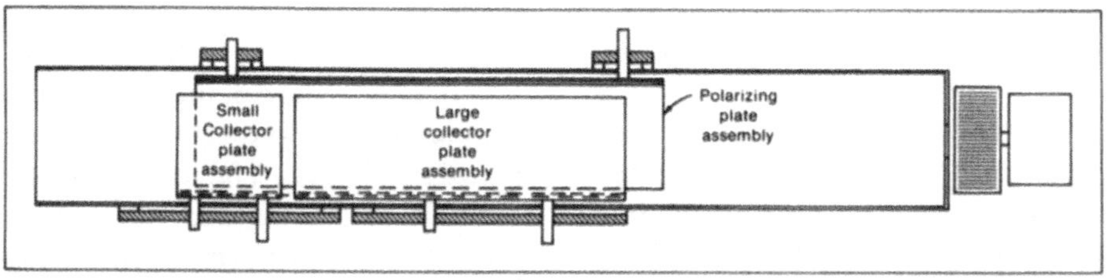

Figure 8-2. Air ion collector viewed in longitudinal section. One of the four collector plates and one of the five polarizing plates are shown. The motor-powered blower is at the right end of the duct.

and negative air ions. For purposes of obtaining relatively high concentrations of ions to be used in laboratory experimentation, or in clinical tests, two energy sources are generally employed: corona discharge or radiation from isotopes such as ^3H or ^{210}Po. Among the former, notable advances in design and performance have been made since the 1930s when equipment was bulky and often produced unacceptable amounts of O_3. There are now available generators operating at only a few kilovolts and capable of maintaining a concentration of 10^3–10^4 negative ions/cm^3 in a room of about 100 m^3. They are conveniently small and have almost no ozone by-product.

For laboratory experiments we have depended upon tritium-powered ion sources. Tritium is a low-energy (0.0185 MeV) pure beta emitter. Fifty mCi is sealed in a foil of zirconium mounted on a stainless steel disc. Unless the disc is charged the ions formed by beta emanation in the plasma next to the foil recombine and no ions emerge. However, when the disc is negatively charged with a direct current of about 900 V, positive ions are neutralized while negative ions are ejected into the adjacent atmosphere. The rate of ion formation is about 10^9 ions/s and due to the nature of radioactive decay, it is constant. An additional advantage is the long half life of ^3H, about 13 yr. We have not found ^3H generators to be an effective ion source for maintaining high concentrations in large, enclosed spaces. For this type of experimental use, corona discharge is a preferred ion source.

Action of Small Air Ions

Microorganisms

Microorganisms offer a convenient means of demonstrating air ion action and over the years many investigators have found that small air ions exert a moderate lethal effect on bacteria, fungi and some bacterial viruses. Beginning in 1933, Tchijevsky and his associates at the Central Laboratory for the Study of Aeroionification in Moscow conducted a long series of experiments with agar plate cultures of *Staphylococcus aureus, Vibrio cholera* and *Salmonella typhus*. In all instances, concentrations of 5×10^4–5×10^6 small air ions inhibited colony formation; negative ions were measurably more effective than positive ions, while electrical fields of 300–1000 V/cm were ineffective. They also observed that small air ions reduced the viable cell count of the normal microflora of air in enclosed spaces and had a similar effect when the air was enriched with artificially generated bacterial aerosols. Lethal effects of air ions on the mold *Neurospora crassa* were reported by Fuerst in 1955. Pratt and Barnard (1960) had comparable results with *Penicillium notatum*. About this same time, Kingdon (1960) observed that negative or positive air ions generated by a corona discharge killed *Escherichia coli*. Biro et al. (1969) and Biro and Svab (1970) reported marked lethal action of small negative air ions on *Staphylococcus aureus* and on four bacteriophages of the T group.

Phillips et al. (1964) used sophisticated techniques to expose *Serratia marcescens* in droplets of less than 5 μm to positive or negative air ions in a chamber holding 365 l. The exponential decay rate was 23%/min for untreated air, 54%/min with positive ions and 78%/min for negative ions. Their data led Phillips et al. to conclude that positive ions act chiefly by increasing the rate of precipitation of aerosol particles. In the case of negative ions, such particulate loss is supplemented by a significant amount of biologic decay.

In 1964 numerous reports on the lethal action of air on microorganisms were presented at the All-Union Conference on Aeroionization in Industrial Hygiene which convened in Leningrad (Boyco et al., 1964). In essence, there was general agreement that the viable cell count of the natural microflora of the air or of artificially created aerosols was quickly reduced by 85%–93% by introducing concentrations of small negative air ions ranging from 5.2×10^4 to 9.5×10^4 ions/cm^3.

Sharp (1967) examined the effect of small air ions on the uredospores of *Puccinia striiformis,* the fungus which attacks wheat, barley and related grasses. He found that small air ions consistently increased the rate of germination.

Elkiey et al. (1977) report on the effects of small air ions on net blotch disease in barley:

> Small positive air ions significantly reduced the number of lesions and percentage of leaf area covered by lesions of net blotch disease of barley. Appearance of disease symptoms was delayed by 3 days when plants were exposed to an atmosphere of positive ions. No such effect was observed for negative air ions. Height and dry weight were significantly stimulated by unipolar ions.

Our first studies in this area involved experiments with *Staphylococcus aureus* suspended in droplets of distilled water sufficiently small to provide a high ratio of surface area to volume. When ion generators consisting of sealed foils of ^3H or ^{210}Po were used to produce about 1.6×10^9 negative ions s^{-1}cm^{-2} on each of these droplets, the death rate was significantly accelerated. This result was obtained only if the droplets were constantly stirred with a vibrating fine quartz fibre in order to bring the cells to the air–water interface (Krueger et al., 1957).

Subsequently we conducted another set of experiments to determine whether the lethal effect was due to a biologically active radical formed when small air ions react with water or to direct contact of air ions with the suspended cells (Krueger et al., 1975). In these tests a corona discharge type of negative ion generator was employed and supplied 3.7×10^{11} small negative ions s^{-1}cm^{-2} at the surface of the suspension opposite the generator. The death rate of *Staphylococci* was not affected by ion-treated water. The lethal action of ions could be demonstrated only when the cell suspensions were stirred sufficiently to bring about direct contact of cells and ions. In both sets of experiments we took steps to avoid certain potential sources of error:

1. Direct action of radiation on the cells in the case of ^3H or ^{210}Po. To avoid this the sealed foils were placed well beyond the maximal ranges of alpha or beta radiation.

2. Agglutination of bacteria resulting in reduced viable cell count. Direct microscopic examination revealed no cell aggregates. Further, the lethal effect was reversed by exposing the cells to intense light of the visible spectrum; this is typical of many instances of cell damage and does not occur in simple agglutination. Finally, the time required to produce a lethal effect was much less than the minimal time of agglutination predicted by the von Smoluchowski equation:

$$t = \frac{1}{4} D \, 2r \, Vo$$

in which: t = time in s;
r = radius of particle;
Vo = number of particles/ml;
$D = RT/N \times \frac{1}{6}\pi\eta r$ (the Einstein equation for the diffusion coefficient).

3. Bactericidal action of ozone produced by the ion generators. Under the conditions prevailing in our experiments, no measurable amounts were generated by either radioisotope or by the corona discharge ion source.

Recently Kellogg et al. (1979) presented experimental evidence that the superoxide ion O_2^- is involved in the bactericidal effects of negative air ions on microorganisms. Cells of *Staphylococcus albus* exposed to a high flux density of negative ions showed a total loss of viability after 5 hr of treatment. Addition of superoxide dismutase to the suspension resulted in full retention of viability after 5 hr.

The data reviewed above, representing

the experimental observations of some 25 investigators, are sufficient to establish the lethal effect of small air ions on vegetative bacterial cells, some fungi and bacterial viruses. The action is not limited to a single species or genus, it can be demonstrated when bacteria are growing on solid media or when the organisms are suspended in small drops with a high ratio of surface to cell volume. Stirring is always necessary to bring the cells to the air–water interface unless the water drops are very small. So far as we are aware, no one working with this phenomenon has obtained other than positive results. While the mechanism underlying the lethal action remains to be clarified, one may hypothesize that it could involve increased oxidative reactions, such as those occurring in other types of cells (Kotaka et al., 1965).

Action of Small Air Ions on Plants

Plants are known to react to the direct application of electric stimuli; for example, Bose (1913) applied small currents to members of the genus *Mimosa* and observed that growth was accelerated by cathode "break" and anode "make" shocks and was retarded by cathode "make" and anode "break" shocks. He also noted definite mechanical effects, such as contraction of pulvinar cells with resultant downward displacement of the leaves. Several investigators have tested the hypothesis that in such cases growth stimulation takes place because of induced translocation of the growth hormone auxin (Brauner and Bunning, 1930; Went, 1932; Kögl, 1933; Koch, 1934). On the whole, the experimental evidence supports this concept although there is doubt that the shifts take place through simple cataphoresis of auxin as an electrolyte. Cholodny and Sankewitsch (1937) concluded that the displacement involves more complex factors in the living protoplasm. Interestingly enough, Kögl's work led him to the idea that some of the results obtained with decapitated coleoptiles of oats could best be explained by diurnal variations in the concentrations of air ions in the ambient atmosphere.

It is not generally recognized that plants respond to less obvious means of administering electromagnetic forces, yet as noted in the introduction, Beccarria, Gardini, and Bertholon long ago theorized that the existence of electricity in the air even in fine weather might influence the growth of plants. Each succeeding generation witnessed exploration of this possibility with results that, taken together, were affirmative, but not entirely conclusive because of lack of available information about some of the factors involved and deficiencies of experimental design. The situation was appreciably changed for the better by two developments: the introduction of a satisfactory method of artificial overhead electric discharge by Lemström in 1885 (Lemström, 1904) and the discovery of air ions 15 years later.

The way was then paved for more definitive experimentation and for improved theoretical analysis of the results obtained. In the first instance, Lemström's innovation led to the development of "electro-culture," a subject well described in Sidaway's paper of 1975. Briefly, a number of scientists, notably in Britain, found that wheat, maize or barley plants grown in pots exposed to electric discharges from overhead wire networks carrying low currents at about 1600 V exhibited increases in dry weight varying from 18%–27%. On a larger scale with field crops of oats, wheat and barley, wires strung 7 feet above the soil carried 40,000–80,000 V and a current of 0.5–1.0 mA/acre. In 18 experiments, 14 showed increases in crop yield; of these, 5 were greater than 30%. Marked physiologic effects on flowering and root dry weight also were recorded. The data from a wide range of experiments were collected by the Electro-Culture Committee of the Board of Agriculture and Fisheries, which was established in 1918. Hicks (1957) made similar observations in the United States. He enclosed pelargonium, rhododendron and bean plants with Faraday cages which served as growth controls. When similar cages were charged with 120 V DC marked stimulation of growth and blooming occurred with a positive electrostatic field and somewhat less stimulation with a negative field.

Our own work in this area began in 1960 and during the ensuing years we have intensively studied the influence of air ions on the growth of seedlings under controlled conditions. Most of our tests were conducted with barley (*Hordeum vulgaris*), although similar results were obtained with oats (*Avena sativa*), lettuce (*Lactuca sativa*) and pea (*Pisum sativum*). Seedlings were grown in washed sand infused with a chemically defined nutrient solution. Our earlier experiments were performed in cubicles of 3.2×10^3 capacity supplied with air purified by passage through an electrostatic precipitator, an activated carbon filter, an absolute filter and an air conditioner (Krueger et al., 1962a). Uniform conditions of lighting, temperature and humidity prevailed. A positively ionized atmosphere was maintained in one cubicle, a negatively ionized atmosphere in the second, while the third cubicle served as a control environment with an average ionic background of 250 positive ions and 200 negative ions/cm^3. The ion generators were tritium powered; ion densities, temperature and humidity were monitored. Electric field controls were included to determine the effects of the positive and negative potential gradients in the absence of artificially enhanced ion densities.

Later experiments were performed in a greenhouse supplied with pure air and the seedlings were grown in tanks of 15-l capacity, filled with chemically defined medium and covered with inverted clear plastic boxes with ion generators mounted in the top (Krueger et al., 1963b). Planter units were rotated in position at least twice a week to avoid effects due to variation in incident light. The minimum temperature was 25°C, but maximum temperature varied with the season. The number of seedlings in each test group varied from 36 to 72; randomization was practiced in sampling and all the data were examined for statistical significance. The methods employed in these experiments are summarized by Krueger et al. (1962a).

It was found that ions of either charge generated in pure air produced statistically significant increases in growth rate as measured by stem length, integral elongation, fresh weight and dry weight. The growth stimulation roughly paralleled the ionic density and was accompanied by a proportional increase in the flow of current to ground. Analysis of ion-treated plants for total nitrogen, protein, total sugars and reducing sugars revealed no deviation from normal composition (Krueger et al., 1963). We also tested the effects on higher plants of exposure to air containing a minimal number of air ions: about 30 positive and 30 negative ions/cm^3. Air ion depletion resulted in a statistically significant reduction in integral elongation, fresh weight and dry weight (Krueger et al., 1965). Chlorophyll production was unchanged, but the plants developed soft leaves and lacked normal rigidity.

We found that unipolar ionized atmospheres exerted a profound effect on the biosynthesis of cytochrome C. This work began with the observation that the development of iron chlorosis in barley seedlings cultivated in iron-free nutrient solution was markedly accelerated by exposure of the plants to air ions of either charge (Kotaka et al., 1965a). Chlorosis is a condition in which the leaves of the plant turn yellow because of interference with chlorophyll formation. As the amount of chlorophyll in the plant tissue falls with the onset of chlorosis, there is a marked rise in the concentration of cytochrome C. In our experiments, when seedlings were grown in an iron-containing medium, and were exposed to positively or negatively ionized atmospheres during growth, the typical ion-induced acceleration in growth rate was accompanied by stimulation of cytochrome C synthesis, but no chlorosis developed and there was no essential difference in chlorophyll content of the exposed and control plants.

Subsequently studies were made on the effect of gaseous ions on the active and residual iron content of barley seedlings during the course of iron chlorosis (Krueger et al., 1964). Active iron is located in the chloroplasts, participates in the biosynthesis of chlorophyll and is soluble in 1N HCl. Residual iron occurs outside of the chloroplasts and is not concerned with chlorophyll production, nor is it soluble in 1N HCl. Positive or negative air ions induced a significant decrease in active iron content of the seedlings as the chlorophyll content declined. At the same time there occurred an increase in residual iron and in the cytochrome C fractions of the seedlings. There is evidence that the rise in residual iron content involves not only cytochrome C, but other cytochromes and iron-containing enzymes as well. In addition, we found that the uptake of exogenous iron (^{55}Fe) was increased during the treatment of very young seedlings with positive or negative air ions (Kotaka et al., 1965b). We developed a hypothesis that incorporates these observations and identifies a major site of air ion action as the regulatory systems responsible for the iron metabolism in the seed and young seedling. It should be pointed out that other investigators have demonstrated the capability of a range of potential gradients to affect the incorporation of iron, aluminum, zinc, manganese and copper (Zhurbitskü, 1958; Zhurbitskü and Shidlovskaya, 1967; Murr, 1963, 1964, 1966b). Zhurbitskü and Shidlovskaya noted that artificially increased densities of light air ions enhance this process.

As our program of experimentation with plants progressed, we found that air ions of either charge increased the oxygen consumption of barley seedlings (Kotaka et al., 1965c). Additional evidence concerning the mode of action of air ions on plants was obtained from experiments with isolated chloroplasts. Chloroplasts shrink when stored in the dark at 4° C and swell again when illuminated. The rates of swelling and shrinking were increased by exposure to air ions. Treatment of the chloroplasts in the presence of ATP with negative or positive air ions at 4°C in the dark increased their swelling/shrinking and ATPase activities, indicating that air ions stimulate ATP metabolism of isolated chloroplasts (Kotaka et al., 1968).

In analyzing the situation which prevails when plants are exposed to atmospheric electricity, both fields and products of corona discharge including air ions must be considered as potential factors. Most of the experimental reports in the literature deal with field and air ion effects separately, neglecting the possibility that they may work in conjunction and are not altogether mutually exclusive forces. When field controls are incorporated in the experimental design of an air ion experiment it is often done without considering the ubiquitous air ions in the background, where their migration in the electric field conveys a current to the plants. For example, Murr (1966b) studied intensively the biophysics of plant growth in electrostatic and electrokinetic fields, and concluded that growth is influenced largely by the magnitude of active current flow. Sidaway (1976) reported that exposure of certain plants to a 1-kV field induced a polarity-dependent inhibition or promotion of flowering. In both these cases air ion concentrations were not measured and there is no way of telling in retrospect to what extent they participated in the observed results.

If the assumption is made that an electric field per se can influence the physiologic processes of a living organism, one would expect it to do so by exerting force on those constituents which carry an electric charge and are located at or near the surface. The direction of the force will be determined by the sign of the charge. Dipolar compounds, because of the positive and negative charges they carry, will become oriented along the lines of the force field. With strong enough fields they will be disrupted, the charges will move in opposite directions and a current will flow with resultant production of corona. That this condition develops in plants exposed to relatively weak electrical fields is demonstrated by the experiments of Bachman et al. (1971), who investigated the extent to which ozone and air ions accompany the biologic applications of electric fields. Using barley seedlings, they observed that in the area close to the leaf tips electric fields may be intensified to a degree sufficient to produce corona even in the low ranges characteristic of fine weather at the earth's surface, about 100 V/m. Applying this mechanism to the tip of a blade of grass, the natural electric field may be increased more than 1000 times, i.e., well in excess of the strength required for production of corona. Barley leaves have spikes along their edges with diameters of about 1/1000 of the diameter of the tip. Intensification of the fields in their proximity will result in formation of corona with applied potentials of only a few hundred V/m. Accompanying these events will be transmission of current to the plant, generation of air ions and formation of gaseous products such as ozone.

Bachmann and Reichmanis (1973) explain the results obtained when barley seedlings are grown in high positive electric fields as follows. There are two field-responsive intrinsic factors, one stimulating growth and one retarding it. At 200 kV/m the opposing forces are in equilibrium. At lower fields the enhancing component associated with by-products of corona prevails, while at values greater than 200 kV/m the retarding constituent predominates. The latter appears to be dependent upon current flow through the plant. This is compatible with the earlier findings of Cholodny and Sankewitsch (1937) and Lund et al. (1974), who noted that current flow from base to apex stimulated plant growth while the reverse flow inhibited it.

During 1976 and 1977 Krueger and colleagues conducted a long series of experiments designed to assess the roles played by air ions and electric fields in stimulating the growth of barley seedlings (Krueger et al., 1978). Basic conditions consisted of adequate populations of seedlings, pollutant-free air, constant temperature and light, relatively constant low humidity and an ion-depleted background of 60–85 small negative or positive ions/cm^3. The essential element of this work was the functioning of tritium foils as bias electrodes at appliied voltages of 67.5–955 V, which provided field strengths of 4.5–90 V/cm and produced concentrations of small air ions of 2.7×10^4 to 1.7×10^5 ions/cm^3. These were sufficient to yield a statistically valid increase in growth rate when compared with seedlings exposed to equivalent electrical field in ion-depleted air. There was no detectable difference in growth rates of seedlings treated with (1) ion-depleted air with no electrical field and (2) ion-depleted air with fields of 4.6–6.8 V/cm.

Apparently the corona discharge resulting from the field intensification on small plant structures did not produce enough air ions to stimulate growth. Growth stimulation occurred only in the presence of a sufficient concentration of small negative air ions and was associated with a flow of electricity through the plants to ground, about 10^{-11} A/plant. This value closely approximates that observed in the experiments of Blackman and Legg (1924) for single plants in pot culture and by Krueger et al. (1962a) for groups of plants. Although these data clearly indicate that the low electric fields themselves did not serve to enhance growth under controlled experimental conditions, it is equally evident that in nature electric fields and air ions are intimately associated because of the ease with which corona forms on the surface of plants.

To summarize this discussion of the effects of atmospheric electricity on plant life, we believe it is fair to say that laboratory, greenhouse and field studies have demonstrated measurable and reproducible physiologic changes associated with exposure to a wide range of air ion densities and potential gradients. The most obvious plant response is growth stimulation, in which air ions presented at low voltages can serve as proximate agents. As potential gradients increase, there is more and more likelihood that the fields themselves will operate to produce corona discharge with inevitable formation of ions. As Bachman et al. (1971) have concluded on the basis of their own experiments and those of others, "Biological effects hitherto attributed to electrical fields must in fact be attributed to the synergistic effects of the by-products of the corona produced by such fields." While considerable progress has been made in elucidating the biochemical mechanisms involved in these phenomena, much remains to be done.

Does the work accomplished so far on a relatively small scale have any wider significance? The answer would appear to be yes. Atmospheric scientists have found that the earth's electric field varies from about 100 V/m in fair weather to about 50,000 V/m under disturbed conditions, e.g., snow and rain carry to earth a precipitation current of fluctuating magnitude along with enhanced ion production (Chalmers, 1967). Worldwide electrification of the atmosphere is a thoroughly documented phenomenon (Shewchuk and Iribarne, 1970), which means that the work carried out so far has broad significance not only geographically, around the world, but biologically as well. All living organisms are subject to atmospheric influences. Thus, even though the mechanisms of action for electric phenomena have not been wholly specified, and this applies for air ions too, some common threads have been cited that tie insect, animal and human literatures to the plant studies.

Action of Small Air Ions on Animals

Air ions, it will be recalled, act on the regulatory systems of plants and affect the ATP respiratory chain through a direct action on cytochrome synthesis. In order to broaden the research on ions via biochemical similarity, an organism whose life cycle also made demands on cytochrome C metabolism had to be found. Such a creature is the silkworm (*Bombyx mori*). In this species, development is dependent on a hormone called "growth and moulting hormone." Its primary function is to stimulate the biosynthesis of cytochrome C and, possibly, related components. If there were an ion effect on cytochrome C, its expression would be in developmental systems. Exposure produced these effects: (1) a marked increase in the rate of larval growth, (2) an increased biosynthesis of catalase, peroxidase, and cytochrome C oxidase, (3) earlier onset of spinning, and (4) an increased weight of cocoon and silk layers (Krueger et al., 1966b). Each of these outcomes was statistically significant and readily reproducible. Other researchers have also reported ion effects in insects (Edwards, 1960; Haine et al., 1964; Helson and Penman, 1970), but the above results are of particular interest because they specifically demonstrate a cytochrome C-linked effect. While such biochemical probings do not constitute a decisive assault on the arcane mechanisms involved in air ion action, they do serve to establish that air ions are responsible for important changes in cellular metabolic processes, and that these changes have the virtue of being reproducible. Moreover, they demonstrate the confirmation in an insect of a hypothesis predicted from plants.

Once the gap between plant and insect effects had been successfully bridged, the transition to increasingly complex life forms was greatly facilitated. For mammals, the influence of ions on the respiratory chain was not directly pursued; however, their influence on respiration was. In one experimental series rabbit tracheal tissue was used. Though other receptor sites could have been examined (Vasiliev, 1960), experience and accessibility predisposed this choice. Furthermore, with this preparation the site and mechanism of action could be studied simultaneously, since ions are probably deposited on tracheal tissue during normal respiration. Experiments were carried out on the tracheal mucosa in situ and in excised strips. The results were clear cut. Positively charged ion complexes decreased ciliary rate, contracted the posterior tracheal wall, induced a state of exaggerated susceptibility to mild mechanical trauma, caused vasoconstriction in the tracheal wall and increased respiratory rate. Negative complexes increased ciliary rate, reversed the positively induced contraction of the tracheal wall, did not alter the mucosal response to mild trauma, did not change the normal vascularity and decreased respiratory rate. All functions were charge dependent, demonstrating on-again-off-again responses to reversed counterbalanced testing conditions (Krueger and Smith, 1960a, b, 1962; Krueger, 1962). It should be pointed out that in this early study the high water vapor saturation required to keep the tissue viable may have, at times, yielded the production of electroaerosols (Andersen, 1972; Krueger, 1962). However, work executed since in laboratories around the world does suggest that a primary site for air ion action in mammals had been demonstrated. Supporting data from influenza experiments may be summarized as follows. When measured doses of influenza virus or of *Klebsiella pneumonia* or the arthrospores of *Coccidioides immitis* are administered by the intranasal route to mice previously treated with small positive ions, there is a significant increase in the cumulative mortality rate. Small negative air ions had the opposite effect in experiments with influenza virus. Air ion depletion, treatment with negative or positive air ions had no observable effect when the challenged dose of influenza virus was administered as an aerosol of 2 μm diameter. Particles of this size tend to clear the nasopharynx and less than 10% are deposited in the tracheobronchial region. An average of 15% of such particles are retained in the lung parenchyma. Influenza virus introduced into the respiratory tract by the intranasal route is subject to all the protective mechanisms of the mucosal surfaces it traverses, and the amount of active viruses persisting in the tracheobronchial tree or surviving to penetrate into the lung parenchyma will be inversely proportional to the efficiency of these functions.

It was then time to study their biochemical mechanisms of action. Interest was centered once again on the cytochromes. That cytochromes play an important role in the biological response to air ions is suggested by the findings that (1) negative ions in vitro enhance the action of cytochrome oxidase in converting succinate to fumarate (Krueger and Smith, 1959), and (2) cytochromes and other iron-containing enzymes are involved in the accelerated growth which occurs in higher plants exposed to abnormally high concentrations of negative or positive air ions (Krueger et al., 1966a). The question posited was whether or not similar mechanisms would be activated in tracheal tissue, where monoamine oxidase (MAO) is responsible in part for the metabolic breakdown of serotonin (5HT) and may be an indicator of overall mitochondrial function. If, like other oxidase systems reviewed, tracheal MAO consisted of a dehydrogenase linked to a respiratory chain that included cytochromes, then ions, at least negative ones, could be expected to have an effect on the cytochrome-linked oxidation of 5HT. This turned out to be the case and the new task became to determine if other empirical demonstrations supported the inference of serotonergic involvement. They did. Smooth muscle contraction, respiratory distress and diarrhea had already been noted in animals exposed to positive ions. Inasmuch as 5HT was a known vasoconstrictor and patients ill with carcinoid syndrome suffer frequent bouts of diarrhea, 5HT emerged as the likely mediator of air ion effects, and the 5HT hypothesis was developed. It suggests that while positive ions induce 5HT release, negative ions seem to speed up the rate at which 5HT is oxidized. Accordingly, intravenous injections of 5HT should duplicate positive ion effects—e.g., decreased ciliary rates, cause vasoconstriction in the tracheal wall, increase respiratory rate, etc.—which is the case. And negative ions should cause a rise in 5HIAA, a 5HT metabolite, which again is the case. Furthermore, experiments in which animal tissue stores of 5HT were depleted by reserpine pretreatment failed to demonstrate typical positive ion effects consequent on normal 5HT release. In corollary, pretreatment with iproniazid, a monoamine oxidase inhibitor, rendered animals resistant to negative ion conditions effected in previous protocols by the accelerated conversion of 5HT to 5HIAA by MAO. There being limited MAO and therefore a reduced medium to accelerate, no negative ion effect was demonstrated. Reversal of conditions in drug/ion or ion/drug combinations were experimentally realized. As the effects of intravenous injections of 5HT are reversed by exposure to negative ions and so forth, indirect confirmation of the hypothesized mechanism of action is tendered. So far, it has appeared that the two charged complexes have disparate mechanisms of action. Positive ions release serotonin and negative ions enhance its oxidative deamination (Krueger, 1974). Whether positive ions operate directly on MAO by reducing its availability for catabolism is a question that has been considered but not subjected to direct testing. Future studies at the University of California, San Francisco, will undertake this assignment.

The effects of small negative air ions on the oxygen uptake of isolated mouse liver cells were studied by Bhartendu and Menon (1978) by exposing the liver cells to varying ion concentrations:

> For concentrations of the order of $1-2 \times 10^5$ ions/cm^3, the oxygen uptake was always higher than in the normal atmospheric conditions of $3-8 \times 10^2$ ions/cm^3. For intermediate concentrations varying effects of activation and inhibition were observed. A statistical analysis showed that the oxygen uptake increased by approximately 14% when liver cells were exposed to ion concentrations of values 1–9 times the normal, by approximately 9% when exposed to 10–99 times the normal, and by approximately 38% when exposed to 100–999 times the normal.

The result of two experiments is shown in Fig. 8-3.

In summary, the observation of tracheal tissue has provided evidence relative both to a primary site of action for air ions and, via analysis of the tissue–ion interface,

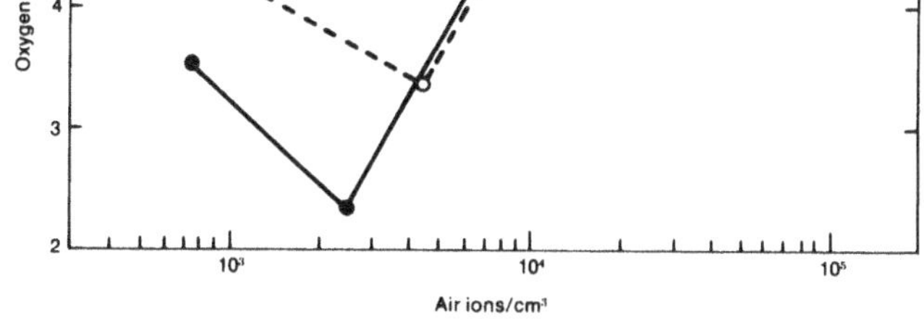

Figure 8-3. Relationship between oxygen uptake of mouse liver cells and small negative air ions in experiments 1 and 7.

their mechanisms of action. The advantages of this preparation over previous studies were at least twofold. First, the isolation of a physiologic system with a short response latency provided an opportunity to study direct and immediate ion effects rather than relying on an increase in death rate, as in the case of bacteria and molds. Second, unlike microorganisms, tracheal tissue is made up of cells that demand a complex intercellular communication system. The network is neurohumoral and the transmitter has already been identified as 5HT. The primary significance of an air ion influence on this system is that it provides a link to other complex organisms which have functions modulated by 5HT and with due circumspection, inferences may be drawn. The basis for this is that during the course of nervous system development, as single cells aggregated they assumed increasingly complex proportions. As they massed, structural limits were imposed and layering, rather than progressive linear expansion, occurred.

When this convoluting occurred, cells once localized on the surface then were pushed toward the interior. From this position they could no longer experience or respond to the environment directly (as can the single-cell amoeba, for instance). Contact with the outside world was quite literally lost. However, the necessity of receiving input about it remained. To fulfill this function, an intercellular information network was required and it became expressed neurochemically. Nervous function, axons, dendrites, internuncials, etc., evolved then as a means for representing the external world internally. Transmitters are among the agents of exchange and 5HT appears to be one of several specific information modulators. Tracheal tissue studies allow for an examination of these events; and while this medium excludes behavioral confounders, its significance is not limited to respiratory function alone.

Given the demands of physiologic economy, it is not altogether surprising that a single, diffuse system evolve to index rather disparate environmental stimuli. Thus 5HT has been implicated in the modulation of several consummatory behaviors, as well as emotion and sleep cycles. While a certain pharmacologic specificity is demanded of a transmitter relative to putative receptor interactions, its behavioral manifestations are in no way bound by the same constraints. This potentially fragments any review of the literature on air ion effects in mammals. However, some order may be discerned if the complexity of behavior is collapsed under the rubric of adaptive significance. In this context behavioral diversity is reduced to a response system attuned to environmental stimuli, e.g., light/dark cycles, seasonal nutrition, or fluctuating ion densities. Its analysis is conducted with respect to these adaptive or selective dimensions. Unfortunately, much behavioral animal testing to date has been carried out under "deprivation" conditions; conditions under which response display may fail because an appropriate target is not in the experimental field. For instance, years ago, when early intracranial stimulation (ICS) researchers failed to confirm topologic correlations with specific behaviors, it was assumed that stimulation had no effect. Only much later did it become evident that "animals don't emit feeding responses if there is nothing to eat" (Valenstein et al. 1970). That is, appetitive responses may fail to surface if there is little probability of their consummation. Such is the message conveyed by a stark experimental environment. A second, and related, difficulty stems from the nature of controlled studies. Precisely because laboratory work exercises control, behavioral variability may be reduced beyond the point at which any relevant response, physiologic or behavioral, can emerge. In the field, variability is a prerequisite for survival. Selection pressures must have a range of expressions on which to operate. Though the general advantages of laboratory work are clear, effects on dependent variables may be totally obscured under some conditions of control. Too stable a field may simply fail to force the expression of other than dominant mode patterns, yet other responses in the repertoire may be the true indicators of an independent variable. In some sense, such is the case with air ions. Evidence is accumulating that demonstrates their particular influence on elderly, infirm or otherwise stressed organisms. But unless testing conditions somehow match environmental demands, such effects may not become manifest. So, in reviewing a behavioral literature in general, and with air ions specifically, the relevance of experimental conditions with respect to treatment and reported outcome must be carefully assessed.

Relevant animal work in the air ion field is scarce, limited further still by the above caveats. However, the work warrants consideration because ion effects *can* be observed in complex organisms, similar mechanisms of action may be operative in humans, animal behaviors are of interest in their own right, and because the old lore of animals as meteorologic indicators may find support in a new wave of investigations. For instance, in a series of experiments where performance was used as a learning criterion, Duffee and Koontz (1965) were able to demonstrate improved maze learning in rats under both negative and positive ion conditions. Negative ion conditions were of greater efficacy than positive in improving performance, particularly for older rats living in a predominantly negatively ionized atmosphere. Jordon and Sokoloff (1959) reported similar findings: old rats (20–27 months) exposed to a negatively ionized atmosphere performed to criterion (escape from a water maze) faster and with fewer errors than a control group. Again, the pronounced effect of ion treatment in older experimentals was noted, though Terry et al. (1969) cite sex rather than age as the significant variable. Others reported gross motor changes in rats consequent to exposure to ions of either polarity (Bachman et al., 1966). Effects on aggression, urine output, defecation, sleeping periods and respiration were recorded. Responses were dose-related. In these experiments, not only are the observed behaviors of interest, several of them (aggression, defecation, and sleep) having been linked previously to 5HT effects, but the notation of dose/response relationships is significant. Serotonin, often considered a behavioral modulator, has been known to effect physiologic changes in a nonmonotonic, often biphasic fashion. This suggests a complex phenomenon, the mechanisms of action for which have not been evinced. Consistent dose/response determinations across protocols could go far in reducing the interlaboratory disparity results for air ion effects and 5HT function.

Relative to and perhaps not altogether distinguishable from changes in gross motor activity are reports of enhanced aggression or emotional responsivity. Frits et al. (1963) administered cholesterol to rabbits and noted that positive ions reversed the cholesterol-induced nervous system depression and increased aggressive displays. Nazzaro et al. (1967) demonstrated differential effects of positive and negative air ions on conditioned emotional responses (CER). Results were suggestive: negative ions reduced anxiety and facilitated learning, whereas positive ions may have increased anxiety and interfered with learning. Strong individual differences, a tendency for positive ions to enhance variability and a small number of experimental animals somewhat confound interpretation. Frey's work (1967), however, supports the general experimental conclusion. In this study, lever-pressing behavior of rats subjected to paired associations of a buzzer sound followed by an electric shock was studied. Animals developed the traditional conditioned anxiety response to the buzzer sound. This CER, indexed by a suppression of lever pressing, was reduced when animals were treated either with reserpine or negative ions. (Reserpine, long used clinically in the treatment of anxiety, essentially depletes tissue stores of 5HT by inducing release of bound 5HT, whereupon MAO terminates its action. Negative ions are operationally similar, inasmuch as they yield a net depletion of free 5HT.) Generally, results such as Frey's lead one to suggest a tranquilizing property of negative air ions. Certainly, the evidence suggests that this is the case, supported again by Olivereau's report (1973) on their anxiolytic effect in rodents. However, this explanation may be somewhat limited. An alternative, though not contradictory, interpretation focusing on pain is available.

Early in the 1960s 5HT had been heralded by some as a new "learning drug." Studies observing performance to criteria in standard testing apparatus and learning paradigms indicated that indeed many animals treated with 5HT learned faster (Woolley, 1967). This work was complemented by studies indicating that p-chlorophenylalanine (p-CPA), a powerful serotonin inhibitor (Koe and Weissman, 1966) retarded learning (Stevens and Fechter, 1969). These claims were challenged by an equal number of reports to the contrary, whereby 5HT inhibited, and p-CPA enhanced learning (Tenen, 1967; Weissman, 1973). Over time, it became clear that it was the interaction between sensory experience and emotional reactivity that determined the experimental outcome for a given animal strain in a given context. If an animal's threshold for pain were lowered, learning appeared to be facilitated as negative reinforcers were responded to as more punishing. But, if emotional responses such as freezing competed with avoidance or escape behaviors, a learning deficit was assumed (Tenen, 1967; Stevens and Fechter, 1969; Wise et al., 1970). As these insights were advanced, 5HT and p-CPA were spared capsulization and commercial exploitation as the newest in a long line of panaceas.

What this suggests relative to air ions is that to the extent that negative ions functionally reduce 5HT, they may be tranquilizing and would therefore logically attenuate conditioned emotional responding. The more proximate locus may, however, be in the modulation of pain. Pain, depending on its perceived intensity and environmental context, may be the primary effector that underlies the varied forms of expression across experimental protocols. Work by Gilbert (1973) indicates that this may be the case. His research has the added advantage of examining the chemical substrate of the behaviors measured. Citing Korn and Moyer's work (1968) that animals reared in isolation were more reactive to handling, Gilbert sought to "determine the effects of both continual and intermittent negative ion inhalation on the emotional level and brain 5HT content of rats experiencing prolonged isolation." He hypothesized that rats exposed to a negatively ionized atmosphere and reared in isolation will display less emotional behaviors and have lower levels of 5HT than will animals subjected to isolation only. Results were in accordance with the hypothesis. Negative ions lessened emotionality and induced a concomitant reduction in brain 5HT. Again, observations reported were consistent with an interpretation broadly relative to pain. That is, animals with higher 5HT levels exhibited signs of heightened emotionality, e.g., startle response, vocalizing and escape attempts, behaviors frequently indexed in the context of pain measurement.

Though pain is clearly a complex phenomenon and studies have been done to raise some fundamental questions about its perception, the work cited here nonetheless supports demonstrations of the biologic activity of air ions. Animals in a free choice situation do select residence in negatively ionized atmospheres (Bonnevie, 1966), a choice presumably based on their positively reinforcing properties.* As is typical, however, along with the advancement of

*Some caution must be exercised to avoid the repetition of classic errors in scientific "insight". In the early 1900s D.L. Gunn studied organisms in another free choice situation. Wood lice were placed in a long tube with a humidified gradient. They were expected to congregate in one of the extremities and indeed they did. However, the site was not based on preference per se. What had actually transpired, as Gunn reports, was that during their explorations, the wood lice meandered into a dense moisture field. Their activity level became substantially depressed by the humidity, the wood lice became quiescent and there they remained, their "preference" determined not by choice, but biologically. The moral of the study may be: a resting place does not a home make.

knowledge of a phenomenon come a new set of unknowns. And here questions on blood versus brain 5HT levels and ion effects arise. While air ions appear to alter brain 5HT under some conditions (Krueger and Kotaka, 1969; Gilbert, 1973), parallel effects are not found in blood 5HT (Krueger et al., 1963a, 1968). Futhermore, each of these effects appears to be dose and time exposure dependent; thus the observation of a zonal response for gross motor activity and air ion concentration (Bachman et al., 1966). Some of these inconsistencies have been attributed to the action of the blood-brain barrier (BBB). But whether or not 5HT actually crosses the BBB has not been unequivocally determined (1971b). Clearly the issue requires resolution in the laboratory. Similarly, laboratory studies may clarify the mechanisms of action of air ions. Though Krueger's work supports hypotheses of 5HT release by positive ions, a finding recently confirmed by in vitro studies of human blood (Tal et al., 1976), questions of the origin of 5HIAA in urine remain. If positive ions release 5HT which must be catabolized, and negative ions accelerate its oxidative breakdown, the net product of either process, in theory, is an increase in 5HIAA. How then does one specify its origin? Empirically, increases in urinary 5HIAA have been linked to negative ion exposure. However, until reports on turnover studies are published, research will have to be guided by these working hypotheses. Pilot studies on air ions and biogenic amine turnover are being undertaken at Rockefeller University.

Gaps in understanding the mechanisms of action of any phenomenon can be problematic for designing experiments, but these do not mean empirical utility need be diminished. Aspirin is an amazing therapeutic agent. Next to morphine, it may indeed be "God's gift to man." But nobody really knows how it works. Air ions too indicate a potential as a physiotherapeutic modality. However, not all of their biologic activity can be inferred solely from known, e.g., serotonergic, mechanisms either.

Vasiliev (1960, 1966) hypothesized a double action of air ions: first, by neuroreflex means, through the afferent nerve fibers originating in the alveoli, and second, by electrohumoral means, through the wall of the pulmonary alveoli and through the blood. According to the Soviet school, chronaximetry (the measurement of the time required for nervous excitation by a definite stimulus) is a method by which functional changes in the central nervous system can be inferred from the changes in the chronaxy of peripheral nerves and muscles. Thus, in rabbits, when chronaxy of the motor neurons of the rear extremities was determined from the contraction of the foot muscles, negative air ions were seen to decrease and positive ions to increase excitation times. Initial values were recovered within 14 min when the ionizing apparatus was switched off. Air ions of either charge do not have a noticable effect on the chronaxy of isolated nerves and muscles, Vasiliev reports. That is, inhalation of positive air ions increased the chronaxy of the skeletal muscles in rabbits, particularly gastrocnemius muscles, by as much as 115%. However, the same dose of air ions was without effect when these muscles were separated from the central nervous system. Rager (1970) also noted that negative ions alter neuromuscular excitability in the intact organism. The reduction of rheobase (the minimal potential electric current required to produce stimulation) and chronaxy by negative air ions are implicated in the modification of the developmental course of experimental diphtheric myocarditis in rabbits. Other examples of ions and neuromuscular excitability will be examined in the human literature when visual reaction time and acuity studies are cited. Meanwhile, it is the Soviet work again that supports the second, humoral, action of air ions. According to Skorobogatova (1962), in crossed-circulatory experiments on paired animals, negative ions induced functional shifts in arterial pressure, rheobase and chronaxy in both donor and recipient animals. Latency periods for recipient responses were in the expected direction: donors were mimicked 10–12 min after initial response onset. Even direct transfusions of blood from subjects exposed to negative ion inhalation have been reported (Vasiliev, 1960) as efficacious in the treatment of ozena (a nasal dysfunction associated with fetid odor). Though these parabiotic experiments are frequently difficult to reproduce, the interpretation of such a series of results is summarized by Vasiliev (1966) as follows:

> Aeroions act on the functional state of the organs, in particular, of the skeletal muscles, partly in a reflex way, by stimulating the endings of the pulmonary afferent fibers which pass into the vagus nerves, partly in a humoral way, changing the electrochemical properties of the blood passing through the capillaries of the alveoli. This, in turn, is primarily expressed in the functional state of the central nervous system, and through it, indirectly, on the peripheral organs. Thus, the electrohumoral mechanism of the action of aeroions on the organism is by no means isolated and autonomous. On the contrary, it is only a link in the chain of the more general and varied mechanisms of a neuroreflex character.

The unfortunate inaccessibility of original resources reduces much of this material to information garnered from English published reviews. The limits of such reports are obvious. However, other studies on the neurochemical and endocrine involvement of air ion action are available for closer examination.

Throughout the 1950s in the United States, modifications of adrenal activity in animals exposed to ionized air were reported. Holloway (1952) reported changes in cholesterol and ascorbic acid content, Rinfret and Wexler (1953) histologic changes indicative of glandular elaboration of salt- and carbohydrate-regulating corticoids, and Nielsen and Harper (1954) reported changes in succinoxidase activity, incorporated as a measure of adrenal activity. The experimental disparity makes it difficult to say what these data collectively mean. Frey (1962, 1965) has commented on numerous experimental confounders relative to early design limitations, failure to ground subjects, and deficient instrumentation, though he did esteem the work enough to organize a "Rational Framework for Interpreting the Behavioral [and presumably neurochemical] Effects of Atmospheric Ions" (Frey, 1961a, 1962). That is, negative ions stimulate the secretion of the glucocorticoids, and positive ions either stimulate the secretion of the mineralocorticoids or inhibit the glucocorticoid secretion. Frey based his hypothesis on information culled from several disciplines. His first line of evidence detailed the aforementioned adrenal work. For his second line, Frey reasoned that since there is some evidence that the adrenal cortex is affected by ions, a differential corticoid secretion (or specificity) must be sought. Support was evinced from Selye's work (1950), stating that stressors increase the production of ACTH. The increase in ACTH secretion will stimulate glucocorticoid production, and though it is not known what stimulates mineralocorticoid production, the two seem to co-vary in some way. While describing putative mechanisms of adreno-cortico-ion function Frey suggests that some relation to stress is implied. In fact, ion effects may not occur at all unless subjects are under stress.

Sufficient conditions for the production of ion effects have not been clearly determined. Though Frey's hypotheses were proferred nearly 20 years ago, accompanied by the recommendation for a systematic research endeavor, such a program has failed to materialize in the United States. The studies of Krueger and his colleagues at the University of California, Berkeley, represent the only ongoing laboratory pursuit of air ion effects over the past two decades. However, recent work carried out in France by Olivereau (1971a, b, 1973) has renewed the interest in Frey's rational framework and has also managed to

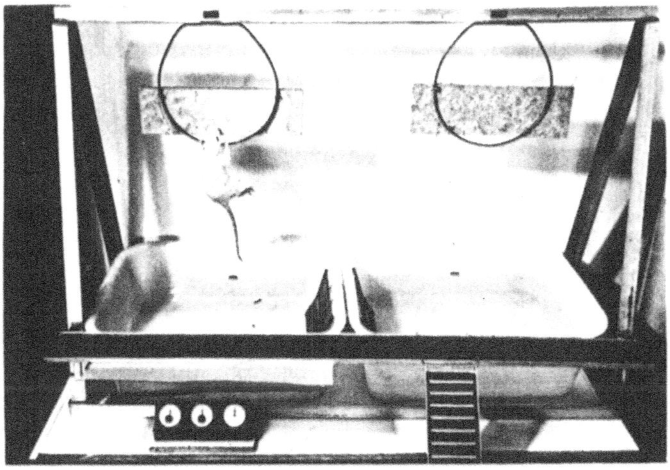

Figure 8-4. Arrangement for testing the decision-making ability of rats subjected to a stress situation. The animals try to keep hold of a ring above a container of water for as long as possible to avoid falling into the water. After Olivereau (1973).

integrate seemingly unrelated behavioral and neurochemical phenomena.

Olivereau (1973), for example, used rats as experimental animals. In a neurophysiologic test of the behavior of male rats, he studied the influence of negative air ions on a fear situation (Fig. 8-4) by means of an adaptive effect. The rats, which hung from a ring suspended above a container of cold water, learned to dive in order to curtail this stressful situation. After a brief treatment with air ions (150,000 ions/cm^3 for 20 min), the frightened animals held onto the ring for a significantly shorter time before releasing it. This effect on animal behavior in a fear situation suggests that serotonin is involved in the action of air ions. Certain paradoxical results, moreover, point to an increase in muscular performance.

To be sure, Olivereau's work is difficult to reduce to a definitive outcome. Readers are strongly encouraged to read original source materials for enhanced clarity of these comprehensive works. Nonetheless, the adrenocortical and serotonergic systems are both fundamentally involved in the mediation of air ion effects, and these systems, acting in concert, account for the wide range of responses to air ion stimuli.

The actual sequence of experimental events is too detailed to recapitulate here. Early on, Olivereau recorded effects of air ion exposure on molar behaviors and later traced them to perturbations at the molecular level. Thus his work on water intake and output by rats and their spontaneous consumption of electrolytes led to the consideration of adrenal hydromineral metabolism and, finally, histologic analyses of neurosecretory granules and back again to behavior. Though historical precedent, which Olivereau briefly recounts, had indicated primary adrenal cortical involvement, he concluded that such explanations were incapable of entirely explaining the phenomena observed. What he sought was a central nervous system connection.* Accordingly, studies of the hypophyseal–hypothalamic axis were undertaken, because this region represents the primary nexus of systemic and central nervous system interface. Olivereau conceived this site as a tissue field for ion, corticoid and transmitter

*These events curiously parallel the historic pursuit of a centrally metabolized toxin in the etiology of mental disorders. Early hypotheses assumed adrenal and/or other systemic dysfunction. However, serendipitous observations noted the structural similarity between mescaline, a psychotomimetic, and the central nervous system transmitters. Researchers immediately shifted focus to the central nervous system, pursuing "the" endogenous neurotoxin, and a course plotted by Thudichum some 70 years earlier was resumed.

interplay and, over time, test results were interpreted as supporting hypotheses favoring such an interaction. Exposure to negative ions did induce changes in hypophyseal–hypothalamic tissue. That is, increases in nucleus and nucleolus size were observed along with diminution of neurosecretory granules. These histologic changes are recognized as classic indices of intense cellular activity. Subsequent tissue studies on other groups of treated animals confirmed the original findings. Again hypertrophy of nuclei and nucleoli were noted. Again the liberation of neurosecretory granules, normally present in the cytoplasm, was observed, this time along with an apparent increase in the discharge of ADH (the antidiuretic hormone). The latter observation was not at all incompatible with the simultaneous behavioral display of a polyurea, and indeed completes response consistency across the varied levels of analysis. The significance of Olivereau's findings is enhanced by the observation that exposure to positive ions reversed these conditions: nuclear volumes were diminished, neurosecretory granules hypoactive, ADH liberation was restricted, and an oliguria was noted.

In sum, whether one cites Frey's hypothetical framework or Olivereau's empirical findings, adrenocortical function is implicated in air ion effects. However, even though recent work has been carried out in an atmosphere of enhanced sophistication in instrumentation, experimental design and biologic measurement, the interpretation of the data does not necessarily have any more finality than in the past. Olivereau readily acknowledges that an explanation of ADH secretion per se is bound to be complex. Any effort to interrelate multifunctional systems is, of momentary necessity, so. He does consider the possibility that a "directorship" is assumed by the corticoadrenal hormone, a status frequently conferred on autocoids of central origin. In this context he suggests the antagonistic actions of the corticoids and ADH on the mobilization of water are regulated via homeostatic feedback whereby the water loss aggravated by the hyperactivity of the adrenal cortex is minimized. The hypersecretion of ADH is not incompatible with the observed polyurea, which may be explicable by concomitant modifications of glucocorticoids and serotonin in the blood. Olivereau relies on Krueger's work to fill in the transition, but in short suggests that serotonin, acting directly at the level of the kidney, probably induces constriction of the afferent glomerular arterioles. This directs a fall in the blood level of circulating 5HT, which then adds its diuretic effect to that manifested by the corticoids and, accordingly, to the manifestation of a polyurea, which ADH tends to moderate, thus closing the loop. Clearly this is a complex set of interactions, the individual sections of which require further investigation. But regardless of the route the reaction takes, the facts remain. Two generations of researchers have now produced and reproduced ion effects. Their action may be direct and/or indirect, humoral or reflexive, operating centrally and/or peripherally, but it exists.

Action of Small Air Ions on Human Beings

Given the reproducibility of these particular effects, it is, of course, of more general interest whether or not these laboratory phenomena have any significance to human populations. Does it matter that ionized air influences animal food and water intake, locomotor activity, copulatory behavior or the adrenals? Probably, for though human behavior need not be reduced to that of other systems, more than enough documentation is in evidence to suggest a relationship between air ions and stress. And that is a phenomenon to which most people can directly relate.

Significant human studies are even scarcer than those on other animals. The same limitations censured earlier are present in force again: deficient instrumentation, weak experimental design, failure to ground subjects. Besides, humans are traditionally less available for longitudinal or even continous periods of study. Finally, these elements are confounded by the already complex animal heritage to which we are heir. Into this precarious area a few have dared. Among the most ambitious are those works engineered by Sulman in Israel. In his studies, subjects are drawn from those "cruel experiments in nature" on whom the Sharav winds appear to wreak physiologic havoc.* Clinical complaints during Sharav conditions range from sleeplessness and irritability to vomiting, diarrhea, respiratory distress, and abdominal pain. None of these ailments constitutes a major somatic insult, but the real danger to health may arise when these minor dysfunctions operate synergistically on an already present or subclinical medical condition. Thus dyspnea may exacerbate respiratory illness; palpitations or pre-cordial pain may aggravate cardiac dysfunction.

Based on extensive sampling, Sulman has grouped recurring patient complaints into three major categories, each characterized by its biochemical and symptomatic profile (Sulman et al., 1970). From the distribution of clinical descriptors it has been deduced that climatic heat stress, or Sharav effect, can be partially defined along the lines of Selye's General Adaptation Syndrome (GAS). Stage I would be an alarm reaction, Stage II, resistance, and Stage III, exhaustion. Over the years, data have been collected from "weather sensitive" subjects to suggest some conformance to this pattern. Predictions of the developmental course for the Sharav syndrome, based on these principles, are as follows. Tourists and new immigrants will not suffer from Sharav, old-timers will. That is, as the hot, dry weather brings about passive dilatation of the peripheral blood vessels and augments perspiration, newcomers will respond to the change with the normal alarm reaction, a hypersecretion of catecholamines. However, while this response reduces the immediate discomfort of climatic heat stress, secondary effects of catecholamine excess may be experienced. With exposure to local weather conditions, however, resistance is developed and an optimal balance of vessel contraction and perspiration is achieved. Long-term exposure to the Sharav winds, on the other hand, will erode the body's capacity to respond effectively to this meteorologic stressor. Failing to balance catecholamine secretion, unable to compensate for peripheral blood vessel dilatation, and least able to resist Sharav effects, long-time "weather sensitives" will gradually enter the exhaustion stage, becoming most medically vulnerable. Urinalyses on over 500 individuals bear out predictions: new immigrants hypersecrete catecholamines and veteran residents hyposecrete (Sulman et al., 1970, 1974). Catecholamine distribution is not the only indicator that Sharav conditions parallel the GAS in some fashion. But since physiologic adjustment mechanisms are generally complex processes, air ion research would probably benefit more from basic research than from the unqualified application of this analogy. Nonetheless, as a link to earlier models of a "rational framework" (see Frey) and for directing future research, the concept has merit. For instance, Sulman describes one reaction to climatic heat stress as the exhaustion syndrome; characterized by catecholamine deficiency. By labeling a diagnostic category as such, a prescriptive for 44% of his Sharav cases becomes intrinsic to the syndrome identification: overcome the catecholamine insufficiency and remedy the problem. Sulman has also examined 5HT, 5HIAA, histamine and hydromineral excretion patterns. Again, metabolic discharges on normal, as opposed to Sharav, days suggest operational groupings; an Irritation Syndrome due to 5HT hyperfunction

*These winds of ill repute are generally characterized by a rapid decrease in relative humidity, values 25%–30% below normal; an increase in temperature of 5°–10°C above the seasonal average, and an elevated positive ion density which precedes the winds by 10–12 hr.

for 43% of the cases and a thyroid syndrome due to hyperthroidism forme/fruste (histamine and creatinine) in 13% of the cases (Sulman et al., 1970).

Investigations on the possibly hazardous effects of artificial air ionization were also conducted by Sulman et al. (1978):

> Five weather-sensitive women and 5 normal men chosen at random . . . were exposed separately during 8 sleeping hours and 8 working hours to the apparatus at 1–2 m distance in a 4 × 4 m room, for 2 months. Thus they were exposed to a daily uptake of 1×10^4 negative ions/cm^3 for 16 h/day during 2 months. Urinary 17-KS, 17-OH, adrenaline and noradrenaline excretion was not affected by the negative ionisation. However serotonin, 5-HIAA, histamine and thyroxine excretion—if increased before—diminished by 50% on an average. There were no changes in body weight, blood pressure, pulse, respiratory rate, oral morning temperature, dynamometer grip strength, routine liver function tests, urinary pH, albumen, glucose, ketones, bilirubin, or occult blood, red and white blood count and ECG records. The EEG revealed the typical changes due to negative air ionisation: stabilising of frequency, increased amplitudes, spreading of brainwaves from the perceptive occipital area to the conceptive frontal area and synchronisation of both hemisphere tracings.

The involvement of serotonergic systems in mediating Sharav effects is of particular interest here. Documentation for an air ion effect on serotonergic mechanisms in nonhuman animals has been reviewed. That is, positive ions release 5HT and negative ions seem to enhance their oxidative deamination. If such mechanisms are the rule for humans as well, then one would hypothesize that Sharav symptoms, induced by positive ion influx, would be reversed by appropriate drug and/or ion treatments. Such is the case. Using both objective and subjective criteria, patients suffering from Sharav effects demonstrate remission of symptoms when treated by negative ions or 5HT antagonists (Sulman, 1971a; Sulman et al., 1974).

Though recent studies have been undertaken at the University of California, San Francisco, Sulman's work is currently the most comprehensive and ongoing air ion research being done in humans. Its value is both general and specific. As an interdisciplinary model it demonstrates the benefit to understanding conferred by a program designed to examine a system in the context of its natural environment. There is an isomorphism between the experiments' complexity and the systems' complexity. Human beings are, at the very least, a mass of biochemical, behavioral and psychological interactions. They operate in all modalities all of the time. The simplicity of this statement obscures its significance, but it is significant nonetheless. For instance, were Sulman to have focused on 5HIAA as a solitary metabolite, it might not have been brought to light that other biochemical events could also be correlated with ion effects, i.e., adrenal or thyroid hyperfunction; or were the euphoria observed by tuberculosis patients treated with iproniazid not recorded, the development of antidepressants might have been considerably retarded. There is clearly long-term value in examining an event in the context of larger systems. This will be considered again when the function of air ions in the total environment is briefly discussed. Of course, working with diffusely or complexly ordered systems is the more prodigious experimental task. Basic research generally requires years of investment to identify and reproduce a given, simple event. The intrusion of confounding variables is more likely. The rules for information transfer across levels of complexity are still nascent. The complaint of deficient characterization of the system and lack of problem resolution is frequently leveled and with enough finality to generally discourage the pursuit of such enterprises in the future. But the facts of complexity remain. This does not denigrate the significance of other work, nor overlook the necessity of pioneering studies in more simple organisms. The power and necessity of meticulously controlled laboratory research is evident. On a grand scale much of modern molecular biology rests on the incisive experiments of J. Loeb exemplified by the investigations he conducted on the behavior of proteins. It is suggested that by appropriately expanding the exploratory field, effective relationships may emerge which otherwise would have suffered exclusion. Again, chickens are not going to peck for food when there is nothing to eat.

Regrettably, it is the general case that funding, space and time constraints preclude the feasibility of interdisciplinary expansion. Many countries already use air ions as a therapeutic agent. The prescription is based on empirical results: it works. Such popular use abroad obviates the likelihood of government investment in a program whose primary aim is to elucidate the mechanisms of air ion action. Clinical efficacy has already been "proved".* But clinical reports generally fail to parcel out the variances in those circumstances in which ion treatment is not efficacious. Clearly a research climate where a priori experimental criteria were established to sight and reproduce effects would be ideal. To this extent, attention may be directed to the United States where public access to air ion treatment has been limited.

In the United States the primary impetus for modern air ion research arose from interest by the business sector. High density living, crowded offices, submersibles and even the prospect of manned space flight all had in common a reliance on, or necessitated a more fundamental understanding of, the air we breathe. As we became more and more creatures of the indoors, occupying inner spaces, "air" warranted investigation. Motivated by economic, if not health considerations, commercial investment represented potentially formidable profit in the burgeoning marketplace of crypto-environments. Much of this demand was actually reducible to a need for increasingly sophisticated heating and ventilation systems since complaints of "dead air" or "stuffiness" were not uncommon in even presumably well conditioned air. The interest by marketing researchers and management specialists, albeit relevant, is much more recent (Robertson, 1978, personal communication). The critical element nonetheless seemed to be ions. During the 1950s, most of the equipment in commercial use functionally deprived the atmosphere of small air ions, and humans seemed sensitive to these atmospheric deficits. Kimura et al. (1939) in Japan, Dessauer (1931) in Germany, Tchijevsky (1933, 1960) in Russia, and Yaglou et al. (1933) and Herrington (1935) in the United States all were in general accord. Excess positive ion or ion-free conditions induced deleterious effects on experimental subjects; negative ion treatments were generally salubrious. Of course, treatment conditions varied and control groups were either nonexistent or varied, as were ion concentrations and exposure times. It is obvious that these early studies are subject to the same interpretive constraints reviewed previously. However, they do form an historical base for modern work, a good deal of which continues to support their general conclusions.

Ironically, the same interest that gave birth to the research boom in air ions was also responsible for its temporary demise, which is to a large extent traceable to reaction to government control of apparatus that were represented as having medical value. In the 1950s and through the early 1960s, ionizers were widely marketed as a

*It is an historic though not isolated curiosity that advances in the clinical domain have limited the pursuit of basic research. Such was the case with schizophrenia and the introduction of chlorpromazine to the U.S. in the middle 1950s. Prior to this, half of all hospital beds in the U.S. were occupied by a person diagnosed as schizophrenic. Following its widespread implementation into the treatment regimen, chlorpromazine operationally became a schizophrenic "curative" in that it was credited with the massive reduction in the numbers of people hospitalized for the disease. The etiology of schizophrenia, however, remains unknown.

therapeutic agent for a multitude of ailments ranging from constipation to cancer. There was no strong data base to support these excessive health claims, and government intervention was necessary and appropriate. A ban on these overt misrepresentations was implemented. Unfortunately, the ban on false advertising and marketing expressed itself as a covert ban on research. Funding sources dried up, as did a once avid commercial interest. So, though there have been scattered reports by clinicians using air ions in therapy, these private investigations are reduced to anecdotal, albeit encouraging, status.

For instance, in a modern Swiss bank, when employees were exposed to negatively ionized air over a test period of 30 weeks, a significant decrease in the incidence of respiratory illness was reported. Working time lost, as compared to a control group, was less by a factor of 16 (Stark, 1974). More anecdotal evidence is supplied by an Oregon school superintendent, who reported a reduction in absences by about 17% in one of four elementary schools equipped with heater/ionizer units. Economic and health factors figured in continued administrative interest in ionizers. That is, in some states, New York among them, fiscal support is based not upon student enrollment, but on average daily attendance. A 17% increase in attendance could amount to millions of dollars over time. Another school system similarly reported a significant 27% reduction in absenteeism following conversion to a ventilation system that restored ions to the atmosphere (Hansell, personal communication). Krueger's laboratory findings on tracheal tissue (1959) bear on these popular accounts since negative ions seem to have a prophylactic effect on the respiratory distress that accounts for a considerable propertion of most school and work absences. The reduction of airborne infections may be reconsidered in light of the destructive effect of air ions on bacteria (Krueger et al., 1957), their enhancement of resistance to influenza virus (Krueger et al., 1971), and influence on the tracheal clearing mechanism (ciliary beat and mucosal flow rate) (Krueger and Smith, 1962).

Clinical applications in the treatment of asthma support the relevance of laboratory findings. Remission of dyspneic attacks (for over 6 months) were reported for 55% of the 830 patients receiving complete treatment sequences (Boulatov, 1968). Palti et al. (1966) report related successes in infants 2–12 months of age. After establishing measurement categories for spasticity and respiratory rate responses, it was determined that "negative ions without any other supportive treatment (including antibiotics) terminate the spastic attack after a much shorter period than that required by the conventional mode of treatment" (Palti et al., 1966). Of added benefit was the absence of side effects with negative ion treatment frequently noted with traditional drug regimens. Jones et al. (1976) failed to find negative ion treatments efficacious in the routine management of asthma. However, a small number, seven, as opposed to the 830 and 19 cases cited by Boulatov and Palti respectively, makes it difficult to evaluate the significance of these negative results. Clearly an understanding of the ion phenomenon would be facilitated by the development of clinical trials for publication whereby peer referees would establish criteria for meritorious work. In Philadelphia, Kornblueh (1968), the pioneer of air ion work in the United States, initiated such efforts in hospital studies with burn patients. Early reports were encouraging. Over a period of 10 years, from 1956 to 1966, over 200 patients with various kinds and degrees of burns received negative ion therapy. For both in- and outpatient populations, with rare exceptions, no analgesics were needed. In a majority of patients, cessation of pain was achieved after 10–15 min of ion exposure. Ion dessicating effects seemed to diminish exudation from wounds, reducing infection and promoting the early formation of eschars. Worden's (1961) work in golden hamsters complements these findings. In the small number of animals tested, healing of surgical incisions and regeneration of several femoral nerves were enhanced by negative ion environments. Again, limits in experimental design placed some necessary constraints on interpretations of results. Kornblueh was aware of these deficits and had, himself, proposed using amounts of sedative and/or narcotics required for pain relief as objective criteria for measuring air ion treatment effects (David et al., 1962). At the Seminar on Human Biometeorology in 1964, Kornblueh said, "Air ion therapy is not a universal panacea, but it is certainly a valuable addition to our therapeutic armamentarium." (Kornblueh, 1967). Though death curtailed full scale pursuit of these ideas, his legacy provided yet another link in the ion chain.

While Kornblueh's work did not directly provide for biochemical analyses, burn patients do represent a bioassay for pain, and serotonin has long been implicated in its production. Recall the historic touting of 5HT as "the" learning drug, only to be recalled as "merely" altering an organism's threshold for pain perception. Though the precise nature of 5HT's functions remains unclear, all the work specifically reviewed in humans supports the 5HT/pain relationship: Sulman's Sharav syndromes, respiratory phenomena based on Krueger's data (1975), Fox's (1976) migraine, reiterated by Sulman, and finally Kornblueh's burn work. Studies in psychiatric populations broaden this relationship to include psychological pain, and demonstrate in yet another way the influence of ions on human systems. In a South American study, some 80% of the people in an anxiety syndrome clinic were reported to have benefited from treatment with negative ions. Psychoneuroses, apprehension and fear responses remitted or diminished considerably in the manifested syndrome. Most conspicuous was the absence of somatic complaints (Ucha Udabe et al., 1968). Attempts to replicate these striking findings in other laboratories appear to be lacking. Psychophysiologic effects of ionized air in a psychiatric population in the United States, however, have been reported to be of sufficient magnitude to warrant further research (Monaco and Acker, 1963). Another psychophysiologic indicator, the EEG, also demonstrated sensitivity to ion influence (Silverman and Kornblueh, 1957). Reduced frequency of alpha waves from 10 or 11 to 9 or 8 Hz, increased amplitude by up to 20%, and an advance of the alpha rhythm from occipital to frontal areas, with a general synchronization of reading from both hemishperes were reported. Slowed alpha rhythm is considered nonspecific and may appear under a variety of circumstances. Subjective reports, however, did include relaxation, alertness and improved working capacity (Assael et al., 1974). Though the accumulated evidence does not amount to a massive assault on the relationships among biochemical, electrostatic and psychological events, at least there are indications that human systems are capable of subjectively describing ionization experiences (McGurk, 1959; Sulman, et al., 1974). Inasmuch as each element of the above set, serotonin, ions and mood is a member of other sets which also have in common at least two of these members, a transitive relationship is suggested. This is particularly strengthened by the detailed and comprehensive work on reward and punishment in animals and humans, linking biochemical substrates to human mental events (Stein, 1971).

Hawkins and Barker (1978) have provided a scientific basis for these field observations. They tested the effects of artificaially increased levels of negative ions on the performance of a number of psychomotor tasks by 45 human subjects. The result was a statistically significant increase in performance, as compared to controls.

The limitations of the medical trials given air ion therapies in the United States, do not allow a proper evaluation. Health issues aside, however, air ionization has aroused interest in another context. It combines those elements already reviewed. It is in some sense economic; in some sense psychological. It is work capacity and work efficiency. Ion effects are of special interest here because of the enormity of the situation. Minority, albeit significantly, sensitized populations are no longer being addressed; rather an entire work force in a given economic system comes into focus. It is an inefficient economy that operates at less than its maximal potential. Working in ion-depleted atmospheres seems one factor in reducing work efficiency. For instance, prior to its installation of negative ion generators, a company processing some £200 million worth of checks daily was experiencing an error rate of about 2½%, or mistakes averaging £5 million/day, over £2 billion a year. Under ionized conditions the error rate dropped to ½%. Overall company morale was reported to be improved as well (Laws and Holiday, 1975). Whether or not the raised spirits were a function of ionization or employee reduction in error rate is of course a moot point. A multibillion dollar savings is cost efficient. Another example, taken from a situation beyond which a price can be put, is reported by a police emergency center. In this station, a problem with maintaining alertness during the "graveyard shift" was encountered. Here, officers working in ion-depleted conditions tended to be overcome by drowsiness. Antiquated ventilation systems aggravated the unpleasantness attributed to a generally foul and fume-laden atmosphere. Upon installation of air ionizers, however, a comfortable ambience was restored, and officers facetiously protested the absence of their customary cat naps (Laws and Holiday, 1975).

There is little doubt that smoke and people and a large surface-to-volume ratio deprive the atmosphere of its normal charge distribution. Independent measurements by Korblueh et al. (1973) and Maczynski et al. (1971) indicate that the concentration of air ions in enclosed spaces may be 2% that of country air. That is, a given room, by the end of a working day, may contain as few as 20 small negative ions/cm^3. Other investigations have shown this to be the case for motor vehicles as well. An increase in traffic accidents during certain atmospheric disturbances has been reported. However, this literature is difficult to evaluate, as ion density counts are generally not factored out in correlating meteorologic changes with an increased incidence of accidents or, for that matter, health related variances (Tromp, 1963). The fact nonetheless remains that ion levels do fluctuate considerably with different weather conditions.

Sporadic attention has been drawn to the same problem in aircraft (Caymaz, 1976). At 11,000 feet there is an acknowledged ion deficit (Robertson, personal communication). This phenomenon should be of particular interest vis-à-vis manned space flight, since an early maneuvering error of a percentage of a degree could cause a target site to be missed by a million miles. The generally low profile assumed by the assumed major sources of interest, i.e., the military and ancillary units, is partly attributable to historical caprice. The periodic replacement of Surgeons General in the United States has interrupted the continuity of longitudinal studies required to establish a meaningful air ion policy. Or, for example, the first commissioned British nuclear submarine was supposed to have been equipped with an ionizer system and effects were to have been reported in the literature. Intervening events obscured the knowledge of whether or not the submarine actually ever had the ionizers installed (Krueger, 1970, personal communication). Russian investigators apparently pursued this interest with a bit more enthusiasm. Nefedov et al. (1966), intending to determine the optimum ionization conditions for a space cabin, exposed subjects to varied ion treatments for up to 20 days, the longest presently known in the literature. No untoward effects were recorded and the recommendation was for use of ions of both polarities. Original source materials are, again, somewhat inaccessible and therefore not subject to detailed analysis. In the United States, NASA and Federal Aviation Agency support is limited, if public research funding is utilized as an indicator of interest. Other studies are available for review, however.

Bisa (1954) studied ion effects on flicker fusion frequency (FFF). This method had the dual advantages of (1) having been extensively studied in other disciplines and (2) being a known central nervous system component, FFF is under cortical control (Wehner, 1962). Essentially, FFF describes a visual response to intermittent stimuli. When presented at low frequencies, stimuli are perceived as relatively discrete or flickering, as were early movie scenes. At sufficiently high frequencies, stimuli hit the retina before the previous image disappears and a continuous stimuli presentation is perceived. No flicker is experienced: images have fused. The FFF method identifies these two critical points in photic stimulation. After elaborate preparations to secure accurate and sensitive readings, Bisa reported that negative ions decrease FFF values. However, FFF values are sensitive to acoustic, odor and thermic stimulations as well (Frey, 1961a; Wehner, 1962). This does not invalidate the empirical findings, though other investigators have since reported effects in the opposite direction (Wehner, 1962). The question is raised rather as to proximate or distal receptivity. That is to say, ions may influence temperature (Winsor and Beckett, 1958), which in turn influences FFF, but sufficient causality is not yet determined even to the extent that it has been for the humoral ion effects.

The same question is raised with respect to other ion-induced effects. For instance, Slote (1962) studied not only FFF, but simple visual reaction time (SVRT) and finger tapping (FT). Though he specifically suggested that FFF is connected with (1) chronaxy,* (2) latent period and (3) refractory period (of the retinocortical system), these concepts can probably be applied to SVRT and FT as well. Accordingly, it would be the general case that the more impulses a center can process per unit time, the better its functional state. Negative ions improved performance across all test conditions; positive ions had the opposite effect. Except for positive ion FT, all effects were statistically significant. Another way of assessing the functional state of a system is in terms of oxygen (energy) economy. The better the oxygen utilization, the more efficient the system, and an improved functional status is conferred. Recall that the negatively charged oxygen molecule presumably enhances the absorption of O_2 in the respiratory system (Guillerm et al., 1967). Though it is doubtful that all of Slote's ion-induced effects could be reduced to this function, results are in accordance with predicted outcomes.

Varga (1968, 1972b) conducted extensive investigations and research on the physiologic action of air ions and their significance as environmental factors. He placed his experimental subjects into a large climatic chamber, where they were exposed to an ion density of 3×10^5 ions/cm^3 at the level of the mouth. The object was to determine the effect of ion exposure on the oxygen consumption of the subjects. Varga found that the subjects fell into two groups according to their response to the ions. The first group displayed an elevated content of oxygen in the blood, while the second group showed no response to the inhalation of air ions. This disparity was unrelated to age or sex, but it was found that smokers invariably belonged to the second group (Fig. 8-5).

The effect of air ionization on the human pulse rate is also highly demonstrative. The pulse rate of subjects in this study declined when the ionization device was switched on (Fig. 8-6). This effect is not surprising, however, when one considers that less blood must be pumped through the body to supply an equal amount of oxygen, since the oxygen content of the blood is increased by air ionization. It follows that the rate of the

*The Russian "reflex action" of ions.

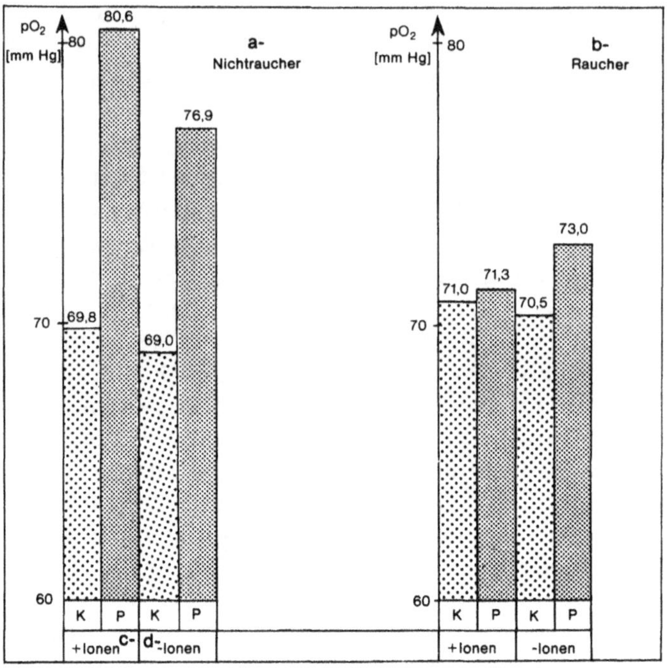

e- f- g-

Gruppe a:	10,7% bei +Ionen 8,2% bei -Ionen	Durchschnitt von 18 V
Gruppe b:	2,1% bei +Ionen 1,6% bei -Ionen	Durchschnitt von 12 V
Gruppe a und b:	6,4% bei+ onen 4 bei -Ionen	Durchschnitt von 30 VP

Figure 8-5. Different effects of ionized air on nonsmokers (left, group a) and smokers (right, group b) in terms of the oxygen level in the blood (pO_2). *Top*: Graph of two typical results for two individual subjects (K = control, P = test group); *bottom:* overall result. After Varga (1972b).
Key: a-Nonsmokers; b-Smokers; c-+ion; d-— ions; e-Group a (etc.); f-percent for + ions (etc.); g-Average of 18 subjects (etc.)

heartbeat must also decline, assuming a constant stroke volume.

Finally, it is reported that a decrease in the CO_2 content of the blood parallels the oxygen increase under the influence of ionized air. This is considered proof of enhanced gaseous exchange during respiration.

Other data support these findings. Knoll et al. (1961) reported that there was "nearly always an effect on RT" after ion exposure. However, results showed a bistability: the directional outcomes for either type of polarized air could not be predicted. Rheinstein (1961), employing varied statistical procedures, confirmed the ion effect, but similarly failed to account for the bistability. As usual, the heterogeneity of dependent variables across research designs confounds the comprehensive assessment of putative ion effects. Thus, Barron and Dreher (1964) studied four sensorimotor measures, but found their pilots significantly differed on brightness discrimination only. This particular finding is of interest vis-a-vis O_2 utilization in that scotopic visual thresholds along with FFF are among the more sensitive indicators of mild O_2 deficiency (Wehner, 1962; Rosenberg, 1972). Whether this phenomenon has anything to do with the increased incidence of traffic accidents at dusk remains to be seen, as it is still difficult to say what the effectors actually are. More parameters are studied, more conflicting reports compiled. Chiles et al. (1962) used a complex mental and vigilance task and found ions of both polarities reduced the number of false responses and errors. Neither result however was considered significant. Bipolar exposures again reduced detection errors in a fast presentation vigilance task, though only the positive ion decrement was listed as significant (McDonald et al., 1967). It is these detection errors and like phenomena that would presumably interest the military, where duties frequently demand radar-oscilloscope vigilance. According to Wofford's results (1966), "negative ionization has a significant behavioral effect on relatively simple tasks, but does not significantly influence more complex forms of behavior" ("under present testing conditions" should be added, in light of the intracranial stimulation studies, Wofford, along with noted Russian investigators, suggests that decreased latencies in RT studies may be a function of enhanced muscle excitability.

It is presumably this effect that motivates the use of ion treatments in some athletic training regimens. There is, unfortunately, limited work to affirm or negate the efficacy of this type of program (Hamburger, 1962). Neither the long-term nor the "staying power" of ion effects are known for humans. Substantive longitudinal research has not been carried out, though some authors have considered these problems. Rosenberg (1972) mentions a few. They are (1) Schultz's ANS activation/deactivation hypothesis; (2) learning theory's performance/activation inverted U; (3) Frey's glucomineralo-corticoid function; and (4) feedback gain deviations. These comprehensive views, while helpful in conceptually organizing complex data, share the common disadvantage of other after-the-fact explanations. Still they do have some predictive value to be applied to the next research series; bistable outcomes. They further have in common a reliance on Wilder's Law of Initial Value (Wilder, 1957), and what becomes obvious in its absence is the need to formally study this denominator. Certainly other investigators have been cognizant of "the state at the time of excita-

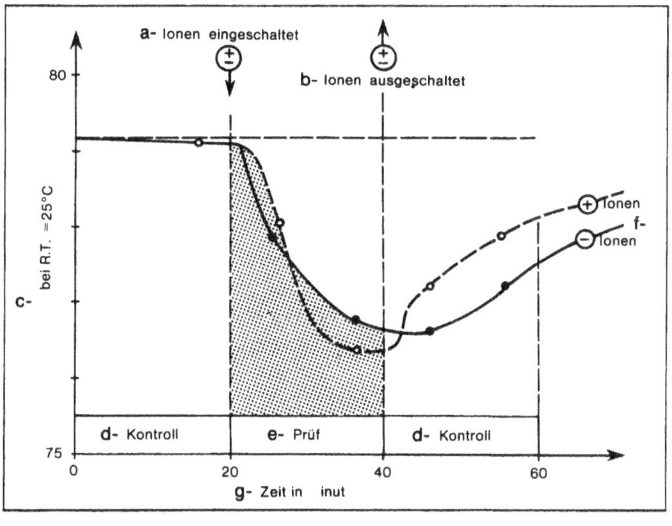

Figure 8-6. Variations in the human pulse rate under the influence of air ions at 25° C room temperatrue (average value for 24 persons and 177 measurements), after Varga (1968a,b).
Key: a-Ions switched on; b-Ions switched off; c-Pulse rate at room temperature = 25°C; d-Control; e-Test; f-Ions; g-Time in min

tion" (Lacey, 1956; Frey and Granada, 1962; Wehner, 1962) concept, but it is generally alluded to in the discussion section of a published paper; proferred as an ad hoc explanation. Rarely is an experimental program undertaken where the determination of "tonus" becomes a design characteristic for testing the main hypotheses—Charry's (1976) air ion work with lability scale determination is a notable exception. Yet an understanding of ion related phenomena demand such treatment.

Final Remarks

As mentioned at the outset, the field of small air ions is exceedingly complex. At least three systems in interface are being studied: electrostatic, biochemical and psychological. A focus on this complexity is essential and will serve to:

(1) remind researchers that knowledge of a segment of function is only a partial, however significant, level of understanding. The wedge of information must be reinserted into the whole operational system in order to not violate or undermine its nature;

(2) establish criteria for nominative or descriptive, ordinal, interval and ratio explanations so that levels of understanding may be made explicit. Laws for the transformation of information may be adduced thereby discouraging naive reductionism;

(3) make the limits of knowledge discriminable whereby what needs to be known is readily defined. The course of future research is likewise directed. In short, to see the whole allows for knowledgable pursuit of the pieces wherein the map is not mistaken for the territory mapped, the content of a message not interpreted by an analysis of its ink.

Much of the seemingly paradoxical outcomes of ion research no doubt could be reduced by adherence to such a perspective, keeping the overall context of the organism in mind. In this vein, ion studies represent a microcosmic view of a macrocosmic dilemma. Proffering levels of explanation inappropriate to the nature of the phenomenon ultimately reduces the value of the original information gained. For instance, that DDT effectively kills an isolated creature is known. However, over time, the species develops a genetic tolerance or immunity to the chemical which means that the annihilant, operating in a larger system, loses its potency. That is, when transposed from a microsystem to a macrosystem, the information is in effect stripped of its explanatory status.

Hippocrates and other Greek practitioners noted this long ago, as did Petersen more recently (1934–1938) in his prodigious works, *The Patient and the Weather*. Their observation and recording of meteorologic phenomena as part of their medical program demonstrates the conception of health and sickness as a dynamic interaction between the individual (the micro) and his environmental (macro) systems. Interestingly, this holistic outlook generally prevailed until the late 1800s when Pasteur and Koch recognized the nature of infectious microorganisms. This monumental medical discovery ironically had the unforeseen consequence of effectively limiting the study of disease to laboratory phenomena, independent of its human host. And though modern medicine has convincingly claimed its greatest triumphs pursuant to the latter model (many physicians will never see a case of typhoid fever, diphtheria, tuberculosis, or polio), modern man may be suffering his greatest ills. Because of the technique inherent in the model, the isomorphism between the level of explanation achieved and the ends to which that information is applied has become minimized, making the original findings of diminished operational value. Though the remarkable efficacy of serum and vaccine treatment should not be underestimated, the conception of illness as an entity isolatable from the milieu in which it operates, confounding variables and all, reduces the phenomenon of man to a passive model. It is doubtful that the cancers, the colitises, the psychoses are the manifestations of invasion by solitary infectious agents independent of their host's status.

Such analytical errors are responsible in large part for our environmental deterioration. No soothsaying about dwindling natural resources will be recited. But it will be observed that such occurrences are the logical indictment of a reductionistic view that exceeds the limits of its applicability. Vis-à-vis air ions: charged molecules are a part of normal, healthy air, though their function in the total environment is not known. However, life did evolve in a matrix of ionized air, and that chemical and electrostatic milieu has been altered. Of course, the presence of the first living organisms also altered that milieu, but now the change is millions of orders of magnitude greater. The microclimate is no longer targeted for change. Rather the climate for macroevolution is. Air quality is being altered on a grand scale. Worldwide, and perhaps for all times, societies not cognizant of the technology required to effect such dramatic change are nonetheless being affected by it. There is no doubt that small air ions are severely, if not totally, absent in polluted, sulfur dioxide- and acid-laden air. This loss in air ions may be a significant environmental factor affecting health.

9

Biometeorology

The subject of biometeorology has been touched upon repeatedly during the discussion of atmospheric electric currents and electric and electromagnetic fields in previous chapters. It would seem appropriate, therefore, to survey this branch of the environmental sciences from a meteorologic perspective.

Meteorologic Aspects*

Meteorologic elements, such as temperature, humidity, wind and infrared radiation from the atmosphere (counter-radiation), are also a part of man's invisible environment. These elements differ from most electric and magnetic phenomena, however, in that man can more or less consciously perceive them via the senses.

But if one considers such a striking spectacle as a thunderstorm, it becomes clear how closely linked the other meteorologic elements are with electricity and magnetism. It can be said, in fact, that all processes and motions in the atmosphere and all natural and vital processes in general are accompanied by electric or electromagnetic phenomena. The utilization of these phenomena as diagnostic tools in medicine (ECG, EEG) and meteorology (atmospherics) is based on this principle. The special science of electro-biometeorology, on the other hand, is concerned with the actions of such phenomena on living organisms (cf. Tromp, 1968, and Chapters 4–8).

History of Biometeorology

Perhaps the most famous explorer-scientist of the 19th century, Alexander von Humboldt (1769–1859), who journeyed extensively in Central and South America, gave the following definition in his comprehensive work on the natural sciences, *Cosmos* (Volume I, 1845):

> The term "climate" in the most general sense refers to all changes in the atmosphere which have a perceptible influence on our organs, namely, temperature, humidity, variations of barometric pressure, wind, the magnitude of the

*This portion of chapter 9 was contributed by W. Sönning

> electrical voltage, the purity of the atmosphere or its contamination by more or less noxious gaseous exhalations...

He further speaks of the "habitual transparency and clarity of the sky," which not only determines the radiation balance at the earth's surface and the "organic development of plants and the maturation of fruits," but is also important for "the feelings of man and his overall mood."

This definition of climate shows that modern climatology has its roots in a biometeorologic way of thinking. The importance that Humboldt ascribed to atmospheric electric phenomena is revealed elsewhere in *Cosmos*, when he speaks of the problems of atmospheric electricity in its "quiet, periodic, daily course" and of the "explosions of the flashing, roaring thunderstorm." He describes electricity as a phenomenon that is closely linked to the behavior of the other meteorologic and geophysical elements, and which "has a profound influence on the entire plant and animal world"—not only through meteorologic processes, but also directly as a "force which stimulates the nerves or promotes the circulation of sap." Such statements, made almost 150 years ago, could still serve as a working program for research into the invisible environment of man. The fundamental problems of biometeorologic research, which are repeatedly expressed in modern works and have yet to be resolved, were also addressed by Humboldt with the following words:

> The meteorological part of nature's tapestry shows that all processes involved in the absorption of light, the liberation of heat,... the hygrometric situation and electrical voltage... are so closely interrelated that any single meteorological process is modified by all other concomitant processes...

Looking back even farther into history, we find that the famous physicians of ancient Greece and Rome described an abundance of bioclimatic and geomedical relationships, which correspond in essence with the approaches and goals of modern bioclimatology. Noteworthy in this regard is Hippocrates (460–377 B.C.) and his text *On Airs, Waters and Places,* as well as the Roman physician Galen (129–199 A.D.), who recommended climatic cures analogous to modern climatotherapy. After studying the ancient writings, von Philipsborn (1949) comes to the conclusion that "a close cooperation existed among the physicians, meteorologists and astronomers of ancient Greece which has remained unequaled even to the present."

Following a sporadic history during the Middle Ages and subsequent periods, the science of biometeorology experienced an upswing in Europe and the United States around the turn of the century owing to improved techniques of measurement and meteorologic analysis. As advances were made in the natural and biologic sciences, the various branches of biometeorology also progressed (effects of atmospheric electricity, radiation, radioactivity, aerosols, etc., on living organisms). A detailed discussion of the history of biometeorology is given by Snorrason (1964) and Assmann (1963), among others.

A number of bioclimatic research centers were founded in Germany in the early 1930s. They provided a basis for diverse biometeorologic activity, which continued after the Second World War and into the 1960s in the form of various medical-meteorologic research groups. Medical-meteorologic and bioclimatologic research and advisory centers in the Germanys are at present supported by the national weather services: the German Weather Service (West Germany) and the Meteorologic and Hydrologic Service of the GDR. The situation is similar in other countries, i.e., U.S. Weather Service in USA. Biometeorology was promoted in general by the founding of research institutes for the various branches, such as medical climatology, tropical and high-altitude climatology, atmospheric electric research, aerosol research, agricultural and forestry meteorology, etc. The growing interest in these problems led finally to the founding of the International Society of Biometeorology (ISB) in 1956. It is active in all aspects of meteorologic, biologic, medical and agricultural research, and has lately become involved in the architectural and engineering sciences as well. The principal goal of the ISB is to stimulate, promote and organize interdisciplinary scientific cooperation (cf. Haufe, 1976). An important aid in this process is the *International Journal of Biometeorology* (Amsterdam), together with other journals that are published by ISB research groups whenever international congresses are convened. Biometeorology is also represented in the World Meteorologic Organization (WMO, headquartered in Geneva) primarily by the Commission for Special Application of Meteorology and Climatology (COSAMC).

Definition of Biometeorology

After various suggestions, the members of the ISB adopted the following definition and classification of biometeorology (cf. Tromp, 1974), which is quoted in part because of its general importance.

> Biometeorology comprises the study of the direct and indirect effects (of an irregular, fluctuating or rhythmic nature) of the physical, chemical and physico-chemical micro- and macro-environments of both the earth atmosphere, and of similar extra-terrestrial environments, on physico-chemical systems in general and on living organisms (plants, animals and man) in particular.

Influence of the Atmospheric Environment on Man

We shall attempt to point out the most important subjects and problems of biometeorologic research from a meteorologic perspective.

Though the question of the effects of weather, atmospheric conditions and climate on organisms has been discussed for many centuries, it appears that man's sensitivity to the atmospheric environment has been increasing in recent times. Contemporary man is apparently growing increasingly "allergic" to weather stresses. Based on extensive demoscopic surveys, Faust (1977) considers it probable that about 30% of the population of Central Europe experiences a disturbance of well-being due to weather. Ranscht-Froemsdorff (1976) obtained the following figures based on 8 years of advisory service: 60%–70% of the persons observed are not meteorosensitive, about 30–35% are "sensitive" and 3%–5% "hypersensitive." The percentages are probably similar in the United States. Both authors make a distinction between "meteorosensitivity" and "hypermeteorosensitivity." The former condition involves an increased overall responsiveness of the organism to environmental stimuli and is manifested in transitory subjective complaints, such as tiredness, headache, scar pain, sleep disturbances or a depressed mood: that is, in disturbances of autonomic function. By contrast, hypermeteorosensitivity can be considered almost an illness. The impairment of the general state of health in such cases is severe (intense phantom limb pains, circulatory failure, respiratory complaints, etc.).

Weather, as a natural environmental stimulus, is not disease-producing in itself. This must be emphasized again and again. Weather-related complaints, rather, are a sign that the organism's response threshold for exogenous influences has been lowered as a result of disease, autonomic dystonia, etc. This decrease of the range within which the organism is still capable of fully compensating for meteorologic stimuli is closely related to man's increasing alienation from his natural environment (living habits, nutrition, etc.).

Problems of Biometeorologic Research

As a branch of the environmental sciences, medical meteorology deals with the action (biotropism*) of the atmospheric environment on man. Thus the effort to understand and clarify biotropism is truly a problem of interdisciplinary research (including biotropic effects in plants and animals).

The measurable expressions of organic processes are in a general sense only indicators of the total complex of biologic activity. In the same way, the total "weather" can be quantitatively grasped neither by direct measurement nor by the computational treatment of measured meteorologic variables. Information on the behavior of this overall complex "weather"—in medical meteorology it is also called "meteorologic impact"—can likewise be obtained only in an approximate manner, through the use of suitable indicators. Owing to the interrelationship of meteorologic elements, almost any can be used as an indicator of common behavior in fairly good approximation.

The notion of the existence of a single biotropic factor that governs biometeorologic relations is difficult to reconcile with the complexity of atmospheric and biologic systems. To be sure, the ever-changing physiologic quantities of the atmosphere, such as temperature, humidity and pressure, are amenable to precise measurement individually and in certain combinations; but all previous studies directed toward identifying one of the meteorologic parameters (temperature, air pressure, etc.) as the dominant biotropic factor have led essentially to a negative result. The biotropism of atmospheric phenomena thus involves a combination of interrelated effects. It should be possible, however, to resolve these effects step by step into partial causal relationships in special areas of medical meteorology (cf. Ungeheuer and Kügler, 1957; and Brezowsky, 1962).

*Biotropism (from the Greek: bios = life, tropē = a turn, turning) refers to the action of meteorologic phenomena on biologic systems. Meteorotropism refers to the ability of organisms to be influenced by the atmospheric environment. Thus, one speaks of "biotropic weather," but "meteorotropic diseases."

Meterologic Impact

The overall climatic or meteorologic impact at a given location or in an area has been divided into various complexes that act specifically on the organism (cf. Becker, 1962).

Thermal complex. The thermal complex encompasses all individual factors that influence the thermoregulation of organisms ultimately through the addition or removal of heat: air temperature, short- and long-wave radiation, wind and air humidity. The duration of action of these elements ranges from climatologic time periods of many years or decades, to seasonal fluctuations, to the application of hot or cold stimuli by the atmosphere on an hourly basis. Even temperature fluctuations on a time scale of minutes and seconds are being investigated for their bioclimatic significance (Jendritzky, 1978b). Among the complexes of atmospheric effects, the thermal complex plays a central role (cf. Hensel, 1973, for example). The level of organization of an organism's thermoregulatory systems is the very trait by which the differentiation between a "higher" or "lower" form of life is made. In the area of physiotherapy and balneology, the use of hot and cold stimuli, applied for the correct duration and at the proper temperature, can evoke profound autonomic changes in patients. This suggests that the thermal stimuli with which the atmosphere "treats" man on various time scales can also produce strong reactions.

For a long time, therefore, efforts have been made to arrive at an objective representation of the thermal sensation and thus at a quantitative evaluation of man's thermal environment. In the United States in particular, various index numbers and biometeorologic indices have been developed for this purpose (cf. Givoni, 1974; Wenzel, 1975; and Fanger, 1970; among others). Jendritzky et al. (1977) recently succeeded in devising a method for the thermal evaluation of man's environment under any outdoor conditions, via Fanger's (1970) "predicted mean value " (PMV). This was done by taking Fanger's original "comfort equation" for optimizing an indoor climate and expanding it into a simulation model by parametrizing the ambient long- and short-wave radiation in terms of the mean radiation temperature (cf. also Jendritzky, 1977, 1978a).

Photoactinic complex. The photoactinic complex includes the short-wave radiation from the sun (UV and visible light) along with the terrestrial thermal radiation resulting from it. The three components of this complex, which have different biologic actions, always occur together under natural conditions, though in varying intensities and proportions depending on the geographic situation, time of day or season. They are further influenced by such purely meteorologic factors as cloudiness and atmospheric turbidity. Most of the energy of the short-wave radiation lies at wavelengths of 0.3–4 μm, with a maximum at 0.55 μm. The terrestrial radiation ranges from 6 to 60 μm, with a maximum at 15 μm (see also Chapter 2).

The factors in this complex act via the visual sense and cutaneous system to produce photochemical reactions within the body, while the autonomic nervous system simultaneously responds to the long- and short-wave radiation. As the most important geophysical "master clock" with its discrete daily and seasonal rhythms, this complex controls highly organized sequences of biologic functions (see below). It thus plays a major role in climatotherapy (heliotherapy) and in physical medicine, as is documented by the medical literature.

Chemical complex. The atmosphere is regarded on the one hand as a permanent gas mixture of natural components, and on the other as a colloidal system of solid, liquid or gaseous impurities of both natural and anthropogenic origin. These include vehicular, industrial and household emissions, and to a lesser extent such natural contaminants as dust, pollen and volcanic exhalations. These factors, which are measured in terms of concentration and partial pressure, not only act on the skin surface (cf. Swantes, 1977, for example), but are also absorbed by the mucous membranes of the respiratory tract, leading to effects on blood chemistry and cell metabolism. The chemical complex thus produces a number of qualitatively diverse effects in the body, ranging from discomfort from dust or odors to allergic reactions to toxic effects. The duration of action covers time scales from hours or days to many years. Consider, for example, the short-term action of toxic substances in polluted air, or the lasting autonomic changes produced by adaptation to high altitudes.

The distribution of injurious substances in the air is dependent on the lapse rate conditions and on the intensity of the air flow, that is, on the horizontal and vertical exchange of air in the atmosphere near the ground. These questions are dealt with by micrometeorology and industrial climatology, on which an immense volume of data has been published.

There is also much interest in the favorable and possibly therapeutic effects of certain trace materials in the atmosphere. In climatotherapy, for example, the iodine and salt in sea air and the essential oils in sylvan air are regarded as "curative agents." In electro-aerosol therapy, medicinal solutions suspended in air are given an electric charge in order to make them more respirable (see Chapters 4–8). The decrease in the partial pressure of the gaseous constituents of air with increasing altitude (alpine regions) is another highly effective therapeutic factor—a subject which is dealt with in high-altitude physiology.

The three complexes discussed so far can be understood quantitatively as partial causative factors (see above) according to their meteorologic and geophysical origin. It appears certain that a direct chain of action exists between these factors and biologic reactions. For example, there is little question that the UV component of solar radiation is the cause of erythema of the skin, or that a high level of heat in the air is the factor that initiates sweat secretion. An intensity-dependent dose-effect relationship is also known in most cases (PMV value, exposure time in heliotherapy, limits for toxic effects of air impurities, etc.). However, little is known as yet about the continuous sequence of physiologic processes that occurs in response to environmental stimuli.

Neurotropic complex. The neurotropic complex includes all processes of atmospheric dynamics that surround man in the form of weather.* The meteorologic bases of its analysis are charts of wind, pressure, temperature and humidity distribution in the upper layers of the atmosphere and at ground level, as well as reports on cloud cover, precipitation and other weather phenomena, as provided by national weather services. All synoptic meteorologic variables are utilized in this analysis, provided they make suitable indicators for describing the weather situation and the dynamic state of the atmosphere. These may include such parameters as the ozone partial pressure at the ground, air-pressure oscillations or atmospheric electric variables, whose function as an excellent indicator of atmospheric processes is known (see Chapter 2).

An orientational survey of the biotropic factors and complexes in the atmospheric environment is presented in Table 9-1. The lower portion of the table also shows two nonmeteorologic factors, radioactivity and cosmic radiation, because their biotropic action is modified by the atmosphere or transmitted to biologic objects (through ionization of the air, for example; see Chapter 8). A detailed survey of the daily and annual rhythmic behavior of each of the factors in the four complexes was compiled by Becker (1962). It should be noted, however, that the sum of the individual effects of the biotropic factors and complexes does not yet explain the total effect (cf. Brezowsky, 1965). The dilemma is that this total effect has been statistically demonstrated for many biologic responses with the aid of medical-meteorologic working hypotheses.

*The description of a complex as "neurotropic" arose from the concept that the factors contained in it act on the human body by way of the nervous system (Pfleiderer, 1952).

Methods of Medical-meteorologic Analysis

Since the Second World War, it has been the task of medical meteorology in Germany to develop a working hypothesis based on previous concepts and to continually review and revise that hypothesis in the light of biologic observations (see the biblography of Weihe and Brezowsky, 1963). Individual research groups, working separately, have developed meteorologic working methods that provide a basis for comparing meteorologic phenomena with biologic data. It is noteworthy, according to Brezowsky (1962), that they all arrived at the same conclusions, even though their investigations were conducted independently in such diverse landscape climates as Hamburg in Northern Germany (Kuhnke and Zink, 1950), Königstein (near Frankfurt) in Central Germany (Becker, 1955), Tübingen (Daubert, 1958) and Bad Tölz (near Munich) both in Southern Germany (Ungeheuer, 1955).

The main question was whether a nonrandom connection could be found between biologic events and simultaneous weather phenomena in all the landscape climates of central Europe. This did not preclude the possibility that biotropism might vary in both magnitude and form under different climatic conditions. Implicit in this question was the further question, whether biotropism is manifested in every

Table 9-1. *Biotropic factors in the atmospheric environment.**

Stimulating factor	Individual elements	Quantitative parameters	Detection of biological action	Complex of effects
Cooling, heating	Sensible heat of the air Moisture content of the air Wind Long-wave ambient radiation (including sky)	Temperature, partial pressure, relative humidity, velocity, radiated power	Thermal sensation (comfort, discomfort)	Thermal
Radiation: short wave (0.3—4 μm) long wave (6—60 μm)	UV radiation, visible light, infrared radiation	Radiated power	Dose (frequency-and amplitude-dependant action of heat), psychological effects	Photoactinic
Natural and anthropogenous constituents of the air	Solid, liquid and gaseous substances	Mass, concentration, partial pressure	Dose, limiting values	Chemical
Atmospheric processes	Dynamic processes and thermodynamics of the atmosphere Distribution of winds, pressure, temperature and humidity aloft Metereological elements at the ground Types of air masses	Synoptic charts, diagrams, numerical meteorological analysis	Statistical correlation	Neurotropic
Air-pressure oscillations	Air-pressure change (short-term)	Pressure (amplitude and frequency distribution)	Statistical correlation, dose, action characteristic	Neurotropic, geophysical, astrophysical
Atmospherics (ELF, ULF), atmospheric electric field, air ionization	Electricity, magnetism	Frequency, radiated power, field intensity. electrical charge		
Radioactivity, cosmic radiation	Corpuscular radiation, X-rays	Intensity, frequency, physical action		

*Factors shown below the double horizontal line would be assigned to a "tellurian" (or geophysical) and a "cosmic" (or astrophysical) complex of effects.

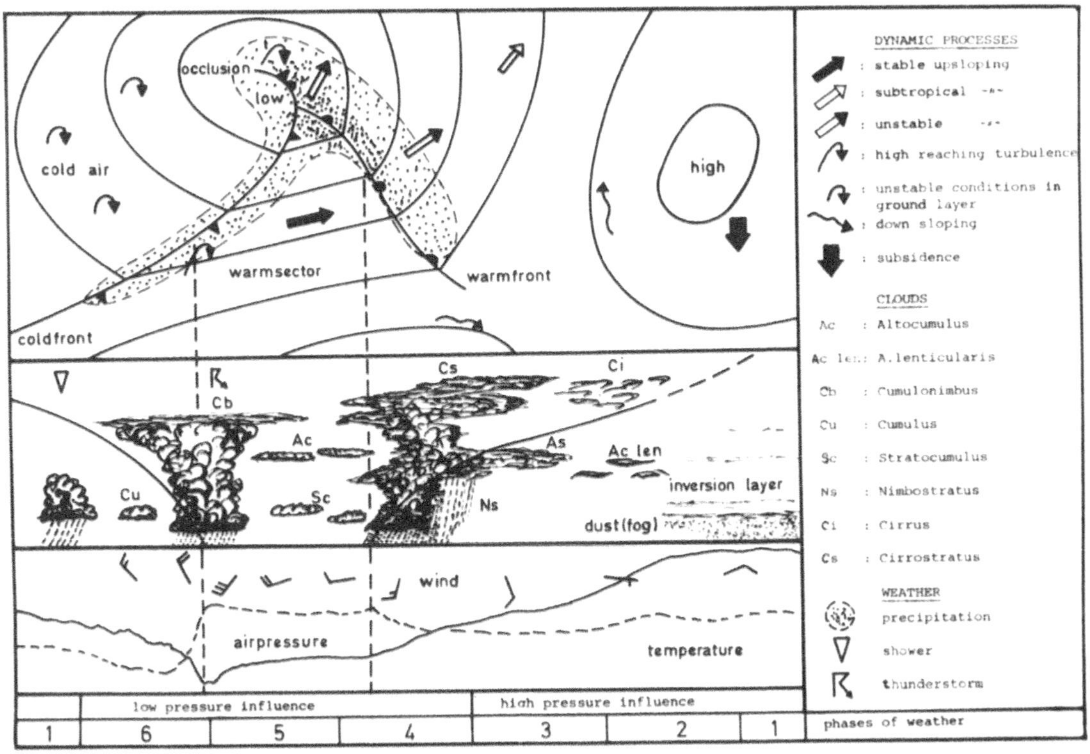

Figure 9-1. The Königstein medical-meteorologic weather scheme. The drawing shows the surface and vertical sections of an "ideal cyclone" together with the major dynamic processes and the progress of certain weather elements (clouds, air pressure, wind, temperature).

biologic process to one degree or another, or only in pathologic phenomena.

The Königstein scheme. Becker (1956) developed in Königstein (West Germany) a method for the precise hour-by-hour analysis of meteorologic developments in a graph called the "bioclimogram." His scheme is based on the "ideal cyclone" of Bjerknes (1862–1951), which is shown in Fig. 9-1 in highly-simplified surface and vertical cross-sections together with the associated processes of atmospheric dynamics and the major cloud forms and meteorologic elements.

A basic distinction is made between cyclonic weather processes, marked by ascending flow, atmospheric turbulence and frontal processes, and anticyclonic processes, characterized by descending flow and subsidence. Ascent occurs when the flow processes in the atmosphere have an upward component. The resultant adiabatic cooling leads to the condensation of water vapor in the air and thus to cloud formation and precipitation. Ascent occurs mainly when warm, usually moist subtropical air at the leading edge of a low advances from a southerly to a westerly direction (in the Northern Hemisphere). Stable or subtropical ascent takes place initially in the higher tropospheric layers, but changes to unstable ascent during the approach of the warm front and in its immediate vicinity.

When the advection of subtropical air takes place aloft, characteristic, high cirrus clouds (fine, wispy structure) are observed at first; these gradually thicken into cirrostratus, until lower cloud layers (altostratus, nimbostratus) appear with the approach of the warm front. Gradually, precipitation in the form of drizzle, light rain or snow begins, which may persist and intensify.

In the warm sector itself, stable ascending or descending flow prevails. Condensation processes are confined to flat layers, resulting in less dense cloud formations (altocumulus, stratocumulus) but seldom leading to precipitation. Stable or unstable ascent may also take place within the same air mass, and can therefore occur independently of frontal processes.

Turbulence exchange occurs if the vertical upward or downward component of flow exceeds the horizontal component. In this case water vapor condenses out of the rising air, and the freed heat of vaporization imparts an additional acceleration to the rising parcels of air. This usually occurs in the region of a cold front, due to the high-reaching unstable conditions there, and in the immediate rear of it. But the unstable conditions are soon confined to the lower layers of the atmosphere, because the inflowing cold air is quickly stabilized by subsidence processes aloft.

Cumulus clouds are typical of unstable conditions of this sort. If the instability extends to high levels, these clouds may build up into large thunderclouds (cumulonimbus), resulting in showery precipitation in the form of rain, sleet, snow or hail.

Subsidence is the characteristic process in the region of the anticyclone. Here the downward component of air motion exceeds the horizontal flow velocity. One result of this is that little or no wind is observed in the central region of a high. The adiabatic heating that accompanies subsidence leads to cloud dissipation. Subsidence inversions (layers with a vertical temperature inversion) develop at various levels and retard or completely prevent vertical mixing. Persistent fog or stratus can occur in the stagnant air near the ground, especially in the fall. If atmospheric pollutants are added to this over industrial or heavily populated areas, conditions are ideal for smog formation.

Descending flow occurs when the air moving aloft possesses a downward component. This process is associated mainly with the boundary regions of a high. Like subsidence, it leads to a drying and warming of the air and is also known as "free foehn." If the descent is strong enough, typical foehn clouds (altocumulus lenticularis) sometimes form. The sky appears deep blue because of the dryness of upper air layers. Unlike the situation at the center of the high, sufficient horizontal air motion is present near the ground to prevent hazardous fog or smog conditions.

A table was compiled according to this scheme showing the meteorotropism of different diseases, health disturbances and acute phenomena, such as infarcts, strokes and deaths, as a function of individual weather processes. It was possible to group meteorotropic reactions into those associated with warm-air processes at the advancing edge of lows (hypotonic reaction type) and those associated with cold advection in the region of a cold front and its trailing edge (spastic-hypertonic reaction type). Between these groups was a broad range of mixed types (cf. Becker and Ströder, 1962) as Fig. 9-2 shows.

Similar methods were employed by the research groups in Hamburg and Tübingen, where weather developments were similarly classified in terms of general meteorologic situation, air masses, radiation parameters at the ground and dynamic daily weather types. Essentially the same phases of biotropic effects were found in these studies (cf. also Brezowsky and Grimmeisen, 1967; Daubert, 1958; Kuhnke and Zink, 1950).

Common to all these techniques is their focus on the dynamic physical processes in the atmosphere up to about the level of the tropopause (6000–8000 m), employing the concepts of synoptic meteorology. Processes aloft were presumed to be the principal origin of biotropic effects. It is interesting that several studies, especially by the Tübingen group, provide evidence of this, as well as support for the theory that the autonomic nervous system is the "point of attack" of the meteorologic impact. How-

AIR MASSES	warm					mixed		cold		
WEATHER PROCESS	⬇	↝	⬈	⇗	wf	⇗	oc	cf	↱	↷
Psychopathy	●									
Excitability		●			○					
Schizophrenia		●	○	●						
Psychic Depression				○	○	●				
Hypotonic Collapse		○			●					+
Neurosis				○	●	○				
Suicide		○		○	●					×
Cold	○	+			○					
Inflammatory Processes					●	○	×			
Pneumonia				○						
Appendicitis	×			×	×	×			×	
Bronchitis		○					○			
Bleeding		●		○	●	×				
Thrombosis		×		●	○	×				
Glaucoma				×	●		×	○		×
Accident Proneness		×	+	●	○	●	×		○	+
Subjective Complaints		●			○	●				○
Heart Infarct	+	○		○	●	●	○	○	×	+
Embolism		×	+	●	●	●	○	○	×	×
Depth of Sleep		●			●	●		○	+	+
Spastic Complaints		●			○			●	×	
Migraine Headaches	×	×		○				×		
Headaches		○		○	○		○	○		×
General Aches and Pains					●	●		●	×	×
Reaction Time		×	+		●		●	○		+
Decompensation of Heart	+				●			●		+
Premature Birth					×			○		
Heart Failure (Death)					×	○	○	●		×
Perforation of the Stomach					○	○	○	●		
Diabetes				×				●		
Rheumatoid Arthritis							○	○	○	
Apoplexy				×	×			●	○	
Epilepsy								○	×	
Colic								○	●	×
Angina Pectoris	+		+			×		●	●	

Figure 9-2. Meteorotropic diseases (summary derived from Becker and Stroeder, 1962). For symbols for weather processes see Fig. 9-1. The symbols for the statistical correlation between the listed diseases and weatherprocesses are as follows:
● = positive correlation at high significance level; ○ = significant positive correlation; × = trend of positive correlation; + = significant favorable influence.
Note: the list is derived under the climatic environment and health conditions of the people in Central Europe in the years 1945-1965.

ever, the causal sequence of effects linking weather processes regarded as biotropic with the reactions of the autonomic nervous system remains a mystery. The "transmitter" of the effects also remains undetermined although even then, as now, atmospherics were often discussed as a possibility (cf. Chapter 2).

Tölz weather-phase scheme. Researchers in Tölz attempted to resolve these fundamental questions with an essentially different approach: by applying biologically oriented standards to the classification of meteorologic activity (Ungeheuer, 1955). These standards were derived mainly from the results of medical biorhythm research then available. The basic idea was that the undisturbed daily and yearly periodicity of meteorologic elements associated with solar radiation acted as a "master clock" for the corresponding endogenous rhythms of the body. A disturbance of the normal (autochthonous) meteorologic periodicity by aperiodic weather developments would thus represent a change demanding adaptation by the organism. These changes could be long-term, leading to acclimatization, or short-term (weather), evoking adaptive responses which mean increased stress for the organism. Thus, several assumptions were made (cf. Brezowsky, 1965):

1. Adaptation is brought about by an aperiodic meteorologic process.
2. This process cannot be represented by a single biotropic factor, but must be regarded as a summation effect.
3. The "meteorological impact" involves all elements and factors that can be monitored in the biosphere by measurement.

With the aid of this working hypothesis, it was possible to compare not only meteorologic and biologic events themselves, but also their patterns of progression. This meant a broadening of the traditional static view (which is still held in most cases) in a more dynamic direction, in that henceforth the *temporal* structures of meteorologic elments were also recognized as having a biotropic value.

The disturbance of the basic pattern of daily meteorologic rhythms is caused by the horizontal and vertical air motions that are a consequence of atmospheric dynamics. A stationary observer witnessing the successive passing of a high and a low will note phases with a weaker and stronger advection of outside air masses. These weather phases are illustrated in Fig. 9-1. Ideally the intensity of advection will show a sinusoidal variation with time, while the autochthonous component of local weather follows an opposite sinusoidal pattern. Ungeheuer (1957) divided this process, with deliberate simplification, into six weather phases (Fig. 9-3).

The connection between this and the Königstein scheme is apparent (see Fig. 9-1; cf. also Sönning, 1974). Besides the characteristic behavior of many meteorologic elements in each phase, a quantity combining temperature and humidity—the so-called temperature–humidity environment (THE)—in the ground layer provides a very sensitive index of advective processes of every kind. Attempts were made to use this objective index as a basis for differentiating between biologically favorable and unfavorable weather phases (Ungeheuer and Brezowsky, 1965). The lower portion of Fig. 9-3 shows the variation of the mean intensity of biotropic stimuli for the favorable and unfavorable ranges as a function of phases 1–6. It is seen that a favorable ("euphoric") effect is present (phases 1 and 2) when the daily periodicity of the meteorologic elements again becomes harmonized to the maximum degree following the disturbed phases, and can thus fulfill its master-clock role for the organism. In this case the organism is, to put it simply, in phase with the meteorologic environment. The situation extending from the middle of phase 3 to phase 6 is biologically unfavorable, on the other hand.

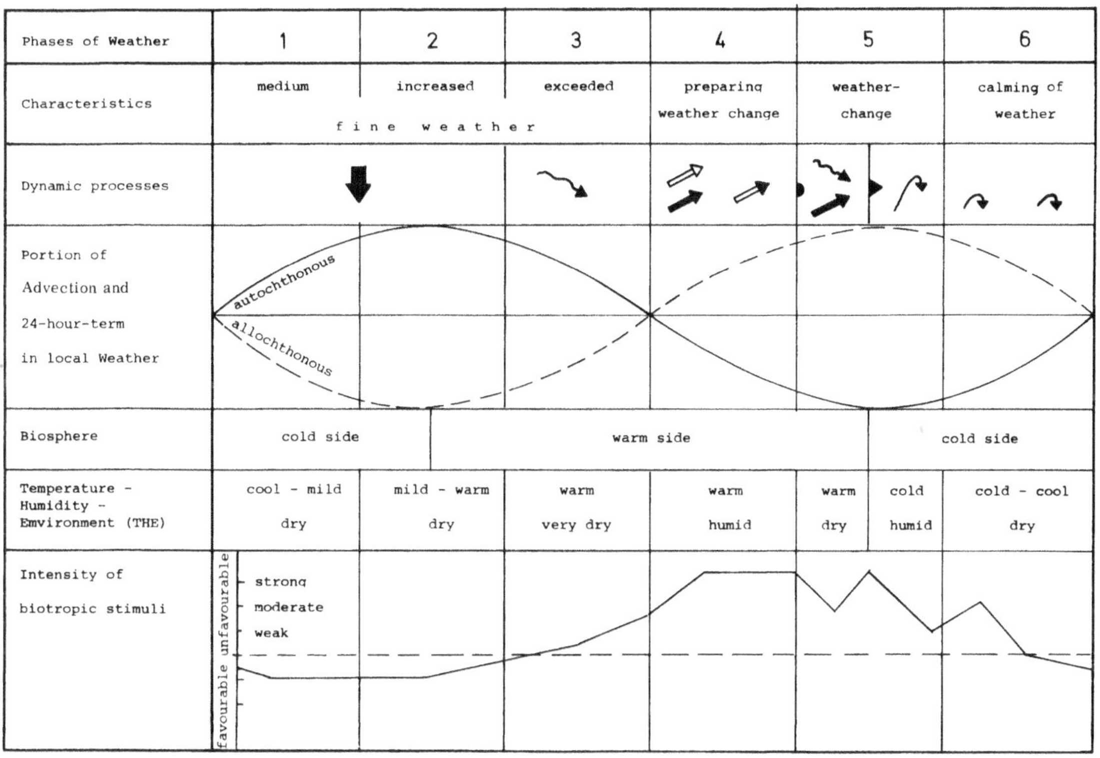

Figure 9-3. The Tölz weather-phase scheme. The ideal weather sequence at temperate latitudes in six characteristic weather phases.

*Biotropic Balance**

With the aid of the Tölz weather-phase scheme and the temperature–humidity environment, the behavior of biologic subjects under the influence of atmospheric environmental parameters could be differentiated on a statistical basis during the course of long-term investigations. A *pathogenetic* biotropic balance was found first, describing the meteorotropic response of individuals with certain clinical conditions—this agreed with the results of the other three research groups. By investigating the spatial distribution of the meteorotropism of equivalent clinical syndromes, it was further possible to find a dependence of biotropic effects on the landscape climate, and thus to determine a *climatic* biotropic balance, which acts in a specific sense to enhance or weaken the biotropism of the weather. A *daily* and *seasonal* biotropic balance could also be established, as well as an *individual* balance, which depends upon reaction type. Summation of all components yielded the normal distribution of biotropism, or the *total* balance, shown in Fig. 9-3. This is the distribution that is obtained on analysis of very extensive space-, time-, disease- and man-averaged groups (e.g., in mortality studies based on the records of statistical bureaus).

It may be that these findings, obtained by relatively simple means, are in need of better substantiation in many cases, or that they will be modified by future research. However, this cannot be the only valid standard for judgment. It is highly probable

**Translator's note:* The "biotropic balance" is a graphic representation of the quality and intensity of biotropic effects with respect to some parameter (such as climate, disease, season, individual reaction type, etc.).

that these results are indeed useful, first because the studies of the various medical-meteorologic research groups in Germany during the 1950s and 1960s can be interpreted as a large-scale blind test (Brezowsky, 1962), and second because more recent research in biorhythms and chronobiology appears to provide a rationale for many of the statistical connections found by meteorologists (Hildebrandt, 1975b). Particular credence is lent the rhythmologic basis of the Tölz method, which may well prove important to the advancement of medical-meteorologic research.

Outlook

In tracing the sequence of biotropic effects in man, the differentiated breakdown of biotropism is nothing more (or less) than a methodologic starting point for further medical-meteorologic research. However, electronic data processing has made it possible to expand previous statistical methods in this area on a multifactorial basis. It has also become possible to lift medical meteorology out of traditional classification schemes and proceed experimentally, that is, search for the most rigorous correlations among the multiplicity of individual factors (cf. Weihe, 1976). It will still be necessary, of course, to find the correct meteorologic and biologic interpretations of events, again proceeding on the basis of working hypotheses.

Despite all the uncertainties in previous medical-meteorologic findings, it must be pointed out that they have aided and continue to aid the practicing physician in the selection of optimal therapies. For example, meteorologic forecasts of the biotropic development of a weather situation are helpful in the prescription of drug dosages, in physical therapy, climatotherapy, dietetics, etc., or in the selection of prophylactic measures. It should also be noted that the prompt prediction of biotropic impairments of the working capacity, responsiveness or mental state of healthy persons is a valuable tool in accident prevention (cf. Reinke, 1977).

However, the practical application of medical-meteorologic knowledge is still no solution to the problem of the relationship of man to the atmosphere or, more generally, of living organisms to the atmosphere. More than ever before, progress in this direction can be made only through close cooperation on an interdisciplinary basis. The "meeting point," so to speak, is the skin surface of the organism. The meteorologist approaches the problem from the atmosphere. The bioscientist (physician, physiologist, biologist, etc.), on the other hand, is interested in the sequences of effects that are triggered by environmental stimuli in the organism itself. He is concerned with causality and examines the thermal, actinic, chemical and other physical environmental factors that might be of importance individually or in concert. Meteorologists and bioscientists thus follow different paths, as they must. It is little wonder, then, that difficulties in communication sometimes arise, as when the biologist prefers to speak of the thermal, actinic or chemical complex of effects, and the meteorologist of the neurotropic complex. But mutual assistance and suggestions are equally possible.

During the last 10–15 years, biorhythm research has revealed not only circadian and circannual rhythms, but also a large number of other rhythmic functions and complex interrelations in man. For example, Hildebrandt (1974, 1975b) speaks of a hierarchic, temporal organization in man, whereby the rhythmic functional order is harmonically structured. The rhythms are also synchronized by external "master clocks," which in turn are coupled to daily and annual geophysical rhythms in the environment. Hildebrandt (1974) states explicitly, "This anchoring of the rhythmic order of the organism in the temporal order of the environment represents the basis of the entire rhythmic organization of man." He also refers in particular to man's reactive periods (which are in turn made up of subintervals), whose significance next to the spontaneous rhythms (daily cycle, for example) has been increasingly discussed in recent times. Interesting surveys of the typical periodicities of rhythmic human functions are also presented by Hildebrandt (1974, 1975b). Sinz (1978), moreover, has attempted to devise a scheme showing correspondences between biologic and geophysical (including meteorologic) periodicities.

The concept of a hierarchic, temporal organization in human physiologic processes could be readily applied to the fields of meteorology and geophysics, which can also be interpreted as a kind of "physiology" (of the terrestrial body). New insights into the extremely complex functional and rhythmic behavior of the (human) organism underscore the importance of the *time factor* in the case of environmental stimuli, quite apart from the "material" effects of such meteorologic factors as heat, cold, aerosols, UV radiation, etc.

We can appreciate how complex the temporal structure of atmospheric phenomena as a whole must be when we consider that besides the daily and annual rhythms imposed by geophysical and cosmic factors, there are other characteristic sequences which we experience and characterize as weather and climate, and which are a particular expression of the laws of atmospheric dynamics. There are, moreover, a great many other less conspicuous periodicities, rhythms and cycles of atmospheric elements of all magnitudes in which an obvious temporal order is at work. It may well account for the qualitative and quantitative variations (with time, location, individual and type of phenomenon) noted in the biotropic effects of atmospheric elements.

Electrophysical Aspects

Electro-biometeorology

One branch of biometeorology is concerned specifically with studying the connections between biologically active atmospheric electric or electromagnetic processes and classical meteorologic factors, such as air pressure, precipitation, humidity, cloudiness, temperature, wind and sunshine. Tromp (1963) presents a survey of the seasonal and annual fluctuations of meteorologically determined electromagnetic processes in the atmosphere and their possible biologic significance. Reiter (1960) also dealt with this problem in detail. He supplemented the extensive literature in this area with a special consideration of the problem of atmospheric electric variables as components of the bioclimate, and their significance with regard to the design of buildings, airconditioning systems, etc. (Reiter, 1970). In the process, he analyzed all relevant atmospheric elements in man's environment, such as fields, charge carriers and electromagnetic impulses of atmospheric origin, from the standpoint of possible effect on biologic systems, with a negative result.

Apparently Reiter (1973) revised this opinion later: "It could be clearly shown that electric, magnetic and electromagnetic fields exert significant biological effects, even at very low intensities." It is this statement that forms the thesis of the present book. If one accepts such a possibility—and based on the evidence presented, any other position is no longer tenable—one must realize its implications. A number of electromagnetic processes in the atmosphere, which are in most cases linked in some way with weather, are measurable and clearly demonstrable. It must be concluded, therefore, that electric, magnetic, electromagnetic and other atmospheric electric phenomena are also responsible for certain biologic reactions associated with meteorosensitivity in man.

The electric field plays a particularly important role in the low-frequency range during weather phenomena (which does not exclude magnetic processes, of course). The physical measurement of electric field fluctuations of several hertz (including higher-frequency signals with pulse repetition frequencies in the same range), which are observed under certain weather conditions, and the complete atmospheric electric passivity seen under certain other weather conditions (foehn, for example), combined with previous results of objective biologic studies, provide clear evidence of influences.

As mentioned earlier, it has been found that different persons react differently to such phenomena. It appears now that one of the main arguments against such speculations offers a plausible explanation for the differences in well-being of persons subjected to identical weather conditions.

Three points, closely interrelated in their effects, may well be responsible for the seemingly confusing phenomenon of meteorosensitivity:

1. Evolutionary theory implies that man normally requires a certain exogenous stress, or "electromagnetic climate," in order to feel well.
2. This stress associated with meteorologic processes varies due to the extremely broad spectrum of field intensities. It can be too strong at times and too weak at other times. It is also frequency-dependent.
3. It is certain that man responds to these external stimuli over an equally large range of sensitivity, probably due to differences in the predisposition of different individuals. Each person requires his own optimal dose of external stimulation, which can also vary as a function of time.

Thus we obtain a picture of how different persons are influenced by weather. It appears that the maintenance of normal cell function requires a continuous external stimulation, which is evidently provided by even very low-level electromagnetic fields. Persons with an excess of "inner tension" are overburdened by the additional stress imposed by external atmospheric electric processes (e.g., the approach of fronts or outbreaks of foehn) and react with typical "weather complaints" (headache, heart trouble, nervousness, etc.). However, if weather conditions are such that all external atmospheric electric processes are largely screened out or absent (after the onset of foehn or under certain inversion conditions), such persons will feel particularly well, because they have no need for this extra electric stimulation (Fig. 9-4).

Persons who possess too little "internal stimulation," on the other hand, are reliant upon external stimulation from the environment and, unlike the type described above, will feel better in the presence of strong atmospheric electric activity. Its absence will result in "weather fatigue" and tiredness.

There are, of course, various gradations between these extremes. According to previous experience, a certain sensitization also appears to be possible during the course of time. This hypothesis is based upon numerous individual observations regarding typical reactions under certain weather conditions (headache, oppressive feeling in the chest, nervousness, fatigue, etc.). But the same subjective complaints were produced experimentally by the use of artificial fields, providing still further evidence of the connection between meteorosensitivity and atmospheric electric fields.

ELF Fields

We shall introduce our discussion of some special biometeorologic contributions with an account of our own experiment, because it must be regarded as highly characteristic in every respect. The experimental subject was seated at a table on which was located a specially designed generator the size of two cigar boxes stacked one on the other (Fig. 9-5). This generator set up an electric field by means of a sheet-metal electrode, about the size of the palm of the hand, mounted above it. The subject was about 1 m from the generator, but within the stray field of the two electrodes, which were about 20 cm apart. The subject not only felt

Figure 9-4. Iron crown used during the Middle ages as a votive crown and as a remedy for headaches. (Bavarian National Museum, Munich.) It is now known that such a device acts on the Faraday cage principle to screen out disturbing atmospheric electric phenomena associated with "weather complaints."

the field, but also gave reports on subjective well-being which varied with the field frequency—without being asked, without knowing when the field was altered, and without any knowledge of the possible effects. These reports agreed with the general findings of other experimenters, as presented here.

Interestingly, similar experimental results were reported by Leventhal (1970), who exposed test subjects to purely mechanical vibrations of the air. Kracmar (1962) also reports personality-specific results in experiments with ELF currents. He performed RC measurements (resistive and capacitive component of the impedance of the human body) in order to investigate the varying responses of the subjects (stimulation or blocking of the vagus and sympathetic). As the frequency (ELF range) of the sawtooth current was varied, varying impedance values were measured.

These findings are reason enough to institute more exhaustive scientific investigations on a broad, statistical basis, for they could yield results of considerable importance to the general public.

VLF Atmospherics

Ranscht-Froemsdorff (1966a) also investigated biometeorologic problems in detail. He explored the effects of VLF atmospherics (see also Chapter 6) and obtained results which fit well into the theses presented here.

Under certain weather conditions there exists a "zero situation," which is typically observed under conditions of descent and alpine foehn. The low pulse activity of the atmospherics in such cases leads, according to Ranscht-Froemsdorff (1966a), to increased incidences of (1) myocardial infarc-

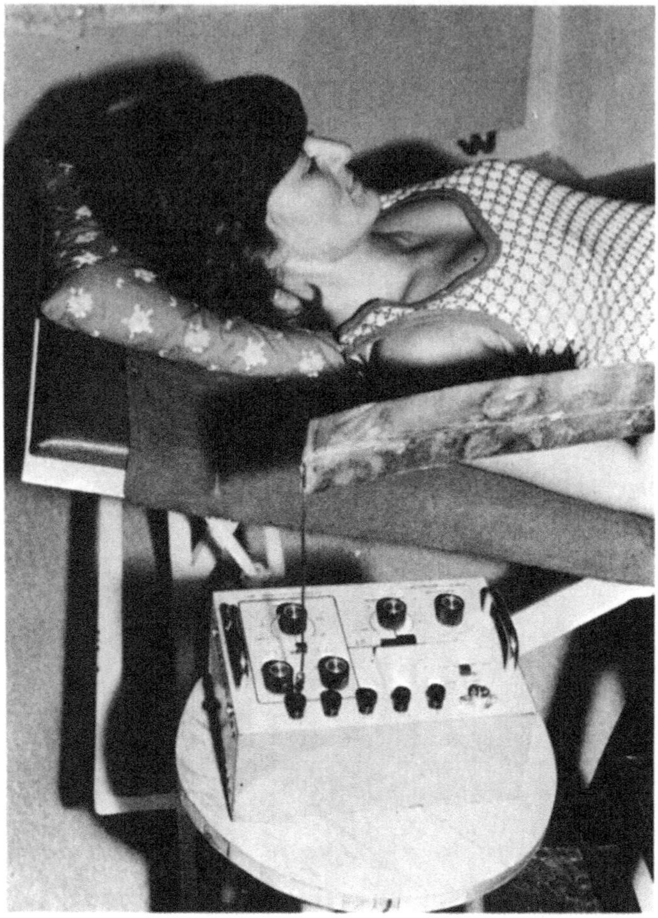

Figure 9-5. Small experimental device for generating an artificial electric field like that measured in nature under certain weather conditions. The "flag" forms one electrode, while the box itself forms the other.

tion or angina pectoris, (2) proneness to hemorrhage, (3) autonomic dystonia, (4) allergic reactions. A higher pulse activity, on the other hand, leads to (1) spasms, (2) colic, (3) embolisms, (4) rheumatic and similar complaints.

Ranscht-Froemsdorff (1966b) also mentions the possibility of designing dwellings (e.g., a metal house) to screen out natural atmospheric radiation in order to create an artificial electric climate. However, such structures should not be considered generally favorable because, as experience indicates, excessive shielding and too much "artificial environment" are advantageous only in a few cases (disregarding specific therapeutic applications).

The significance of electromagnetic energies in the ELF and VLF ranges in biometeorologic problems was also recognized by Persinger (1973) (see also Chapter 4).

Atmospheric Electric Factors

The importance of air ionization in connection with biometeorologic problems was the principal concern of Sulman (1971b). He investigated the biologic effects of hot, dry desert air primarily in the Southeastern Mediterranean (Sharav in Israel) and found that the native population in particular suffers from such hot, dry winds. Sulman (1971a,b) and Sulman et al. (1974) advance a biochemical explanation for the warm-air stress, stating that the potassium excess that accompanies the water loss under such conditions exerts a toxic action on the cardiac muscle. This would account for the specific susceptibility of cardiac patients under foehn weather conditions. Based on various investigations, Sulman (1971b) then formulated biochemical working hypotheses and attempted to find a causal relationship between all the reactions observed in the human body and the special air-ionization situation that precedes a weather change. The strong shift toward positive ionization at such times is suspected of playing a particular role (see also Chapter 8). However, it is questionable whether these relationships found under the climatic conditions of Israel can be generalized or applied to similar phenomena in other countries (e.g., the foehn at the northern foot of the Alps) due to the essentially different bioclimatic conditions.

Sulman (1972) points out in another paper that besides the marked change in such meteorologic data as air temperature and humidity, which is observed in the presence of warm, foehn-like winds, there is a high level of negatively charged "hazardous" ions combined with the simultaneous occurrence of 1500 small ions/cm^3.

Although the natural ion concentrations mentioned are considerably lower than those employed in laboratory experiments to which Sulman (1971b, 1972) refers in documenting the biologic activity of such ionization conditions, this does not rule out all connections between weather-related complaints and special conditions of atmospheric ionization. Nevertheless, the problem remains that certain foehn-type complaints occur even before the onset of the foehn wind itself, and thus before any related changes in ionization conditions. It then becomes necessary to consider the electromagnetic field energies emitted by fronts and other meteorologic processes when seeking the active factor.

Gualtierotti (1968) deals with the same problem in great detail. Based on his own investigations and drawing on the international literature, he confirms the importance of air ionization in connection with questions of bioclimatology and biometeorology and in its application in the form of aeroionotherapy. Mountain climates undoubtedly play a special role in this regard (Gualtierotti, 1977).

A comprehensive body of material and extensive bibliography are presented by Faust et al. (1977) on the problems of biometeorology and their relation to synoptic meteorology. In agreement with earlier investigations, these authors hold that meteorogenic fluctuations of well-being and diseases associated with certain weather conditions are not true "weather sickness": weather effects are, rather, only an indicator of sites of diminished resistance.

10

The Divining Rod Phenomenon

This chapter is not intended to present a full or even partial solution to the divining rod problem. Its primary purpose, rather, is to present a survey of the results of recent investigations in this area as they pertain to electromagnetic fields. It is also hoped that a closer look at the question of the existence of "geopathogenic," or location-dependent, effects will shed further light on the factors underlying the volatile nature of this controversial subject.

The divining rod phenomenon has generally provoked an attitude of opposition on the part of the scientific community. The main reason for this is the behavior of many dowsers, who feel it necessary to advance a pseudoscientific rationale for their apparently indisputable abilities. They have borrowed impressive-sounding terms from classical science and physics without clearly understanding them, and have used them to explain the divining rod phenomenon. The reaction of the exact sciences to this assault on their theses and nomenclature, as well as their basic view of the world, is hardly surprising. Dowsing was relegated to the realm of the occult or ascribed to human foolishness. The situation was not improved by the touting of "radiation neutralizing devices" and similar costly apparatus which was worth only a fraction of its purchase price and seldom delivered what was promised. If serious scientists did address themselves to this problem, their actions were viewed with suspicion.

Be that as it may, we shall attempt to present a certain overview of what has been learned from research into the dowsing problem by responsible scientists. This research has focused on two main areas:

1. The divining rod phenomenon as such; that is, the ability of certain persons, called dowsers, to react to certain conditions with a movement of the divining rod. The question arises whether this movement is based on some external influences or whether a more subjective reaction is involved which cannot be objectified and is thus of no evidential value. A further question is whether the dowser represents a kind of detector that responds to certain conditions, and how this response could be explained in physiologic terms.

2. The other area of research pertains to the question of what causes the movement of the divining rod. Here the place where the reaction occurs appears to play a dominant role: namely, along narrow paths called "dowsing zones" or "crossings" (in-

tersecting zones) on the earth's surface. Normally the dowser will react with a particular movement of his rod (or pendulum) when he crosses such spots. This talent is an important one, because it supposedly enables the dowser to locate underground water as well as oil and gas deposits, which modern science is only occasionally able to do with a far greater technical effort. This fact is made more problematic, if not more significant, by medical and health-related aspects. Dowsers have always maintained that dowsing zones, and especially crossing zones, give off "rays" which are injurious to health, particularly if persons spend a great deal of time in such spots (place their beds there, for example). Dowsers thus speak of "geopathogenic zones" that affect human health and well-being. If it is indeed proven that places exist which have a long-term pathogenic effect, then this is a problem which clearly merits closer study in the interest of public health.

For completeness, we should also mention the practice of long-distance dowsing, or "teleradiesthesia." This refers to the alleged ability of some dowsers to obtain information over great distances with the aid of a rod or pendulum. If we look for possible explanations of this phenomenon in the realm of classical physics, we must conclude that it involves the direct action of certain energies, and thus rays, on the practitioner, thereby lending credence to the dowsers' claims. However, such explanations are not applicable to phenomena separated by great distances, for then the radiations present in one area can no longer account for effects in the other area—especially if the propagation of such radiations in the form of electromagnetic waves, for example, can be ruled out. At this point we again enter into the realm of parapsychology and all its variations. But this is not the subject of the present discussions, which must be limited to the two problems mentioned earlier.

The divining rod problem is, in any case, an extremely complex one which cannot possibly be judged, let alone solved, by a doctor, physicist or biologist. The many facets of the problem are illustrated by the fact that both physicists (Brüche, 1960) and physicians (Hartmann, 1967, 1976) have dealt extensively with this phenomenon in recent years. Hartmann, drawing on decades of experience, affirms the existence of such geopathogenic zones which can severely impair human health. He is supported in this by Diehl and Tromp (1954), who studied the geographic and geologic distribution of cancer mortality in Holland. These authors found a significant correlation between cancer incidence and geologic conditions.

Experimental Investigations

A large volume of observational data has been gathered to date on the phenomenon of dowsing. Although these data are for the most part based on individual case reports, which are important as such and have a certain statistical significance when viewed as a whole, they still leave much to be desired from a purely scientific standpoint. And while it is true that some large-scale investigations have led to successful, generally valid results, they do not meet strict scientific standards. But perhaps this was not intended, for whatever reasons.

Historical Review

The divining rod problem should be resolved with assistance from representatives of the classical sciences as well, for the results of previous investigations already have profound implications. What is meant by this? According to Hartmann (1973, 1976) von Pohl reported on diseases associated with "ground emanations" as early as 1930. To prove this connection, he conducted a search for underground water flows with a divining rod in various small towns in southern Germany. A comparison with cancer cases occurring in houses located over these underground water currents revealed astonishing connections between illness and subterranean influences. In 1934, Rambeau (1969) published similar results based on studies in villages. Instead of a divining rod, however, he used geophysical apparatus which were used at that time for the detection of fractures and faults in the earth (portable medium-frequency transmitters and receivers for field-intensity measurements). He concludes his research report with a remarkable statement: "We searched in our statistical study for a house that was located on geologically undisturbed ground and was nevertheless occupied (on a long-term basis) by persons suffering from cancer. Such a house was never found."

Two main tasks emerge from our look at the divining rod problem. First, it is necessary to determine the extent to which there are zones or points on the earth's surface which play a role in connection with human disease, especially cancer, and what physical factors might be involved. Second, it is important that a better understanding be gained of the dowsing phenomenon as such, for this will allow inferences to be made reagarding the properties of geopathogenic zones.

The main purpose of this discussion is to show, within the scope of the problems posed, what role electromagnetic fields might play in these phenomena. The two main problems outlined above will be investigated by a look at comparative studies of the special biologic significance of geopathogenic zones, and by an analysis of location-specific measurements of various physical parameters.

What is the status of our knowledge in this area? Hartmann made an important contribution to this question by publishing his own investigations, which are supplemented by a survey of some historical studies on the divining problem. Even the layman will appreciate the great deal of time and effort that have been devoted by idealists to the solution of this problem. Nevertheless, the critical scientist must occasionally raise objections which, while not belittling the general value of the studies performed, should still be taken as a warning that, due to deficiencies of scientific method, these matters should not be regarded as conclusively resolved.

Hartmann (1967, 1976) reports on studies done in 1937 by Henrich and Dannert with a sawtooth-wave transmitter that emitted low-intensity electric signals at frequencies of 1–10 Hz. It is reported that spontaneous cancer was induced in mice when a frequency of 1.8 Hz was administered. When such fields were applied to humans, a variety of sensations were produced at signal frequencies from 1 to 20 Hz. One frequency range caused headaches, another caused palpitations, while a third produced mild visual disturbances (measurements by the author confirm this). The studies were conducted because some dowsers apparently reacted to such artificial electric fields with a turn of the divining rod, and a connection was presumed to exist between these low-frequency fields and the so-called geopathogenic zones.

Biologic Effects in Dowsing Zones

Most dowsers believe that in some places dowsing zones form a grid-like pattern on the earth's surface. These parallel zones are said to be about 2 m apart. Dowsers also claim that the points where zones cross are particularly injurious to human health. Hartmann (1967, 1976), for example, reports on many amazing cures, especially in chronic cases, which he effected as a doctor simply by moving the patient's bed slightly. Many studies have been done in an attempt to demonstrate the effect of these geopathogenic points on man and other organisms. For example, it was found that the infrared emission from the human body changes when a person stands on one of these critical points. Petschke (1953a,b) demonstrated a connection between blood sedimentation and such points in numerous experiments. He picked out three points only a few meters apart for his experiments: one was a neutral point, one was in a simple dowsing zone, and the third was at a crossing of such zones. A total of 62 series of tests were conducted, each at least 8 hr in duration. The 72 hourly results obtained in each series were converted to 24 mean values, so that finally the results of more than 5000 readings were evaluated, including half-hour values. The following findings were reported:

1. The sedimentation rate of blood samples located over the dowsing zones and crossings was altered in relation to the sedimentation rate over neutral ground. The deviations of the two altered groups followed a parallel course, which tends to rule out pure chance.

2. The sedimentation rate was not always accelerated over the dowsing zones and crossings; in some cases a retarding effect was observed.

Hartmann (1967, 1976) reports on similar experiments. He placed blood sedimentation tubes into a rack that was mounted crosswise over the bed in a house where two generations had died of a similar cancerous disease. The point above the floor which corresponded to the location of the affected organ in the bed had been called geopathogenic by dowsers. Since the tubes were spaced 6 cm apart, only a few were located over this critical spot, while most were outside it. This experiment, which has been repeated many times, showed that blood sedimentation took a markedly anomalous course in the tubes located over the geopathogenic spot. Similar reproducible, location-dependent deviations were found in germination experiments with cucumbers, beans, peas, radishes, corn, etc., in which germination was suppressed over such points. All observers of blood sedimentation tests like that described above report an additional variation in the course of the sedimentation which appears to be associated with weather.

If there are certain points that differ markedly from their immediate surroundings with regard to their ionizing radiation, this would be of considerable importance to persons who sleep in such places. Scheller (1969) developed a technique for detecting the biologic effects of "injurious rays." Location-dependent blood studies showed that an elevated level of incident radiation leads gradually to the formation of granules, globules, vesicles, fibers and various other particles in normal blood which were not there previously. Such changes can be detected in the erythrocytes by the so-called Scheller test, in which ultramicroscopic structures of the mitochondrial type are revealed by dark-field examination. This technique could prove that irradiation of the blood has a visible, demonstrable biologic effect in the most radiosensitive cells, the erythrocytes.

According to Scheller (1969), ionizing radiation acts on the erythrocytes and alters them (contrary to previous beliefs) and finally cancer can develop. If this hypothesis is borne out, it would be evidence that geopathogenic points play a role in carcinogenesis. Measurements lend support to this hypothesis, for they reveal peculiar conditions of ionizing radiation at such points (at least for a certain type of geopathogenic zone).

Hartmann (1972a) provided further evidence of the peculiar biologic properties of geopathogenic points by means of reaction time measurements. These were carried out at two points about 1.2 m apart. The subjects were seated either directly on a "crossing," or at a neutral point. The subjects on the critical point showed a 1%–10% slower reaction time than the subjects at the undisturbed point.

The problem of the possible effects of geophysical factors on biologic responses (geobiologic effects) is treated by others as well. For example, Bortels et al. (1970) reported on a study of the intensity of both aerobic respiration and anaerobic fermentation of *Saccharomyces* yeast under "constant" experimental conditions. They found variations in measured biologic values with both time and location. Several parallel tests were conducted in the thin-walled attic and the cellar of a very massive building under otherwise identical conditions. Relative maxima and minima of the yeast respiration coincided with a strong rise and fall of air pressure, respectively. These results were the same qualitatively for both test sites. However, the average intensity of respiration was lower in the cellar than that measured simultaneously in the attic. Relative maxima of fermentation intensity in the attic coincided on the average with air pressure minima immediately following a strong pressure drop, that is, with conditions in the warm sector of a cyclone (low). Conversely, relative minima coincided with air pressure maxima immediately after a strong pressure rise, that is, with conditions in the region of an anticyclone (high). In the cellar, however, these correlative relationships were reversed.

The investigation of biologic effects in several 20 × 20 cm areas located close together yielded interesting results. For example, three cages, each containing 4 mice (1 male and 3 females), were kept in a certain room for experimental purposes. One group of mice was located over a "crossing" (critical point), one over a neutral point (only about 0.5 m from the first), and the third was kept in a Faraday cage in another location 2 m away. Within 6 months, according to Hartmann (1970), the 3 females in the cage over the neutral point bore 124 offspring and those in the Faraday cage bore 118, while those over the critical point bore only 56. The mice on the crossing were always restive, bristled, and continually devoured their young. Their behavior differed markedly from the control animals in other respects as well.

In order to objectify further the possibility of location-dependent biologic effects, Hartmann (1972b), also conducted experiments with rats at the same locations. He sought to answer three questions:

1. Do rats inoculated subcutaneously with Yoshida tumors exhibit different tumor growth at certain (geopathogenic) points in relation to the nearby ("undisturbed") environment?

2. Does holding in a Faraday cage and the associated shielding effect have any influence on tumor-inoculated rats?

3. Does a high-frequency radiation with low-frequency modulation influence tumor growth?

A total of 132 animals were tested in 12 series of experiments, each series lasting 3–4 weeks. In the first 4 series of tests, 3 cages with 3 animals each were used. One cage was on the critical point, and the other two were located about 1–1.3 m away. The rats in the Faraday cage were treated with a 21-cm high-frequency radiation with 1.75-Hz modulation twice for 15 min. The power density was on the order of 1 mW/cm^2. In each test the irradiation was done immediately after the tumor inoculation.

In the remaining 8 series of tests, a fourth cage was used as well. This cage was placed about 30 cm from one of the two control cages in order to check whether differences in tumor growth could still be observed over relatively small distances. This was not the case, and so this particular differentiation did not have to be made. In contrast to the initial series of tests, the rats in the Faraday cage received no further treatment after their inoculation.

The animals were inoculated with tumor suspension, care being taken that they were all from the same population. After 10 days' exposure at the points selected, the tumors were palpated, molded in plastiline and weighed by a known method (by a University of Heidelberg veterinarian). The progress of tumor growth is shown graphically in Fig. 10-1. The curves for the first series of tests represent the mean value of 4 test series with a total of 9 rats per series. Curve 1 (values for the "cancer point") clearly shows the sharpest growth increase. Curve 2 shows the mean values at the undisturbed neutral point. Curve 3 shows the mean values for the irradiated animals in the Faraday cage. Compared with the exponential increase that is typical of biologic growth processes, the latter curve shows practically a retardation of tumor growth.

The results of the second series of exper-

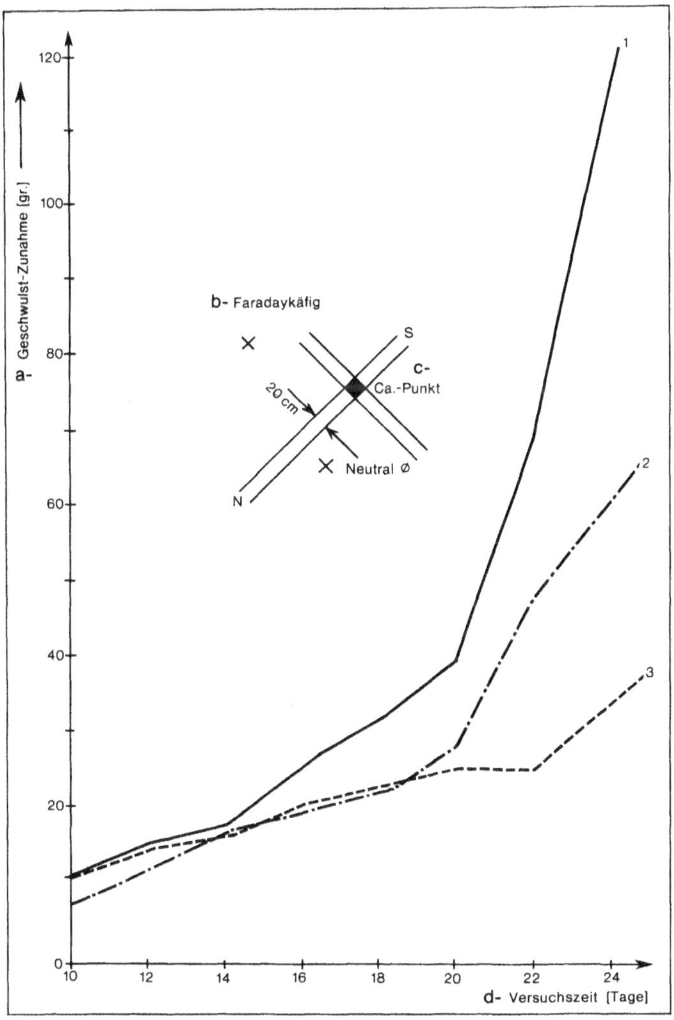

Figure 10-1. Tumor growth in rats as a function of location: (1) growth on a "crossing" of dowsing zones (cancer point); (2) growth at a "neutral point" (about 1 m from the "crossing"); (3) growth in a Faraday cage in which the animals were also irradiated with ELF-modulated high-frequency energy. After Hartmann (1972b).
Key: a-Increase in tumor weight (g); b-Faraday cage; c-Cancer point; d-Time (days)

iments, showing the mean values from 8 series of tests with 96 rats, are presented in Fig. 10-2. In contrast to Fig. 10-1, the animals in the Faraday cage, which were not irradiated this time, displayed the greatest increase in tumor weight. The other two curves show a trend analogous to that in Fig. 10-1. The test results indicate that the rate of tumor growth is apparently dependent on various previously unrecognized locational and environmental factors.

The spot designated as the cancer point in these experiments was surveyed with respect to the static terrestrial magnetic field. A probe traversing this spot in one direction indicated a drop in magnetic field intensity to about one-fifth its normal value (Fig. 10-3). Because earlier measurements also showed increases in gamma radiation levels at such points, sometimes by as much as 10%–20%, it is assumed that the peculiar circumstances of oxygen ionization and oxygen concentration there may be responsible for the magnetic field anom-

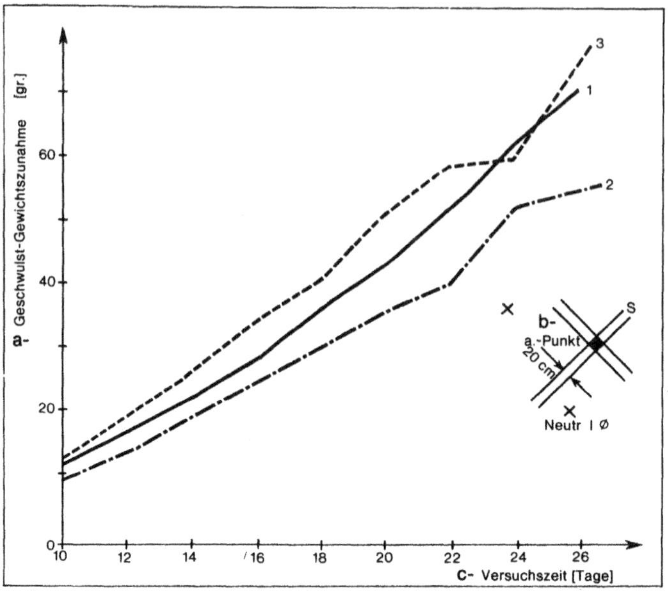

Figure 10-2. Tumor growth in rats as a function of location: (1) and (2) same conditions as in Fig. 10-1, (3) progress of tumor growth under pure Faraday conditions. After Hartmann (1972b). Key: a-Increase in tumor weight (g); b-Cancer point; c-Time (days)

aly, owing to the diamagnetic properties of oxygen.

Hartmann (1972b) considers this result a major contribution to the realization that there are certain circumscribed areas in dwellings that may disturb normal biologic processes given a sufficiently long period of exposure. Apparently locational and environmental factors are relevant with regard to the cancer problem as well (Fig. 10-4).

Besides experimental investigations of the places which produce the dowsing reaction, efforts are also being made to verify the dowsing phenomenon itself. Fadini's (1971a) statistical evaluation of experimental results supplied by numerous dowsers independently of one another appeared suitable for this purpose. Approximately equal numbers of men and women of different ages and occupations used identical divining rods in the same manner. In addition, the divining rod reactions were compared with the response of a VHF field intensity meter (more about this below). To evaluate the tests, Fadini (1971b, 1972) developed a special statistical approach: the area to be investigated, which measured 2×2 m, was divided into 8 equal sectors. The dowsers were given the task of determining which sector was the critical (geopathogenic) one. A large number of these tests proved statistically (probability $> 99\%$) that the detection of specific geopathogenic points by dowsers is not determined by chance alone.

Measurement of Physical Parameters

Experimental results confirm the biologic significance of spots that produce the dowsing reaction (Fig. 10-4). But other experiments have been conducted by both interested experts and laymen in an effort to find measurable physical parameters which might undergo anomalous changes in such places. According to a partial survey by Hartmann (1967, 1976), magnetometric measurements with a dual compass showed magnetic peculiarities at the spots indicated by dowsers. Wüst (1953) used a field balance and local variometer to show distinct magnetic field anomalies at such points.

Wüst and Petschke (1954) investigated the electric ground resistance in the vicinity of dowsing zones and found marked changes there. Tromp (1954) reports that ground conductivity is generally maximal in dowsing zones, but that resistance peaks are also observed. As early as 1931, Lehmann (1960) demonstrated a significant decrease in the potential gradient (up to a factor of 2) of the electrostatic field over such points. He also found an increase in the conductivity of the air over spots

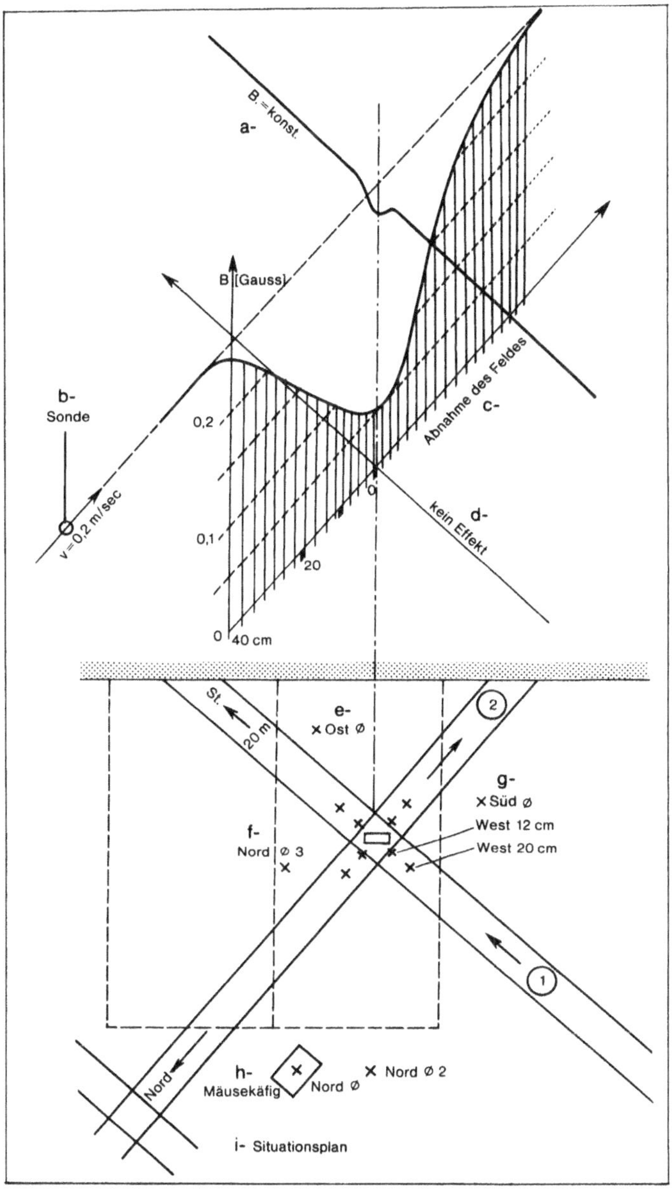

Figure 10-3. Measurement of the intensity of the geomagnetic field at a "cancer point." In the north–south plane (2) the field intensity decreases from 0.025 mT (0.25 G) to about 0.005 mT (0.05 G) when the Hall probe is moved through the area at a speed of about 0.5 m/s. No such effect is observed in an east–west direction (1). After Hartmann (1972b) and Varga.
Key: a-B = const.; b-Probe; c-Decrease of field intensity; d-No effect; e-East; f-North; g-South; h-Mouse cage; i-Site plan

Figure 10-4. Damp walls are apparently typical sites which produce a movement of the divining rod. In this case visual cues may be involved.

which, according to dowsers, marked the location of an underground "water vein." Brüche (1960) reports that the radioactivity in such places is also different from that of the surrounding environment; it may be increased or decreased, depending on weather conditions.

Tromp (1972) reports on the use of divining rods to locate groundwater. He also associated measurable physical parameters with the spots that indicate "water veins" for dowsers. According to Tromp, even EEG changes can be observed.

Wüst (1959) introduced a special measuring technique in 1958. He showed that the VHF field intensity of conventional VHF radio transmitters undergoes marked changes over dowsing zones. This effect is seen in open terrain as well as indoors. Hartmann (1967, 1976) applied this principle to his medical practice. A considerable number of investigations showed that the VHF field that can be measured in the space around sick-beds and its anomalies (particularly extreme field-intensity minima) bear a certain relation to the occurrence of diseases. While such results do not necessarily establish a causal relationship between diseases and the VHF field, they do show that this field may be important as an indicator. In any event, investigations and theoretical considerations from a purely technical standpoint have shown an order-of-magnitude agreement between room dimensions and the wavelengths of VHF fields. Thus, standing waves can form in parts of the room due to reflection, leading to the formation of the maxima and minima measured there. It remains unclear whether the VHF fields are responsible for the observed pathogenic effects, either alone or in combination with other physical factors, or whether the VHF fields are merely an indicator, implying that other physico-technical peculiarities, which are involved in the pathogenic process, exist in such places.

Aschoff (1967, 1976), for example, reports on his experience as a physician with VHF field-intensity measurements in more than 125 cases. When 85 of them were moved to an area in the room with a uniform VHF field intensity, more than 28 patients, some of whom had not responded to prior treatment, showed an immediate improvement of chronic complaints that had been treated for years without success.

Prompted by these findings, Hartmann (1967, 1976) developed a portable device that corresponds in principle to the apparatus of Rambeau (1969): a VHF transmitter with a vertical rod antenna (basic frequency 32 MHz), and a receiver (tuned to triple the transmitting frequency) with a horizontal rod antenna. Both antennas are about 30 cm in length, and both devices are mounted on a board about 60 cm apart. This arrangement, carried horizontally at waist level, made the VHF field measurements in rooms independent of external radio transmitters. This apparatus confirmed previous results on VHF fields that had been obtained without the use of a transmitter.

According to Hartmann (1967, 1976), amazing results can be obtained simply by measuring the resistance to the passage of a direct current between the hands of test subjects. If the measurements are properly carried out, the measured resistance values cannot be influenced (at least directly) by the subjects.

It was found that the resistance of the body exhibits a significantly different time course at a geopathogenic point than at a neutral point. The difference in resistance values is also clearly dependent on the state of the weather. According to Eckert (1971), the curves obtained by Hartmann (1967, 1976), called a "georhythmogram," show not only resistance values that deviate from the mean, but also unusually strong short-term fluctuations of resistance in subjects standing on a critical point.

Endrös (1969) conducted more elaborate investigations of the physical effects of underground water flows, focusing his attention on the infrared and microwave emissions from such spots. The complex diversity of the phenomena associated with the dowsing problem was once again confirmed.

Measurements by Wüst and Petschke (1954) showed that anomalies of ground conductivity are always present in spots that produce a movement of the divining rod. They found two basic types of anomaly: the first is a sizable increase in ground conductivity over a short distance of about 2 m across the zone, and thus a discontinuity of resistance (Fig. 10-5); the second is the presence of a boundary zone between two adjacent areas of different conductivities (Fig. 10-6). While the absolute value of the ground conductivity is apparently of primary importance in the first case, the gradient is a more important factor in the second case. These anomalies of ground conductivity have a significant impact on electric and magnetic fields, air

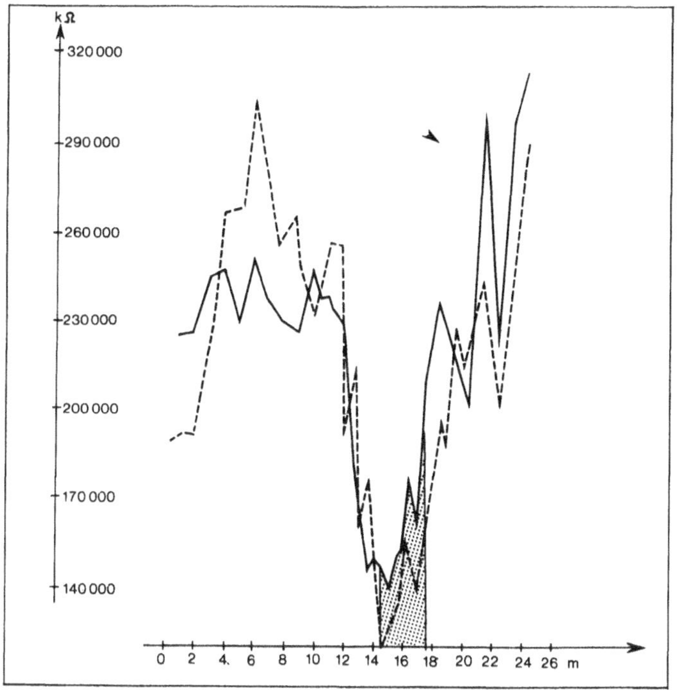

Figure 10-5. Variation of ground resistance across a "dowsing zone" (shaded) at a depth of 1 m. A pronounced resistance minimum is seen. Both measurements (solid and broken lines) were made along parallel paths 2 m apart. After Wüst and Petschke (1954).

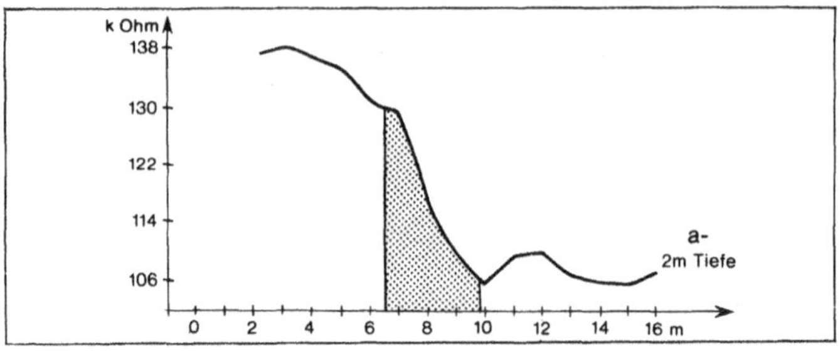

Figure 10-6. Variation of ground resistance across a "dowsing zone" (shaded). The boundary between two regions of different resistance is apparent. After Wüst and Petschke (1954).
Key: a-2 m depth

ionization, the vertical atmospheric electric current, as well as currents flowing in the earth, which suggests a link with the dowsing phenomenon. The biologic activity of all these physical parameters must also be taken into account, however.

According to Hartmann (1976), measurements of the intensity of VLF atmospherics at different locations showed that the incidence rate of VLF signals varied at a pathogenic point by up to 30% in relation to the surrounding environment. This suggests that reception conditions can vary considerably within an extremely limited space, an effect that could also be explained theoretically on a physical level (e.g., by the resistance anomalies mentioned above).

Bürklin (1965) reported on investigations of the increased frequency of lightning strikes at low-voltage power transmission poles. All poles that attracted lightning and thus had a lower-than-average life expectancy were located on critical points, according to dowsers. Grounding the ground wire at the crossings indicated by the dowsers apparently provided excellent lightning protection.

It was also discovered by acident that paraffin–graphite blocks stored in a cellar evoked a response from dowsers on the floors above directly over the point where the blocks were stored. Measurements over these paraffin–graphite blocks revealed a radiation from the ground that produced readings similar to neutron radiation that is slowed by about a factor of 100. Paraffin–graphite blocks have also been used by dowsers for experiments. A certain time delay observed in the dowsing reaction suggests that the high-energy radiation produces an ionized column of air over the blocks, which takes a certain amount of time to form or disperse when the experimental situation is changed. The presence of these ionized columns of air would also account for the high incidence of lightning strikes at critical points.

The tendency of lightning to strike overhead transmission lines has also prompted studies to determine whether peculiarities in the geophysical properties of the ground and the associated geoelectric structure of the subsoil might not be the cause of this. Kretschmar (1967) points out the existence of "nests" of lightning, determined by the morphology of the earth's surface but also by the electric properties of the subsoil. It must be noted that the absolute conductivity of the subsoil is of no importance with regard to lightning strikes as long as it is completely homogeneous. However, the electric structure of the subsoil becomes important if it is not homogeneous, for then it influences the starting point of the counterdischarge and thus the course of the main discharge. Zones in which good and poor geologic conductors adjoin have proved especially prone to strikes, indicating that contacts, faults and zones of geologic disturbance are in greater danger than electrically homogeneous subsoil. Such findings have implications for other geophysical parameters as well. Thus, location-dependent intensity variations are to be expected for electric fields and electromagnetic waves in particular.

Stängle (1973) made an important contribution to the objectification of the divining rod phenomenon by analyzing underground water flows with the aid of scintillation counters. He used sodium iodide as the scintillator, because it is generally preferred for the detection of gamma rays. The moderator mounted in front of the crystal influences incident radiation from below, enabling it to excite the scintillator material. Stängle (1973) mounted his apparatus on a wheeled frame suitable for cross-country use. The apparatus converted incident gamma radiation into measurable impulses, which were then automatically recorded and plotted continuously in the form of a curve. This arrangement has proved a reliable means of locating sites for water wells. Mention has already been made of the earlier studies of von Pohl cited by Hartmann (1975, 1976) indicating that the beds of cancer patients were all located in dowsing zones in certain German towns. Stängle (1973) used his scintillation apparatus to recheck these old investigations in an attempt to answer two questions: (1) Is radiation from the ground distorted over these zones such that its intensity is increased? (2) Do these changes in ground radiation behave in a manner typical of water flows, according to Stängle's experience?

The results prove that both questions must be answered in the affirmative (Fig. 10-7). The radiation intensity is nearly twice as high within the dowsing zones as over adjacent ground. Based on his experience in locating water, Stängle (1973) assumes that corresponding underground water flows exist in these places.

Disregarding for the moment the divining phenomenon, which is still considered ominous by many, it must be said that the importance of these investigations can no longer be disputed. If it can be shown by physical measurement that there are zones on the earth's surface that have a higher level of radioactivity than their surroundings (regardless of whether an underlying

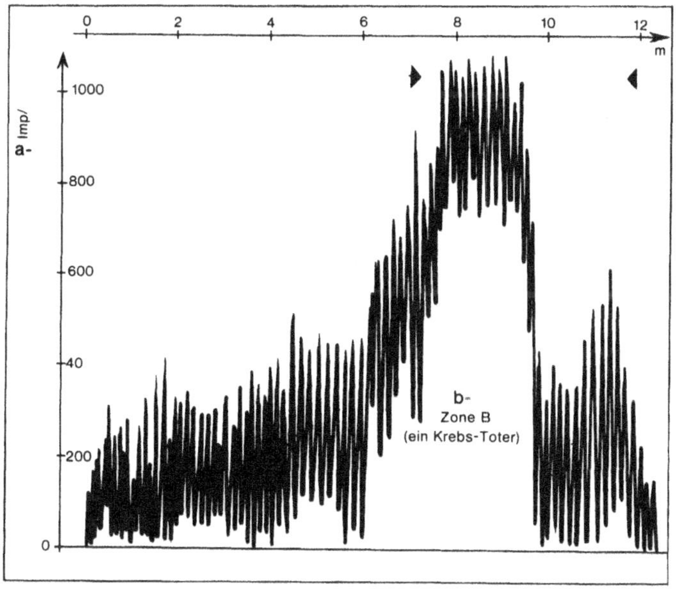

Figure 10-7. Measurement of ground radiation with a scintillation counter designed by Stängle. Von Pohl's divining rod experiments were repeated by Stängle in Vilsbiburg, Germany. He confirmed the existence of geopathogenic zones with a significant number of cancer deaths which coincided with zones of radiation anomalies. After Stängle (1973).
Key: a-Impulses/sec; b-Zone B (one cancer death)

water vein is the cause or not), and such striking connections are found between these zones and cancerous diseases as the ones described here, then the orthodox sciences must regard these findings as sufficient cause for exact and precise investigations in the interest of all mankind. Even large-scale statistical studies would be helpful in collecting important data. They would also provide a basis for determining whether possible protective measures should be seriously considered.

Williams and Lorenz (1968) report on gamma radiation measurements with a highly sensitive instrument for the detection of geologic fractures. Fractures at depths to 1700 m have been localized by this technique. The increase in radiation accompanying a fracture apparently stems from the sedimentation of radioactive minerals in springs or groundwater, or possibly from the migration of radioactive gases that rise to the earth's surface through geologic fissures. In any event, it has been established that the main source of the radiation is the uppermost 2 m of earth.

To explore further the significance of gamma radiation in the "cancer point" phenomenon, Hartmann (1970, 1976) used probes to compare the radiation at such a point with that of the immediate surroundings. He found that the critical point almost always exhibited a different (usually higher) radiation pulse count than the surroundings (at a distance of 12 or 25 cm, measuring simultaneously with two probes). The measurements were carried out over a period of several months. On a scale of hours, a higher average gamma radiation was measured at the critical point, as mentioned. However, the pulse difference does not arise from a more or less constant pulse deficit, but occurs in a rhythmic manner. Periodicities of about 15 min and 3 hr are clearly seen (Fig. 10-8).

To rule out the possibility of any radioactive source in the floor of the experimenting room, the experiments were repeated at a height of 1.75 m above the floor with the aid of a wooden frame. The result was the same. The increased radioactivity thus appears to occupy a column-like space extending vertically upward from the floor. Observations also point to a certain dependence of the gamma radiation on weather conditions, that is, on the general meteorologic situation.

In further experiments to investigate the effect of lead shielding, equal amounts of lead were placed beneath three measuring probes. Surprisingly, the effect of the lead was different at the critical point than at the other measuring sites, suggesting a different type of radiation at the critical point than at the other sites. In any case, the radiation pulse count generally decreased in different degrees, depending on the measuring site, when the lead was put down. When paraffin–graphite blocks about 10 cm thick were interposed, the pulse rates generally increased, again in different degrees depending on the location.

The experiments show that marked differences in gamma-radiation pulse counts (10%–30%) are observed over relatively small distances of 20–40 cm, which evidently have nothing to do with the immediate flooring material. The correlation with weather, on the other hand, suggests a certain relationship with the localized escape of radioactive gases (radon), which could create this peculiar condition over geologic fissures due to the influx and efflux processes that accompany air pressure changes. Since we are dealing here with ionizing radiation, such findings are of profound importance. If "ionized columns" do exist in the space above certain points on the ground, they would have considerable effects on all types of electromagnetic fields, the natural atmospheric electric current, air ionization, etc., as well as the phenomena associated with these parameters. Direct ionization processes in a localized tissue region could also be of significance. Hartmann (1970) does not rule out the possibility that such processes could result in cell degeneration, particularly if persons place

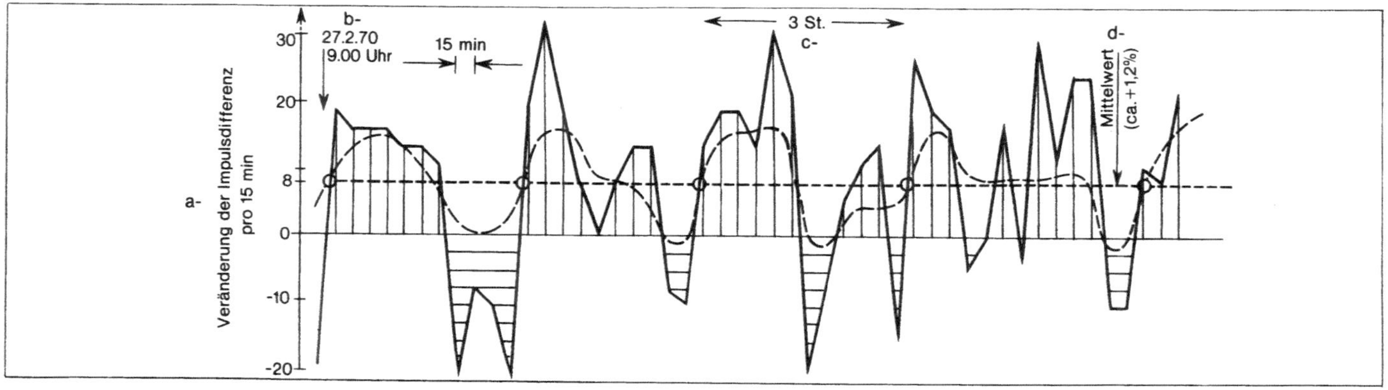

Figure 10-8. Differences in the gamma radiation of two sites only 30 cm apart (one of which is a "crossing"). Superimposed onto the higher average radiation count (~ 1.2%) at one site is a rhythmic rise and fall with periodicities of about 15 min and 3 hr. Data from Hartmann, (1970, 1976), correspondingly reworked.

Key: a-Change in impulse difference per 15 min; b-2/27/70 at 9:00 A.M.; c-3 hr; d-Mean value their beds in such spots, and a single part of the body is exposed to such disturbed conditions over long periods of time.

According to a report by Rocard (1964), anomalies of the geomagnetic field must also be taken into account in the divining rod phenomenon. He reported that the dowser shows a reaction if he walks at a uniform speed through an area in which the geomagnetic field is not completely uniform, that is, where field anomalies are present. He presents the following findings:

1. A field intensity gradient of 24–40 (mA/m)/m (0.3–0.5 mOe/m) can be detected by dowsers with a time delay of about 1 s.
2. An increase in the gradient to 0.16–0.24 (mA/m)/m (2–3 mOe/m) improves the locating ability of the dowser.
3. A gradient less than 8 (mA/m)/m (0.1 mOe/m) cannot be detected.
4. Weaker field gradients can be detected in a moving automobile or airplane, probably due to the greater speed with which the anomaly is traversed.

Moreover, dowsers are apparently incapable of detecting still water in a pond or flowing water in a river, as opposed to water that is filtered through a porous medium or penetrates permeable strata. The well-known "electrofiltration potentials" created by filtering water may be responsible for the anomalies of the static magnetic field (Fig. 10-9). Measurements at earth dams, for example, showed variations of the magnetic field over a range from $+4 \times 10^{-8}$T to -6×10^{-8}T(+40 gamma to -60 gamma). Thus, all factors that influence the magnetic field in any form must also be considered possible causes of the dowsing reaction.

To support this hypothesis, Rocard

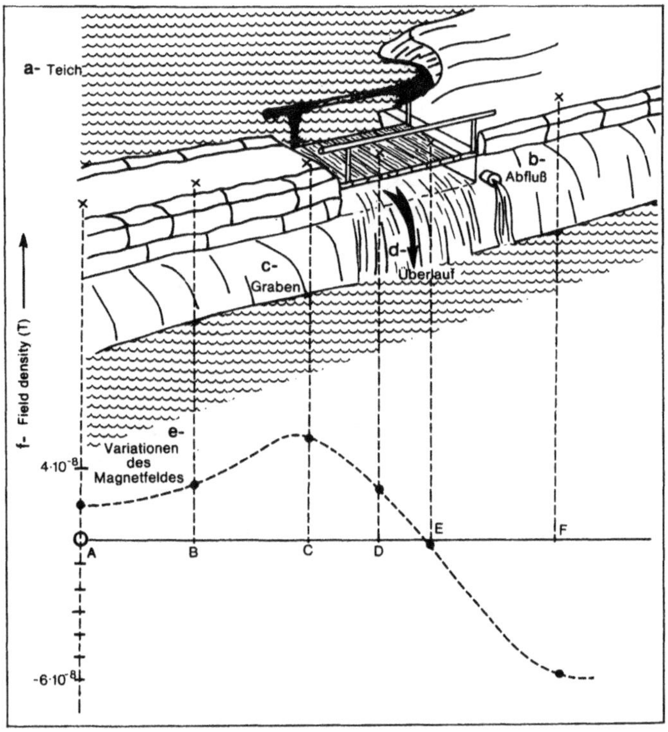

Figure 10-9. Anomaly of the static geomagnetic field along an earth dam at a lake with various drainage points. The field variations are given in teslas (1 T = 10^4 G = 10^9 gamma). After Rocard (1964).
Key: a-Pond; b-Drain; c-Ditch; d-Spillway; e-Variations of magnetic field; f-Field density (T)

(1964) conducted experiments with artifically produced static magnetic fields. He had dowsers walk past a 50 × 100 cm frame around which was wound 100 turns of wire. A direct current of several milliamps produced a magnetic field of about 0.1 A/m, or several millioersted (mG in air), at the center of the frame. Walking at a speed of 1–1.2 m/s, the dowsers thus traversed an anomaly of about 10 mA/m (several tenths of a millioersted or mG in air) in 1.5–2 s. After appropriate training, good dowsers were able to run through 5 or 6 experiments with 30 to 36 trials and indicate correctly in almost every instance (excluding cases of fatigue) whether the coil was carrying a current or not, that is, whether a magnetic field was present. The dowsing reaction was enhanced when two such frames were set up about 3 m apart, and the dowsers had the opportunity to pass both frames in one transit. More detailed statistical data on these experiments are not available. It was reportedly found in parallel tests that the electric resistance between the palms of the hands is apparently a good index of a person's suitability as a dowser. According to these studies, the resistance of good dowsers is only one-third to one-fourth that of persons unsuited for dowsing.

We conducted our own experiments to test the response of dowsers to magnetic fields. Our main concern was not to prove that magnetic fields are responsible for the dowsing reaction, but rather to check the genuineness of the reaction itself. Such experiments with artificial fields can be conducted basically in three ways: (1) the dowser locates an artificial field at a certain spot within a prescribed area; (2) an artificial field is produced intermittently at a known location, and the dowser must decide whether the field is absent or present; or (3) field-generating devices are set up at several locations, but only one is operating. The dowser must then decide which of these devises is generating a field. The latter two methods are decidedly superior, for they require a clear yes/no decision by the dowser. These experiments can meet the most rigorous scientific standards, because (1) the experimental situation is reproducible; (2) the experiments can be conducted in double-blind fashion, where neither the experimenter nor the subject is aware of the solution to the assigned task; (3) the test can be statistically evaluated with clarity and relative simplicity.

The experimental procedure was therefore based on the guidelines developed by Fadini (1971a,b; 1972): The correct solution had to be selected from among eight possibilities. In an initial test, a single coil with 24 windings and a diameter of 0.5 m was attached to a wooden stand so that the coil axis was at a height of about 1.5 m above the floor. Neither the current lead nor the coil itself was shielded to prevent stray electric fields. It must be assumed, therefore, that the current that generated the magnetic field in the coil also produced an electric field of roughly 10^{-6} V/m at a distance of 0.5–1 m about the coil.

Each subject was first exposed to a known "field-on" and "field-off" situation. If a subject thought he was able to feel the difference, he was given the task of selecting the "field-on" situation from among 8 situations produced by a selector switch. Thus, his chance of success was 1 in 8. Subjects who showed no reaction to the artificially produced magnetic field in the pretest were eliminated from the experiment with no evaluation.

Most of the tests were conducted initially under single-blind conditions in order to gain experience. The initial results: At a current of 70 mA through the coil and a frequency of 6 Hz (rectangular current), producing a magnetic field of 4×10^{-6} T (0.04 G) at the coil center, 6 of 30 trials were successful. Under the same field conditions but at a frequency of 200 Hz, 6 of 15 trials were successful. One test was noteworthy in that the subject was able to detect the field at a frequency of 59 Hz and current of 30 mA (1.7×10^{-6} T or 0.017 G at the coil center), even under double-blind conditions.

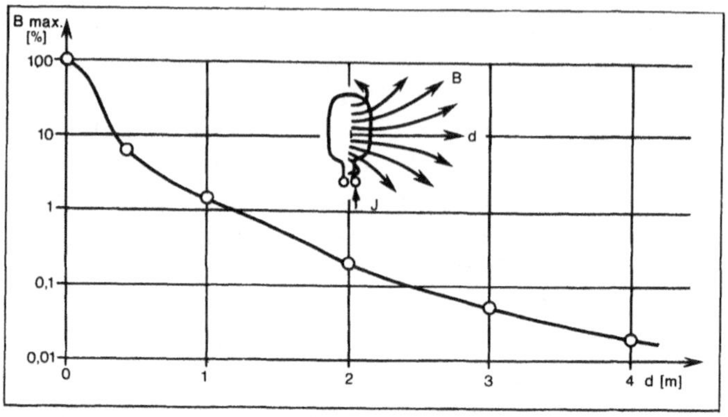

Figure 10-10. Percentage decrease in the maximum value of the magnetic field B_{max} produced at the center of an idealized coil 0.5 m in diameter, as a function of axial distance d in meters from the coil center.

The magnetic field intensities indicated pertain only to the center of the coil and decrease very rapidly with distance (Fig. 10-10). The dowsers approached the coil in an axial direction or passed by it to the side at a distance of about 0.5–0.75 m, at which point the maximum field intensity at the coil core has already fallen by more than 90%. However, the movement of the divining rod itself occurred in most subjects at a distance of 1–2 m (or more) from the coil center, an area in which the field intensity is several orders of magnitude lower than that of the terrestrial magnetic field.

One disadvantage of our procedure was the need to approach the coil repeatedly, which was unfavorable for the dowsers. Therefore, in a second series of tests with different subjects we arranged 8 identical coils in a row, and the dowsers were given the task of determining which of the coils was carrying current (see Fig. 10-11). In this way the dowsers could make their determination in a single run, which corresponded better to normal dowsing conditions. Basically only one frequency, 30 Hz, was employed in this experiment, while the current (i.e., the magnetic field intensity) was variable. The results of this test series (again done under single-blind conditions) are summarized in Table 10-1. The subjects identified the correct coil in 23 of 46 trials. A statistical evaluation of the results shows a 99.9999% probability that the success rate achieved in these tests was not a matter of chance. The qualification must be made, of course, that the experimenter knew in most cases which coil was carrying current, so that only single-blind conditions were present. But this detracts little from the significance of the results. If one also considers the positive results of other experimenters in similar experiments, it becomes evident that we have here a promising method for shedding further light on the divining rod phenomenon. The result of this experiment is at least a contribution to the realization that the movement of the divining rod in the hands of the dowser is not simply a voluntary or random act, but can also be attributed to the presence of physical quantities.

The Divining Rod

Based on the size and shape of the divining rod, it would not be misleading to regard this instrument as a kind of tuned high-frequency antenna. After exhaustive investigations, Schneider (1973, 1978) took the view that the divining rod held by a dowser must be excited by high-frequency energy like a dipole-type electromagnetic radiator. The properties of the divining rod as a radiator or resonator can be described with ac-

Figure 10-11. Dowsing experiment using artificially produced electromagnetic fields. Female dowser carries vertical rod past eight test coils, attempting to locate the one coil which is carrying a current.

curacy if it is regarded as a V-dipole, taking into account the shortening factor which is customary in high-frequency engineering. Experiments with artificially produced high-frequency fields show highly promising results in terms of the reactions of dowsers to such energies. High-frequency engineering principles were employed, in fact, to develop a divining rod designed to operate in the manner of a "Lecher wire" (see Fig. 10-12). The Lecher wire functions as a type of resonator based on the reflection of electromagnetic energies within the system. Schneider's experiments with dowsers indicated a selective response of the dowsers to certain frequencies. However, more precise investigations are needed to determine whether this has to do with

Table 10-1. *Results of a test of dowsers.*

Artificial magnetic field with a constant frequency, f = 30 Hz			Recognition of field-generating coil (number of cases or trials)			
Coil current (mA), rectangular pulse	Induction (in 10^{-7} tesla = 1 mG)					
	at coil center	at 1-m axial distance	Correct	Incorrect	No response	Total trials
80	263	1.1	7			
32	105	0.46	2			
25	84.2	0.37	11			
16	52.6	0.22	1			
4.8	15.8	0.07	2			
		Total	23	16	7	46

Figure 10-12. Dowser with a special divining rod during an experiment with artificial high-frequency fields. The rod is designed according to the "Lecher wire" principle in order to study resonance-specific reactions.

frequency-dependent reflections of high-frequency energy, which are determined mainly by the geometry of the experimenting room, or whether this effect is in fact produced by the resonance tuning of the Lecher wire. In any case, experiments with the aid of a signal generator at a frequency of about 1.6 GHz showed that a type of resonance-dependent reaction was evoked in some dowsers: a slow frequency sweep of the signal over a bandwidth of about 150 MHz produced a turn of the divining rod at 2-s intervals in synchrony with the sweep frequency.

Preliminary measurements of the high-frequency properties of such a rod have not yet answered all questions. True resonant step-ups were observed at about 570 MHz, 1.2 GHz and 1.8 GHz for a particular rod design. The resonant properties of the instrument were not entirely eliminated by holding it in the hand, but were significantly impaired (leaving a step-up of only about 10 db).

In any case, the following has been learned about the mechanism of action of the divining rod. The rod, whether made of metal or an insulator, can be viewed as a type of (dipole) antenna in the high-frequency range and is thus capable of transmitting absorbed energy to the dowser by some means, thereby causing the familiar movement of the rod. Experiments with low-frequency fields show that this is not the only possibility, however. The successes obtained in such experiments with metal as well as plastic rods demonstrate that the divining phenomenon cannot be explained by the antenna action of the rod itself where low-frequency electric and magnetic fields are concerned. It is evident that the dowser himself must be the "receiver" in such cases; the rod is merely held in such a way that slight movements of the hands will produce a large, visible movement of the rod, which thus serves an "amplifier" function.

Neutralizing Devices

The practice of setting up certain objects or devices to neutralize (render harmless) radiations believed to be injurious is probably as old as dowsing itself. By their very association with the divining rod, such devices have become an object of suspicion and ridicule (and probably with good reason in many cases).

Despite the fact that such neutralizing or anti-disturbance devices are sometimes employed with an observable effect, if not success, there is seldom any scientific evidence to indicate that the results are anything other than random or subjective in origin.

Hartmann (1967, 1976), for example, developed "bioresonators" which he used for a variety of experiments and studies, and with which he reportedly achieved amazing results. This prompted König (1968a) to conduct independent studies of their properties as ultrahigh-frequency resonators from a purely technical standpoint.

The heart of the bioresonator is a 3-mm-thick wire of copper-, brass- or zinc-plated iron, which is shaped into a coil (Fig. 10-13) with a diameter of 105 mm and a length of 120 mm, resulting in a total wire length of about 2 m. The coil is wound in a single layer of 6 windings, with a spacing of about 20 mm between windings. The striking feature of the device is that it comprises two such coils wound in opposite directions. However, this only appears to create a disparity, for it can still be shown theoretically that such a coil has a resonant frequency in the neighborhood of 1500 MHz. More precise laboratory investigations showed that one such coil exhibits numerous resonance peaks ranging from 1.3 GHz to 2.0 GHz. If two coils are mounted a small distance from each other, this arrangement actually differs from a single coil only in the greater number of measurable resonance peaks within the same frequency band (Fig. 10-14). Thus, from a technical standpoint these coils represent a type of broad-band resonator whose properties are improved when a pair of coils is used.

The coils are fabricated from elastic wire material and can therefore be bent to a degree, much like a mechanical spring. This change in the coil structure simultaneously effects a change in the electric values of the coil, and thus produces a frequency shift of the various natural coil resonances. The natural mechanical resonance of the coil constructed of elastic material is on the order of 10 Hz, thus it occupies a frequency range in which electromagnetic processes are considered to be biologically active.

If we take into account the floor vibrations that would be present if such a device were installed in the home, it is evident that the high-frequency field always present in

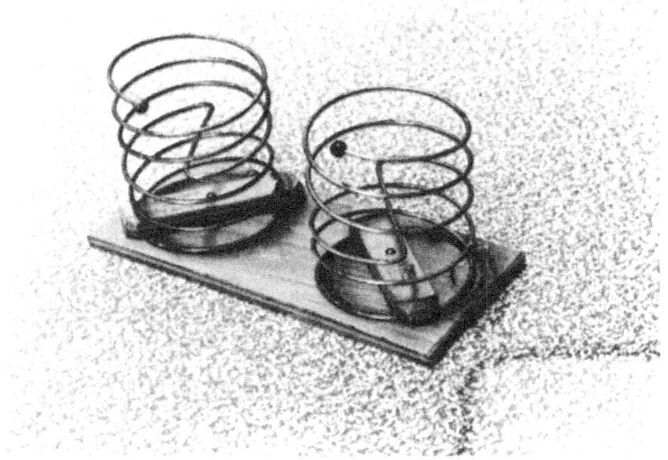

Figure 10-13. "Anti-disturbance coils" designed by Hartmann. This "bioresonator" consists of two identical coils, each composed of two half-coils wound in opposite directions.

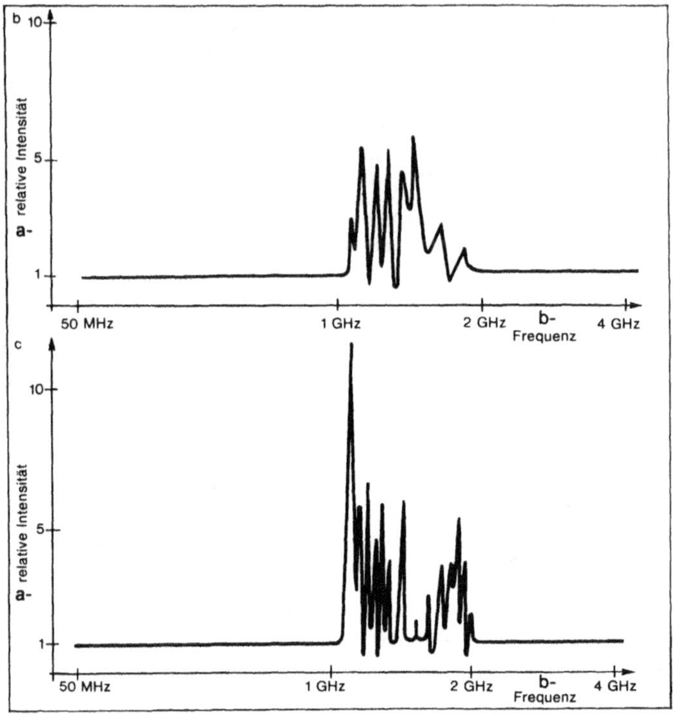

Figure 10-14. Characteristic frequency responses of "anti-disturbance coils" designed by Hartmann (see Fig. 10-13). *Top:* Frequency response of a single coil; *bottom:* response of a coil pair. Numerous strong resonance step-ups are observed in the range between 1 and 2 GHz. Key: a-Relative intensity; b-Frequency

such places would be influenced by this high-frequency resonator, especially with regard to the possibility of low-frequency modulation discussed above. Naturally such results do not resolve the problem of neutralizing devices. However, they do show how little the objective, scientific search for truth is served by working with preconceived judgments without examining the material more closely.

It should be noted that the terms anti-interference and neutralization in connection with such devices are misleading. Previous experience shows that electromagnetic energies can act over the entire spectrum of frequencies, while anti-interference devices can at best claim only a frequency-specific effect. It is inconceivable that they could influence configurations of any other frequencies. Any possible effect of such devices can therefore only be of a therapeutic nature. But such findings would have profound implications if adverse health effects were associated with such "radiations," as is claimed and will be discussed below. It would then be impossible for these devices to screen out or neutralize such radiation (Fig. 10-15). The users of the devices would be deluded into a false sense of security. In the long run, the physical avoidance of such potentially dangerous radiation would be the only sure remedy.

The Geopathogenic Cause of Cancer

Certain compelling arguments led Hartmann (1968, 1976) and others (Diehl and Tromp, 1954) to assert that there is a geopathogenic cause of cancer, that is, a locational factor that predetermines or at least contributes significantly to the development of cancer. According to these authors, a large number of cancerous diseases in man can be linked to a location factor that encompasses an area usually no larger than 50×50 cm, called a cancer point. This location factor can be reproducibly demonstrated by physical, biophysical and biologic methods. This cancer point coincides much too often with the location of the beds of cancer patients to be dismissed as pure chance. Measurements and experiments have shown:

1. In blood sedimentation tests done at these geopathogenic points, zones or their crossings, the sedimentation rate generally follows an entirely different course than at points in the vicinity. Germination and growth experiments with various species of plants have repeatedly shown the absence of germination in these "disturbed" zones. Mice in boxes placed in such zones leap out of the box to escape. Smaller litters, weight loss, increased spontaneous tumor formation and an increased susceptibility to artificially induced cancer (tar brushed onto the skin) in comparison with control animals are all typical of such zones.
2. Direct-current resistance values (between the hands) monitored in human

Figure 10-15. The horseshoe is well known as a good-luck charm (shown here in the living room of a cottage in Lower Bavaria). However, it also constitutes a high-frequency resonator whose natural resonance occupies a wide range around 1 GHz and thus lies within the range of hydrogen resonance at 21 cm. A striking parallel exists between the horseshoe and bioresonators, as a certain type of "anti-disturbance" device used by dowsers is called.

subjects over long periods of time follow a different course if the measurements are carried out in such spots.
3. The electric currents measured in the human body by galvanometer are markedly different.
4. VHF field-intensity values show pronounced anomalies (maxima or minima) relative to the environment.
5. The electric conductivity of the ground is altered.
6. Anomalies of the static geomagnetic field are present.
7. Radioactivity varies significantly within a small area. The radiation of certain terrestrial neutrons is markedly different.
8. The incidence or intensity of VLF atmospherics deviates from that of the immediate surroundings.
9. The flicker fusion frequency of test subjects is reduced by 0.5–2.5 Hz within 1 min after the subject is placed on such a spot.

These effects are observed in various combinations, depending on the location. According to Hartmann (1976) and Aschoff (1973), these geopathogenic zones and points exhibit biologic and physical properties which should be sufficient, based on previous experience, to make even the greatest skeptic more cautious in his remarks on the possible existence of geopathogenic effects. Indeed, one should expect an open and unbiased science to show an active interest in finally resolving the grave and profound questions that have been raised here.

Conclusions

The discussions in this chapter are hardly sufficient to provide exhaustive information on the divining rod problem. But this is neither possible nor intended within the present framework. There is already a large body of scientific literature dealing with various points of view on this subject, and there is an even greater subjective and unscientific literature. It is certain that dowsing will continue to be associated with charlatanry, and rightly so, especially where so-called neutralizing devices are concerned. The detractors of dowsing have published very comprehensive and critical works contending that the divining rod effect is of little relevance. Undoubtedly the questions addressed in these works have always been the subject of attempted unscientific explanations by dowsers, and it is certainly useful in the scientific scrutiny of such questions when competent experts adopt a critical attitude toward particular topics. However, the interests of objective inquiry are not served by approaching a subject from a negative standpoint, or simply doubting experiments whose results were properly obtained. The objection must be raised that such works are attempting merely to document and verify a preconceived opinion, instead of reaching conclusions in an objective manner.

In any case, evidence gathered to date makes it clear that when viewed from a realistic, neutral perspective, the divining rod phenomenon can no longer be dismissed as nonsense. An objective appraisal of this problem from the outside requires not only the necessary physico-technical expertise, but also a considerable degree of knowledge and experience in the field of biology. All too often, classical physicists and engineers make the mistake of treating biologic systems like apparatus during measurements, particularly in the case of the divining rod problem. But by their very nature, biologic systems, and man in particular, do not lend themselves to clear and reproducible measurement. Every biologic system must be judged with regard for the principle of individual diversity. Objective results and information cannot be obtained by the study of individual systems, or individual persons, especially in critical areas. In this case, only investigations that can be evaluated statistically can produce scientifically acceptable results; and here, too, it is often only the experienced researcher who can avoid environmentally determined errors in his investigation. External physical, psychic and psychological influences can lead to considerable difficulties as well as erroneous conclusions, especially in experiments with humans.

The phenomenon of the turning of the divining rod in the hands of the dowser must be regarded as the real product of physically determined environmental stimuli. As in the case of meteorosensitivity, more than just a single agent is responsible. Electric and magnetic fields from the static range to the highest frequencies, ionizing radiation, air ionization and the vertical atmospheric electric current represent the possible physical parameters to which the dowser can respond. And because every person is unique, and thus a nonreproducible individual, every dowser will also react specifically, that is, in different ways to different parameters. Added to this are such influences as his attitude toward what he is doing, his momentary frame of mind, his constitution, and other environmental stimuli. These many variables explain why such marked contradictions continually arise in dowsing experiments, and why they in fact must arise and will continue to do so. One can thus appreciate the urgency of the desire to develop a physical method of measurement that can at least confirm the dowser's indications, if not replace them altogether. In the light of previous findings, however, it appears doubtful that an ideal solution will be found, particularly with regard to replacing the human dowser with instrumentation.

11

Electromagnetic Fields: Biologic Stress or Therapy?

General Considerations*

Lang (1977b) dealt specifically with the question posed in this chapter in a comprehensive study. Orientation by electric and magnetic fields is an ability that has been demonstrated in many animal species, from protozoans, worms and insects to higher animals such as birds and small mammals. This includes orientation for homing purposes over great distances, as well as simple avoidance behavior or flight responses to escape from disturbing geophysical environmental influences. Observations that man is also capable of consciously perceiving such fields, which could not be substantiated initially according to scientific criteria, have recently proved to be statistically valid (Gossel, 1977).

The most important geophysical environmental factors from a biologic standpoint are summarized in Table 11-1. The conscious perception of these factors or their changes with time and location by the human organism is often nonspecific and may be manifested in general discomfort (e.g., the "stuffy" feeling experienced when humidity is too high relative to room temperature); without prior experience, the person would be incapable of accurately identifying the parameter causing his discomfort. On the other hand, numerous examples in the animal world prove that organisms have evolved a sensory apparatus that is appropriate for the environmental factors affecting them, and that enables them to adapt to external conditions.

Table 11-1. *The most important geophysical factors which influence the organism. After Drischel (1975).*

Temperature
Humidity
Air motion
Air pressure
Oxygen partial pressure
Radiant energy
Gravity
Mechanical vibrations
Air ions
Electric fields
Magnetic fields

*The concept of this part of Chapter 11 was developed originally by Dr. Siegnot Lang, who died on January 30, 1978, at the age of 36. Dr. Lang was one of the small group of scientists who deal specifically with the biologic effects of atmospheric electricity. The end of his highly active and successful career leaves a gap in this special area of science which will be difficult to fill.

Working from this premise, Presman (1970) and Lang (1972) formulated the so-called evolutionary hypothesis regarding the action of electric, magnetic and electromagnetic fields on organisms. According to this hypothesis, the average intensities, as well as the rhythmic and arhythmic intensity fluctuations, of the atmospheric electric and magnetic fields at the earth's surface were largely the same during evolutionary history as they are today. Every living creature has adapted to these electric and magnetic climates, or to the microclimate predominant in its own biotope, so that it preceives the intensity pattern and time course of the geophysical factors affecting it as adequate. Now if these environmental stimuli are absent or act on the organism in unnatural patterns (time course, intensities), such a milieu represents a biologically inadequate state which, as expected, finds expression in various metabolic and behavioral changes. The classification of the currently known effects of electric, magnetic and electromagnetic fields as orientation behavior, frequency-specific responses, stress responses and field-specific adaptive responses will be explained in the sections that follow.

Orientation Behavior

Golden hamsters, mice, rats, bees, goldfish, birds, eels and may beetles "notice" electric fields and, at higher field intensities, apparently perceive them as a disturbance of their environment that is to be avoided (Schùa, 1953; Zahner, 1964; Warnke, 1975b; Altmann and Lang, 1973; Lang et al., 1977; Brinkmann, 1976). During periods of animal activity, electric fields at more or less adequate intensities (and frequencies) are accepted (Altmann and Lang, 1973; Lang, 1976b). Inadequate field intensities, on the other hand, lead to heightened activity and aggressiveness, assuming that the animals are unable to avoid the field (Schùa, 1953; Warnke, 1975b). A relation between the response of the animals and the field intensity and duration of exposure cannot be clearly deduced from these findings, because the corresponding parameters were determined primarily from technical rather than biologic considerations. Bees consistently displayed an extremely sensitive response, abandoning their nest within 30 s on exposure to a field intensity of 800 V_{pp}/m and frequency of 50 Hz (Warnke and Paul, 1976). Resynchronization of the circadian rhythm at a field intensity of 2.5 V/m, as described by Wever (1967, 1968, 1973, 1976), also appears to represent a very sensitive response. Small mammals display a certain tenacity in the presence of such fields; i.e., they tend simply to become accustomed to their surroundings. As the free-choice experiments of Altmann and Lang (1973) prove, the action of fields is not just a matter of exceeding a constant stimulus threshold of a certain value; rather, the level of this threshold is determined in large measure by the well-being of the animals and their acclimation to their environment. The finding of Dowse and Palmer (1969) that circadian rhythms can be influenced both by a static field and by a spectrum of alternating fields is subject to various possible interpretations. On the one hand, certain frequencies (the static field also acts as an alternating field in the range of 1–20 Hz, as described below) may trigger rhythmic physiologic processes from outside; on the other, the electric stimulus may effect a completely nonspecific arousal of the organism's regulatory systems.

To be sure, the sensitivity of organisms to such environmental factors as temperature, humidity and light also appears to be much more differentiated than was previously assumed. Warnke (1975a,b; 1976a,b,c) further pointed out the significance of the ability of diverse body tissues to acquire an electrostatic charge.

Conductivity measurements performed on body surfaces by Becker and Speck (1964), Athenstaedt (1972, 1974) and Ellenby and Smith (1968) showed that the humidity of the surrounding air, the concentration of small air ions, air temperature and illumination are all capable of influencing the electric conductivity of insect cuticle. In the short-wave light region, a photoelectric effect could be demonstrated for insect tissues. Pyro- and piezoelectric effects showed that electromagnetic vibrations induce mechanical microvibrations at body surfaces, especially in the resonant range. Conversely, electric impulses are produced in the skin by air-pressure fluctuations or sound waves. The full significance of these effects is not yet clear.

Arendse (1976) reported on experiments with mealworms that were bred in various artificial biotopes and then exposed to magnetic field-free conditions (the terrestrial field was offset) or directional magnetic fields. The animals normally displayed a directional preference that corresponded to the direction of darkness in their breeding containers. However, they were able to maintain this orientation only if the direction-indicating effect of a magnetic field was present.

Proceeding from experiments done in 1958 by Merkel and Fromme (1958), Wiltschko (1973) and Wiltschko and Fleissner (1976) were able to show that red-breasted robins can perceive the polarity of the horizontal component of the earth's magnetic field. However, this orientation is useful only if combined with concurrent orientation by the force of gravity. The magnetic compass of robins functions only within a very narrow range of intensities. If the total magnetic field intensity is reduced by 25% or increased by more than 50%, the birds are no longer capable of magnetic orientation. Wiltschko and Fleissner (1976) tested robins in equal-energy alternating magnetic fields with frequencies of 50, 1 and 0.8 Hz. While the 50-Hz field had a disorienting effect on the birds and sinusoidal fields had no apparent orientational value, they were able to orient themselves in the 1- and 0.8-Hz fields with a rectangular time course.

Termites are influenced by atmospheric magnetic fields with respect to their feeding activity. According to Becker (1975a, 1976a,b,c), the daily wood consumption of these insects is correlated with atmospheric magnetic disturbances associated with the 27-day rotation of the sun. Effects from weather conditions, and especially atmospherics, were also determined.

A comprehensive survey dealing specifically with the orientation behavior of animals in the earth's magnetic field was presented by Martin and Lindauer (1973). No adequate explanation has yet been found for the alignment of animals, or of general plant and animal growth processes, with respect to the anode or cathode of an electric field (Kemmer, 1967) or along the vectors of a magnetic field. It can be stated in general, however, that animals, and probably humans, possess some form of receptors for electric and magnetic fields, through which information on these environmental factors can be supplied to the central nervous system, thereby enabling such responses to occur.

Frequency-specific Responses

Besides the essentially frequency-independent orientation behavior described above, some animals such as the robin display an orientation ability in magnetic fields that is clearly dependent upon the pulse form and frequency of the field. In this section we shall examine the biologic effects of electric and magnetic fields that are clearly a function of the frequency employed.

Animal Experiments

Though his report received little attention, Petrow pointed out in 1935 that a neuromuscular preparation could be excited both in vitro and in vivo by an electromagnetic vibration of 1 kHz, while vibrations at 33 MHz produced no effects. Forty years later, Lott and Linn (1975) showed that the action potentials of the frog sciatic nerve could be increased in vitro by means of a

10-Hz square-pulse electric field. These authors suggest that this phenomenon may be due to a physical change (pore deformation) in the membrane or a field-induced alteration of cell metabolism leading to a change in ion flux.

However, the findings of Altmann et al. (1975) on the increase in the abdominal-skin potential of the frog by a 10-Hz square-pulse field, or of Adey (1975) on increased calcium efflux from the chicken brain under the influence of 10-Hz alternating electric fields cannot yet fully explain how the increase in brain and hypothalamic activity under the influence of an extremely low frequency electric field, as measured by Lott and McCain (1973), is brought about. The investigations of Persinger (1969) on the prenatal effects of alternating magnetic fields with a frequency of 0.5 Hz and magnetic induction of 0.05–3 mT were recently supplemented by the findings of Ossenkopp and Ossenkopp (1975). They found that the exposure of pregnant rats to the field led to reduced activity and increased defecation in both male and female offspring. If the pups continued to be exposed to the field during early life, the males showed both increased activity and increased defecation. In the females, defecation increased while activity levels were indistinguishable from those of the controls. By comparison, Persinger and Lafrenière (1975) reported on thyroid gland studies in rats exposed to fields of 10 G, 1 G, 0.05 G or 10^{-4} G (controls). While the thyroid weight was significantly increased in the animals exposed to 10 G, the follicle count in the thyroid tissue was elevated only in the rats treated with the 0.5-G field. The other field conditions produced varying effects.

Natural geomagnetic storms (July 5–6, 1974) also produced an activity increase in rats, according to Persinger (1976). As in the termite studies of Becker (1975a, 1976 a,b,c), Persinger's measurements showed a certain time delay in the onset of the effect, so that the activity increase was not evident until 1–2 days after the increase in magnetic activity was registered.

Bawin et al. (1973) incubated the brains of 500 newly-hatched chicks (2–7 days old) in a physiologic medium with $^{45}Ca^{2+}$. Afterwards the preparations were washed and exposed to artificial field conditions: 147 MHz carrier frequency, modulated by sinusoidal oscillations at 0.5–35 Hz (1–2 mW/cm^2 maximum). The increase in calcium efflux from these brains as a function of the modulation frequency is illustrated in Fig. 11-1. The effect demonstrated in these experiments could not be blocked with cya-

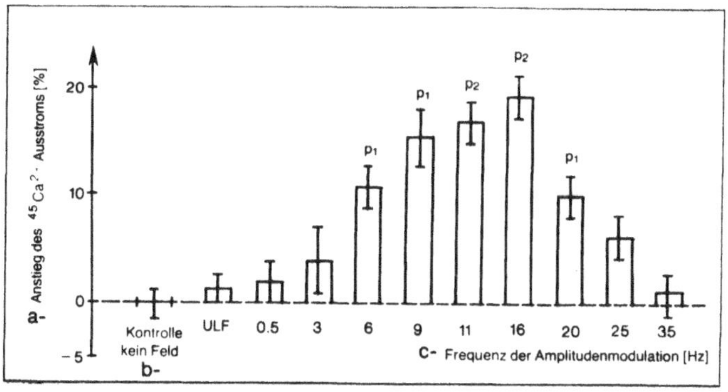

Figure 11-1. Effects of 147-MHz fields with varying amplitude modulation on the efflux of $^{45}Ca^{2+}$ from the isolated forebrain of newly hatched chicks. Results are given as a percentage of the increase of calcium efflux, in comparison with control conditions without a field, $p_1 < 0.05$, $p_2 < 0.01$. After Bawin et al. (1973).
Key: a-Increase of $^{45}Ca^{2+}$ efflux (%); b-Control without field; c-Frequency of amplitude modulation (Hz)

nide ions, indicating that it is independent of any metabolic processes.

Human Experiments

The frequency specificity observed by König and Ankermüller (1960) in the effects of electromagnetic fields on the responsivity of human subjects finds a possible correlate in these results. Maxey (1976) also observed this frequency-specific effect in subjects exposed to alternating magnetic fields at frequencies from 1.6 to 11.4 Hz (at 1000 $\gamma = 10^{-6}$ T). Ten percent of his subjects reacted to 7.8 Hz with a direct coupling of the respective alpha waves in the EEG as long as the field influence was present. About 4 s after the start of exposure, theta waves with a frequency of 4–7 Hz appeared in the EEG. The same effect occurred at 10.8 and 11.4 Hz, but was not observed at 1.6 or 4.7 Hz. Maxey (1976) compared this effect to the known individual-specific sensitivity of certain persons to extremely low frequency light pulses and pointed to correlations between atmospheric magnetic storms and aircraft accidents ("pilot-error weather") with regard to the impairments of concentrating ability and responsivity due to this effect.

Ludwig (1972; 1973a,b,) and Ludwig et al. (1973) distributed 700 ELF generators (10-Hz Schumann-resonance frequency range, harmonic content, field intensity 100 A/m at close range) and 220 sham devices among 860 patients with psychosomatic and rheumatic complaints. Healthy subjects hardly reacted to the devices, while the patients each responded best to a certain frequency, depending on their type according to Curry (1946); frequencies of 1–6 Hz had a calming, sedating effect, while higher frequencies of 8–20 Hz had a stimulating and analgesic effect. While a very precise frequency setting was required for some patients ($\pm$ 3%), others were less critical ($\pm$ 20%). Apparently the "W types" (vagotonics) are responsive to fields of higher frequencies, while the "K types" (sympatheticotonics) react to frequencies above 12 Hz with nervousness and irritation. According to Ludwig (1972; 1973a; 1976a,b) and Ludwig et al. (1973), the necessary exposure time lies between 1 min and 1 hr. The mode of action of the small devices was found to be strictly local. The harmonic magnetic vibrations produced by them caused no increase in the thrombocyte adhesiveness of the subjects' blood, as was achieved by Jacobi and Krüskemper (1975, 1976) with 10-kHz electric field oscillations modulated by low-frequency alternating fields.

After Jacobi et al. (1973) had established a statistical correlation between thrombocyte adhesiveness and the general meteorologic situation, weather phenomena, atmospheric dynamics and electromagnetic radiation, they simulated artificial sferics in steel chambers (Faraday cages) in order to investigate its effect on various parameters in experimental subjects. They found that pulse repetition frequencies of 10 Hz at 0.4 V/m (10 kHz carrier frequency) produced a highly significant increase in thrombocyte adhesiveness after 3 hr in comparison with shielded, zero-field conditions or pulse frequencies of 2.5 Hz and 20 Hz (Fig. 11-2). This effect could be prevented by the administration of 75 mg dipyridamole and 300 mg acetylsalicylic acid. A 10-Hz pulse repetition frequency impaired responsivity. Mentally labile subjects showed stronger responses than mentally stable subjects (Jacobi and Krüskemper, 1975, 1976).

Bein (1976) conducted a comprehensive study of 110 subjects exposed to an electrostatic field, taking special account of meteorologic activity during the course of the experiment. He found an increase in the mental activity of his subjects which, as mentioned above, must also be regarded as the effect of an alternating electric field, according to the average frequencies of respiration and movement exhibited by the subjects.

Persinger et al. (1975) exposed 70 male and female subjects to a 40-min field program in which electric fields with frequen-

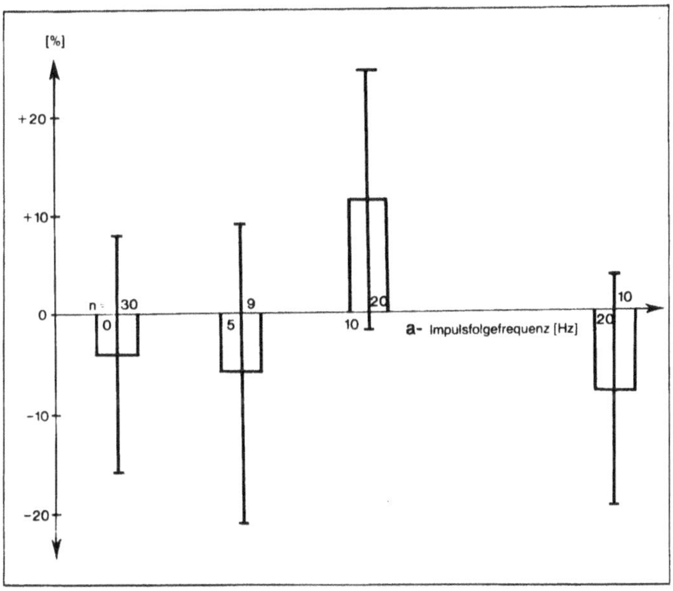

Figure 11-2. Percentage change in the thrombocyte adhesiveness in the blood of subjects (n) exposed for 3 hr to simulated atmospherics with a carrier frequency of 10 kHz and pulse repetition frequencies of 0 Hz, 5 Hz, 10 Hz and 20 Hz at a pulse amplitude of 0.4 V/m (width of columns corresponds to scatter). After Jacobi and Krüskemper (1976).
Key: a-Pulse repetition frequency (Hz)

cies of 10 Hz (sinusoidal) and 3 Hz each acted upon the subjects for 10 min. This program was implemented at three different field intensities—3.0 V/m, 0.3 V/m and Faraday conditions—and the reaction times of the subjects were measured throughout the course of the experiment. The mean results of these measurements showed no differences between the field programs employed, the field intensities, the field frequencies or the sex of the subjects. Statistical evaluation based on the standard deviation, which represents a certain measure of the consistency of concentration performance, showed a continuous functional capacity on the part of the female participants during the 40-min program, assuming a constant readiness to perform. During exposure at intervals, both the weaker and stronger field produced a significantly greater regularity in the performance required of the male participants.

In this context Persinger (1974, 1975b) called attention to the many sensations that are reportedly felt by the human organism during disturbances of the terrestrial magnetic field. Persinger and Janes (1975) found a significant correlation between the anxiety of students and magnetic storms which had occurred before and after the births of the subjects tested. A valid correlation was also found between daily self-descriptions of current emotional state and meteorologic parameters such as temperature, pressure, relative humidity, duration of sunshine, wind velocity and geomagnetic activity. "Bad moods" tended to be associated with a short duration of sunshine and a high relative humidity, and "good moods" with low relative humidity, with a correlation coefficient of 0.27.

Stress Responses

The Stress Syndrome

More or less strong stimulations of the human organism generally evoke a standardized, stereotypical response that is phylogenetically very old: if the individual finds himself unexpectedly subject to attack by an aggressor, the cerebral cortex together with the limbic system activates the hypothalmus, which triggers the immediate release of catecholamine from the adrenal medulla. The result is a drastic increase in the cardiac minute output. The attendant rise in systolic blood pressure and simultaneous collateral peripheral vasoconstriction ensure an optimal supply of nutrients to the muscles. This is reinforced by the acceleration of glycogenolysis and lipocatabolism by epinephrine in the liver. This sympathetic emergency response represents the phase of the alarm reaction of the general adaptation syndrome with which the organism counteracts the stress-producing agent. The special feature of this stress response is the stereotypical pattern of physiologic mechanisms that occur, and the fact that the organism responds to any change in his environment in the same manner: catecholamine secretion is increased during great joy as well as great fear.

Electric Fields as Stressors

As mentioned, stimulation of the hypothalamus triggers the release of catecholamine from the adrenal medulla. Over 100 papers have been published to date on the ability of electric fields to produce symptoms of the first stage of the stress response, the alarm reaction, with no recognition of the contextual significance of these findings. Only Becker (1975b) and Marino (1975) pointed to the importance of the electric stresses produced by the extremely high field intensities near high-voltage transmission lines in connection with the Common Record Hearings on the Health and Safety of 765 kV Transmission Lines before the New York Public Service Commission in 1975.

In another context, Verheijen (1958) pointed out that the intensity of artificial environmental parameters acting on an organism should not deviate too strongly from natural intensity values. This condition is necessary to ensure that the stimulus intensity lies roughly within the control range of the regulatory mechanisms that evolved in the organism during the course of its phylogenetic development. Aside from considerations of biologically adequate and inadequate environmental parameters (discussed above), this means that any electric, magnetic or electromagnetic field condition is perceived by the organism as an unnatural stimulus as soon as the field intensity exceeds the maximum value occurring in the organism's natural biotope at a particular frequency. The release of catecholamine and its consequences—heightened activity, the mobilization of energy reserves (fat metabolism) and cardiovascular changes—are the most important symptoms by which the course of the alarm reaction can be monitored. The increase in motor activity is the physiologic correlate of the biologically important fight or flight response to the acting stressor. Mice, birds and insects all exhibit this behavior in electric fields, as reported by Altmann and Lang (1973), Fischer (1973), Becker (1975a) and Moos (1967). The increased energy supply necessary for this activity in the form of energy-rich fatty acids and triglycerides was confirmed in rats and mice by Beischer et al. (1967), Lang and Reuss (1975) and Klingenberg et al. (1976a,b). Only Kröling (1976b) reported negative findings on the motor activity of mice in an electric field.

The rise in blood cholesterol caused by catecholamine secretion enhances the clotting ability of the blood, which was observed by Brezowsky and Ranscht-Froemsdorff (1966) under artificial atmospherics and by Boenko and Shakhgeldyan (1968) under the influence of an 8-kHz, 1500-V/m electromagnetic field. Jacobi and Krüskemper (1975, 1976) recorded an increase in thrombocyte adhesiveness as a measure of blood coagulability under conditions of natural atmospherics, and could confirm this effect in a laboratory experiment with simulated atmospherics. A direct action of the stress-induced increase in catecholamines on the circulation is reflected in the results of Blanchi (1973), whose ECG studies showed a lengthening of the P-R and R intervals and the QRS complex. His rats survived a dose of 100,000 V/m and 50 Hz over a period of 1000 hr, during which time a pattern of 9 hr exposure followed by 3 hr recovery was employed. In experiments performed by Solov'ev (1967) with half this intensity (50,000 V/m at 50 Hz) and full-day exposure, all the mice and insects died within a few hours. Fischer and Richter (1976) reproduced the findings of Blanchi (1973) with rats exposed to 5000 V/m, 50 Hz, over a 50-day period; they found a continuous decline in the heart rate of the animals for the full duration of the experiment.

Silny (1976) monitored the heart rate, respiration rate, systolic blood pressure, skin and rectal temperature, and ECG and EEG of rats exposed to a field at 80 kV/m (60 kV/m for cats), 50 Hz, from the periods of the fourth to eighth and twelfth to sixteenth hours of a 16-hr session. The heart rate of the animals increased by 21% during the first 2 hr, while the body temperature rose by 3%; the increases in respiration rate and systolic blood pressure were not significant.

Studies in persons exposed occupationally to high-intensity fields at 50 Hz have shown the full gamut of possible cardiovascular disorders, ranging from unstable pulse, arrhythmias and diminished cardiac blood flow, to bradycardia and disturbances of potency (Lyskov and Emma, 1975; Asanova and Rakov 1966a,b; Sazonova, 1967; Korobkova, 1971, 1972; Strumza, 1970). The associated danger of infarction was recognized as early as 1966 by Brezowsky and Ranscht-Froemsdorff in connection with the much weaker field intensities of natural atmospherics.

The alarm reaction of the general adaptation syndrome is concluded by a rise of the diastolic blood pressure as the cardiac minute output returns to normal. Before this, however, the hormones thyrotropin, corticotropin and somatotropin are secreted by the anterior pituitary gland.

Related findings are presented in other studies. According to Persinger (1969), Ossenkopp and Ossenkopp (1975), Ludwig (1972) and Ludwig et al. (1973), the thyroid gland weight was increased in rats exposed to 0.5-Hz electromagnetic fields, indicating an increased production of thyroxine. The thyroidal response of organisms to electric fields is expressed in increased oxygen consumption. According to Damaschke and Becker (1964), this effect was observed in termites on intermittent exposure to a 20,000-V/m electrostatic field, as well as under the influence of labile atmospherics. The same observations were made by Lotmar and Ranscht-Froemsdorff (1969) with regard to the cellular respiration of rabbit skin, as well as in mouse liver cells exposed to corresponding artificial atmospherics programs (Ranscht-Froemsdorff, 1969). According to Altmann (1969a) and Fischer (1973), the oxygen consumption of guinea pigs and mouse liver cells, respectively, increased just as much under constant field conditions as that of mice exposed to a strong 10-Hz square-pulse field (Lang, 1972a). The thyroid hormones act directly at the mitochondria, whose function is governed by these hormones. Riesen et al. (1971) found that the mitochondrial activity of brain and liver cells ceased completely within 40 min in a 60-Hz field at 155 V/m. Threshold values of 6.3 V/m and 60 min were determined (cf. Ng and Piekarsky, 1975).

The secretion of somatotropin by the anterior pituitary leads to the activation of m-DNA and thus to an increase in the rate of cell synthesis. Growth effects of electric fields have been observed in bacteria (König and Krempl-Lamprecht, 1959; Busch, 1961), in *Physarum* (Goodman et al., 1975), in insects (Mittler, 1972), in birds (Durfee et al., 1975; Krueger and Reed, 1975; Watson et al., 1975; Becker, 1975b; Marino, 1975), in small mammals (Marino, 1975; McElhaney and Stalnaker, 1968; Mamontov and Ivanova, 1971; Bassett et al., 1974; Norton, 1974; Knickerbocker, 1967) and in man (Marino, 1975;

Becker, 1975b). Plants also display a significant average growth increase of 40% under the influence of electric fields, given an adequate supply of water and minerals (Lemström, 1902; Stetson, 1957; Sidaway, 1966; Hicks, 1957). The secretion of corticotropin by the pituitary stimulates the glomerular zone of the adrenal cortex to step up its hormone production, and the glucocorticoid level increases. The decrease in the number of neutrophilic and eosinophilic leukocytes in mice exposed to a 50-Hz, 100,000-V/m field for 42 days (Blanchi et al., 1973) is a clear index of this reaction. The lymphocytes are also reduced. The same effect could be observed in rats after only 6 hr exposure. While De Lorge (1975) detected no changes of this kind, Marino (1975) found an elevated corticoid level in the blood serum of rats, as well as an enlargement of both the pituitary and adrenal glands, as might be expected. The serum glutamate and oxalate transaminase were also increased in the blood of these animals. Seeger (1968) claims that he has observed the suppressed inflammatory response that accompanies increased glucocorticoids in human subjects as well. The alleviation of pain in patients covered with an electrically shielding blanket during natural atmospherics (Ludwig, 1973a) may also be attributable to the action of glucocorticoids.

Fischer (1976a) reports on the effects of a rippled electrostatic field on the metabolism of mice and rats. In the intensity range from 1000 to 5000 V/m with a ripple (residual hum voltage) of 1%, the motility, food and water consumption, oxygen consumption, brain serotonin level and antibody production were increased in comparison with the corresponding metabolic indices of animals kept in Faraday cages in which the ELF spectral components of external electric fields were screened out by more than 99%. Fischer (1976a,b) and Fischer and Richter (1976) made the supplementary finding that only about one-third of the effects, on the average, could be attributed to the action of the electrostatic field, while two-thirds were due to the 50-Hz ripple in the field (about 10–50 V/m). Of particular interest is a comparison of the heart-rate measurements of Fischer and Richter (1976) in rats with the results of Silny (1976). At a low 50-Hz field intensity of 50 V/m, the heart rate of the rats was reduced by about 14% after 15 min and remained low during the duration of the 50-day experiment (Fig. 11-3). A field intensity of 5300 V/m produced an even greater reduction of about 25% after 15 min, which

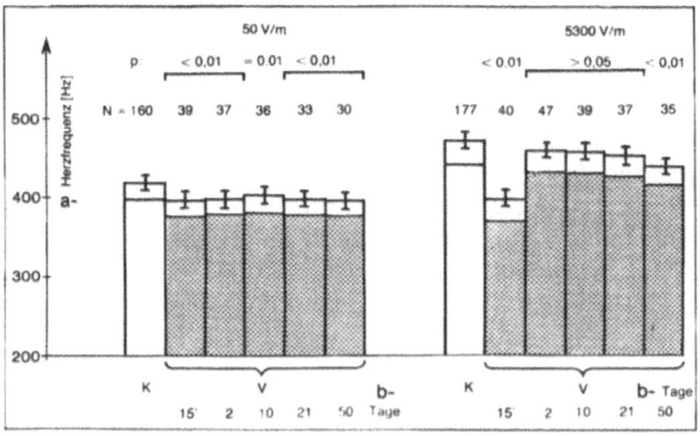

Figure 11-3. Effect of 50-Hz fields of various intensities on the heart rate of rats, as a function of exposure time (K = control, V = experiment). After Fischer and Richter (1976)
Key: a-Heart rate; b-Days

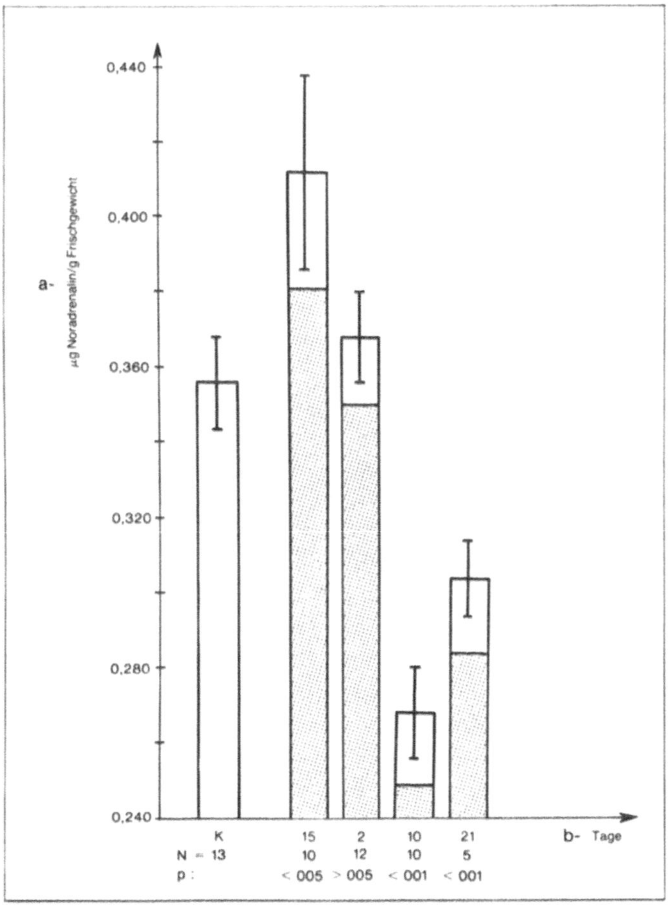

Figure 11-4. Norepinephrine level in the brain of rats during prolonged exposure to a 50-Hz field (5300 V/m) as a function of time (K = control). After Fischer and Richter (1976)
Key: a-µg norepinephrine/g fresh weight; b-Days

measured 4% after 2 days, and then gradually fell back to 11% by the fiftieth day. The body temperature of the animals was also reduced at 5 kV/m, unlike the reaction observed in a field at 80 kV/m. The norepinephrine content of the brain rose from 0.360 µg/g fresh weight to 0.415 µg/g fresh weight in 15 min, representing an increase of about 15% (Fig. 11-4). Only a slight 3% increase was present after 2 hr, and a 25% decrease was measured after 10 hr; after 21 days the decrease still amounted to 15%. Fischer and Richter (1976) suggested that the animals may have been able to sense alternating electric fields through microvibrations of the fur and microvibrissae. Later experiments with hairless rats (result of degeneration) refuted this hypothesis, however.

It was primarily the increased epinephrine level in the brain which then led Fischer to speak of an electric stress. Due to the thermoregulatory mechanisms peculiar to the rat and mouse, the secretion of epinephrine in these animals causes a stronger stimulation of the cold center than the heat center (unlike man), which would account for the reduction of body temperature.

Using much weaker electric fields in the same frequency range, Krueger and Reed (1975) were unable to obtain corresponding results in mice. They exposed 3-week-old mice for 25 days to a week (5 V/m) and strong (100 V/m) sinusoidal 45- and 75-Hz field. They investigated their weight gain, blood and brain serotonin levels, and susceptibility to influenza viruses during and after the experiment, as compared with the same indices in control animals. No differences could be found between the experimental and control animals with regard to any of the parameters tested.

In rats exposed continuously for 1 year to a 50-Hz, 100-kV/m electric field, Brinkmann (1976) investigated motor activity and startle response to acoustic stimuli (as a correlate of reaction time in humans), as well as heart rate and blood status. After the exposure period, the animals were sacrificed and subjected to histopathological study. Compared with the controls, the field-exposed rats showed a 23% lower total activity, although it must be noted that the comparability of the activities of the two populations had not been previously checked. The number of white blood cells was 30% lower as well. No differences were found with respect to growth, activity rhythm, histology or the other blood studies. When the animals were given the choice of building their nests within the field or outside it, they showed a clear preference for the field-free zone. The lowest field intensity at which the animals could be driven from their nesting place lay between 10 and 20 kV/m.

In a follow-up to the investigations of Hauf (1976a), Rupilius (1976) exposed 10 subjects to a combined electric (20 kV/m) and magnetic (3 G) field with a frequency of 50 Hz. The duration of exposure was 3 hr. In accordance with the previous experimental program of Hauf, the studies performed included ECG, EEG, pulse, blood pressure, blood status, BSR, thrombocytes and reticulocytes in the blood, Quick test, changes in serum triglycerides and cholesterol, and reaction behavior. No changes could be detected in the persons exposed to the combined field relative to the control groups. Not even the mild stimulative effects reported by Hauf (1974) in a 50-Hz electric field at the same intensity could be observed.

Meda et al. (1974) exposed mice, rats and pigs in 3-hr intervals to a 50-Hz, 100-kV/m field for a total of 1000 hr. The differences observed by other authors regarding the concentration of blood cells (see also Hauf, 1974) and the ECG (see also Silny, 1976) were confirmed under these conditions.

Cabanes (1976a,b,c), Gary (1976), Hauf (1976b), Kornberg (1976), Cerretelli (1976), Malboysson (1976), Silny (1976), Le Bars (1976) and Rivière (1976) debated evidence on the biologic effects of 50-Hz fields and judged the apparent inconsistency of the various results to mean that artificial 50-Hz fields had no significant effect on man.

De Lorge (1975) has gathered much evidence on the reaction of monkeys to sinusoidal electric, magnetic and electromagnetic fields at frequencies of 7, 10, 15, 45, 60 and 76 Hz and a variety of intensities. On the whole, he has found no consistent effects with regard to activity, body weight, various behavioral parameters or metabolic indices.

Ng and Piekarsky (1975) experimented with rats immobilized in cage arrangements. A static field was applied to the shaved backs of the animals via an insulated metal electrode connected to 7 kV. After 8 days at 3 hr/day exposure, the animals exhibited an increased respiration rate, increased hair growth, and a diminished rate of skin epidermic-cell mitosis in comparison with the controls. These effects could be achieved with both a positive and negative field. The authors attributed these findings to the dependence of the mitosis rate on the level of the transmembrane potential (Cone, 1971), which was influenced by the exogenous electric field. At such field intensities—at an assumed electrode spacing (on which no data were given) of 1–5 cm, the field intensity would be approximately 140–700 kV/m—it is entirely possible for alignment effects to occur in dipole structures in the physiologic substrate. A more important factor, however, would appear to be the mechanical vibrations of the animal body caused by respiratory movements, and the resulting alternating fields of the same frequency which are set up in the static field. Assuming a mechanical rise and fall of the animal's dorsal surface by about 0.5 cm during respiration and a respiration rate at rest of about 2 Hz, the animals (or at least the dorsal areas investigated) were exposed to an alternating field of the same frequency, and to periodic changes in field intensity on the order of 250 kV_{pp}/m.

Rivière et al. (1964, 1965) and Pautrizel et al. (1966, 1969, 1970, 1975) first reported in 1964 on rats that showed a complete remission of tumor growth when irradiated with alternating magnetic and electric fields. The field conditions employed in these studies consisted of alternating fields at frequencies of 375 MHz–100 GHz or 15 MHz–300 MHz. The maximum field intensity was 620 G, the duration of exposure 40 min/day. Whereas the infected control animals all died between the twenty-second and thirtieth day, the experimental animals survived and recovered completely. If the electromagnetic therapy was not initiated until 14 days after the start of tumor growth, a daily irradiation of 40 min was no longer sufficient. In this case, only full-day irradiation could still ensure the survival of the animals. The same authors later obtained identical results in rats and mice with lymphosarcomas. In further experiments, control mice infected with trypanosomes dies within 4 days. The experimental animals were each exposed to the aforementioned field conditions for 12 hr/day and survived; by the fifth day, no more trypanosomes could be detected in their blood. The same results were obtained with rabbits. When maintained on a high-cholesterol diet, these animals showed a substantially lower hyperlipemia under the influence of the field than under control conditions with no field.

A 6-hr exposure daily to the conditions described above nearly doubled the animals' production of antibodies. At the same time, the authors were able to show that the multiplication of trypanosomes in the so-called diffusion chambers that had been implanted in the experimental animals remained unaffected by the field treatment, thereby ruling out a direct action of the field on the parasites in the body of the host. Even though the field pattern employed in these experiments was of a considerably higher frequency than the other alternating fields discussed here, we are still dealing with nonthermal, nonionizing effects of electromagnetic fields, which produce an effect strikingly similar to the steady fields used by Fischer (1976a).

Field-specific Adaptive Response

During the course of the stress response, the alarm reaction (catecholamine secretion) is followed by the secretion of glucocorticoids from the adrenal cortex. This marks the beginning of the adaptive response, which represents a kind of convalescent phase in which the consequences of the severe stresses associated with the alarm phase can be relieved. The glucocorticoids have a mineralocorticoidal side effect in this process, resulting in an extracellular–intracellular shift of electrolytes. According to Krück (1976), sodium ions flow into the cell during this process, while the intracellular potassium concentration is reduced. However, Altmann et al. (1976c) and Lang (1976b) reported that the potassium concentration in the erythrocytes is actually increased, while the sodium concentration is reduced, in rats during prolonged exposure to an intense 10-Hz field (Fig. 11-5). This increase in the activity of the sodium pump may be analogous to the aforementioned finding of an elevated membrane potential in frogs (Altmann et al., 1975).

Urine analysis in rats and mice under

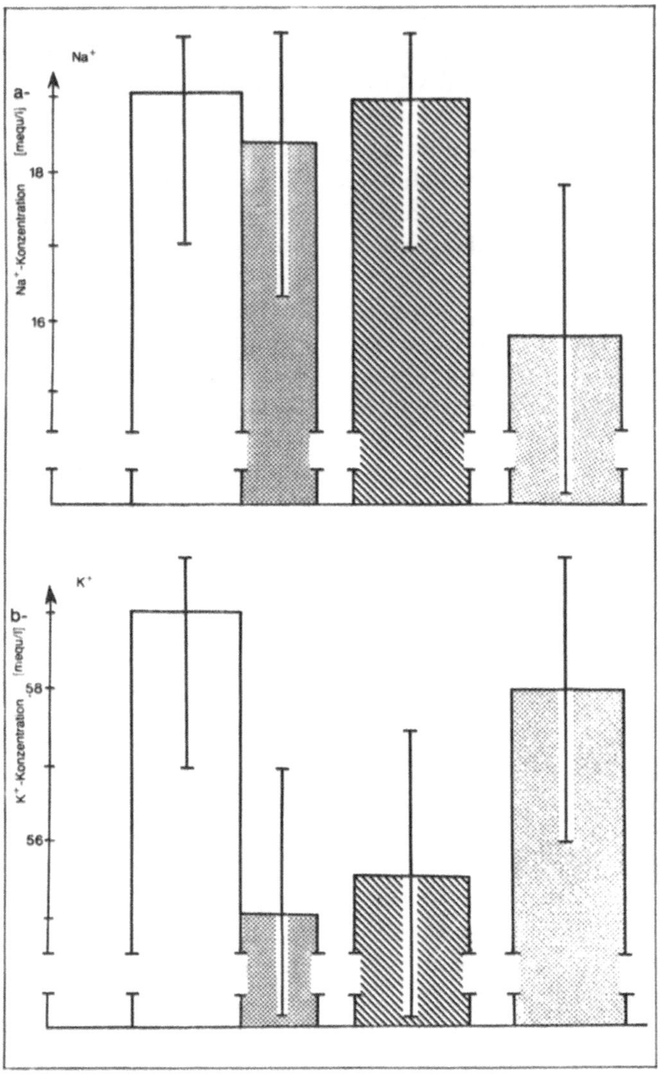

Figure 11-5. Changes in the sodium and potassium concentration in the erythrocytes of rats. Changes were measured under normal conditions with low biotropism (unshaded), with high biotropism (shaded), under Faraday conditions (hatched), and in a 10-Hz square-pulse electric field (lightly shaded) with an intensity of 200 V/m, after an exposure time of 8 days. After Altmann et al. (1976c).
Key: a-Na^+ concentration (mEq/1); b-K^+ concentration (mEq/1)

electric field conditions directly confirms the intracellular–extracellular electrolyte shifts: the urine of rats in strong and long-lasting 10-Hz electric square-pulse fields shows a high sodium concentration, while potassium is retained. These results can be explained only by a diminished blood concentration of mineralocorticoids, which should increase in order to compensate for the aforementioned mineralocorticoidal side effect of the increased glucocorticoids. We are thus dealing with a response of the mammalian organism that is independent of the general adaptation syndrome of the stress response, and which can be considered a specific response to the action of electric fields. Another specific response is a decrease in the secretion of ADH (antidiuretic hormone). The water content in the blood and tissues of mice and rats is diminished, and the excretion of urine is increased (Lang, 1976b; Altmann et al., 1976c; Altmann and Soltau, 1974). The weight decreases described by Marino (1975) and Knickerbocker et al. (1967) also represent nothing more than increased clearance in animals exposed to electric fields. An elevated osmotic pressure accompanying the diminished water content of the blood was confirmed by Klingenberg et al. (1976a,b). Lang (1976b) suggested earlier that this behavior may stem from a reaction mechanism that would enable the organism to adapt to varying exogenous electric field conditions.

According to this hypothesis, the biologic organism represents a good electric conductor in an electric field and therefore attracts lines of force. The electric force acts first on the outermost portions of the skin. From the outside inward, there follow the various layers of the epidermis. The outer layer of the corium, the stratum papillare, lies at an average depth of 300 μm. The total thickness of the corium, including the stratum reticulare, is 1–2 mm. Below this is the subcutis. If we are seeking skin structures that are capable of perceiving electric field effects, we must first examine the macromolecular structure of the tissue in more detail.

The cellular structures which underlie the layered structure of the skin are surrounded by the basic substance of the interstitial space, which has a relatively large volume in the connective tissues in comparison with other tissues. Free nerve endings, nerves that terminate in the interstitial space, are still found in the region of the stratum lucidum. At these endings the intracellular region of the nerve axon is separated from the extracellular, interstitial milieu only by the cell membrane.

Pischinger (1969) made an early reference to the significance of this interstitital peripheral region as the site of important regulatory processes. The role played by the skin as the most important sense organ for exogenous influences can thus be correctly interpreted only if the connective-tissue cell-milieu system is regarded as a "synergetic system of cell, capillary and nerve with the common field of action of the intercellular fluid" (Pischinger, 1969).

Thus, a physiologic action can hardly be ascribed to vibrations excited in electrically charged particles in the interstitial fluid by exogenous fields, if this additional motion of the ions or charged molecules simply competes with Brownian movement as an additional motional impulse.

For electrical oscillations lying in the VLF range of frequencies, however, Ludwig (1974) points to the effect of quasi-resonance absorption by the collagen molecules of the interstitial peripheral region. Exogenous fields excite the π-electrons of the peptide bonds in the filamentous polypeptides to vibrations whose frequencies lie in the Debeye range of the crystalline phases of these macromolecules.

Busch and Busch (1965) found that the slightest stimuli, whether in the form of pressure, moisture, chemicals, temperature or electric fields, elicit local responses in the skin areas affected. Their investigations also showed that the electric conductivity of the skin is reduced in areas to which an electric field is applied. Taking into account the calculations and measurements of Ludwig (1971) on the antenna effect of collagenous fibers, the observed increase in tissue conductivity can be explained as follows, based on the molecular structure of the peripheral interstitial space.

The skin is capable of compensating for local field effects. Measurements by Reiter (1960, 1962) of connective-tissue pH in electric fields confirm this. The polysaccharides of the interstitial connective tissue are characterized by strongly polar side groups such as SO_3H and others, which may be in a free state (due to dissociation) and exert an acidotic action. But the polymerizing effect of exogenous electric fields reduces the number of free acid radicals, and the pH increases. This alkalization of the tissue was observed (Reiter, 1960,

1962) as early as 1953 during the exposure of guinea pigs to artificial square-pulse fields, even though no explanation of this phenomenon was possible at that time. The impairment of free diffusion in the extracellular region of the skin tissue leads to an increase in the membrane potential at the free nerve endings in the connective tissue, either via the indirectly resulting decrease in the chemical activity of extracellular sodium ions, or through direct concentration changes in the clusters of the polymeric network. The further conduction of the resulting impulse activates the central regulatory mechanism (see also Lang, 1976b).

Under natural conditions, atmospheric electric field forces act on the organism without restraint, and the equilibrium of the polysaccharide–glycoprotein network is shifted toward polymerization. This situation represents the normal state of affairs, to which the organism has become accustomed. But natural or artificial Faraday conditions evoke a local skin response that results in increased dissociation of the polymeric structure of the basic substance in the peripheral interstitial space and, with it, an increase in tissue conductivity as a compensatory effect. The excitation of the free nerve endings is diminished at the same time, however. Preliminary investigations show that the blood vasopressin concentration is increased in rats subjected to Faraday conditions. It must be presumed, therefore, that the diminished excitation of the free nerve endings results in a higher secretion of vasopressin. This leads to the significant water retention observed in mice, guinea pigs and rats under Faraday conditions. The sodium concentration in the blood of the rats is significantly increased by 5.5%, and the potassium concentration is reduced by 7.3%. These concentration shifts are also found in the erythrocytes, according to Lang (1975).

The accumulation of water in the tissues results in a general "dilution effect." The erythrocyte count per unit volume is decreased, and the hemoglobin level and protein content of the blood plasma are strongly reduced by the increased water content. As a result, the colloidal osmotic pressure in the plasma is insufficient, and a water efflux from the intravascular region to the interstitial region takes place. But the resulting increase in the water content of all the tissues promotes the dissociation of weak electrolytes, as represented by the mucopolysaccharides of the interstitial fluid with their polar side groups. This loosens the polymeric network of these structures, and the electric conductivity of the tissues is increased throughout the body. One effect of this process if a sensitization of the organism to exogenous electric factors, perhaps due to a resulting change in field distribution within the organism, but due primarily to the increase in molecular mobility. The net result is a compensation of the Faraday shielding, with the consequence that attenuated exogenous electric fields may have a stronger effect than they would normally.

Survey

Although it is very time consuming to become thoroughly informed on current scientific evidence regarding the biologic effects of electric, magnetic and electromagnetic fields, such a survey is nonetheless essential if we are to properly evaluate available results.

Problems in Experimental and Measurement Technique

Problems in measurement technique often arise from uncertainty with regard to experimental conditions. This is especially true for field experiments with mice and rats in cages. It is likely that the true field intensity acting on the laboratory animal has been accurately measured in only a very few cases. In most cases electrodes were mounted in cages, the voltage applied to these electrodes was balanced against the electrode spacing, and the resulting field intensity was computed in V/m. Aside from various sources of disturbances such as feeding racks, drinking-water containers and partitions in the cage, as well as furniture and apparatus in the laboratory, which all cause more or less strong inhomogeneities in an electric field, depending on the materials used, precautions always had to be taken to prevent the animals from touching the charged electrode. Depending on the frequency of the field used and the electric properties of the protective grate mounted between the specimens and the voltage-carrying electrode, such precautions prevented a considerable portion of the electric field from entering the actual experimental chamber. Moreover, the Faraday shielding for screening out atmospherics had to be grounded and thus tended to divert field lines from the measuring site or cage interior, thereby diminishing the effective field intensity in these areas.

The electric charges acquired by the animals through friction with wall surfaces and other animals are also important in this context. These depend, in turn, on the population density in the cage and the cleanliness of the cage walls, as well as the conductivity of the flooring. If the animals are standing directly on a well-grounded electrode, their bodies represent a "bulge of the bottom electrode" due to the high conductivity of their skin surface (cf. Brinkmann, 1976). If they are standing on dry, poorly-conducting material scattered over a grounded electrode, they represent conducting bodies in a capacitor field, and the field lines will tend to concentrate on them. In addition, the movements of the animals set up alternating fields, whose dominant frequencies are shown in Fig. 11-6 as a function of the number of animals present. If the animals rub against plastic walls, the walls acquire a negative charge, while the animals' fur acquires a positive charge. If the plastic walls are replaced by metal walls, only the mutual electrostatic charges of the mice will form electric fields. The same will occur with humans, depending on

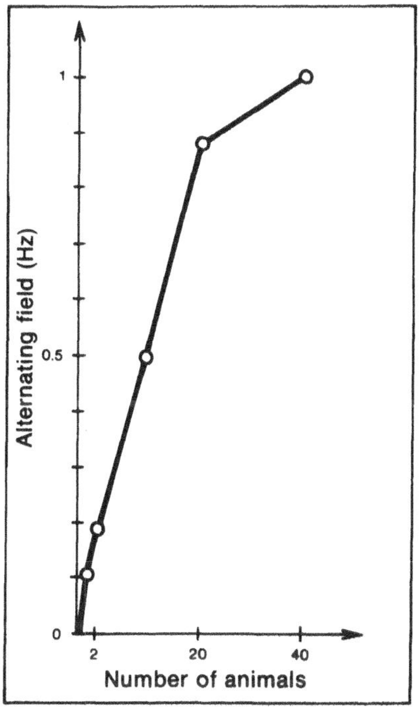

Figure 11-6. Frequency of alternating electric fields caused by the build-up of electrostatic charge due to friction associated with the movement of mice, as a function of the number of animals per cage.
Key: a-Number of animals; b-Alternating field (Hz)

the electrostatic charge on their clothing or surrounding furniture (Ludwig et al., 1976).

Besides these field imponderables, the reason for the conflicting results often published on the measured biologic effects of electric fields may also lie in the different responses of dominant and subordinant animals to the fields. According to Henry et al. (1974), dominant rats respond to stress situations with a much greater blood pressure increase and catecholamine secretion than animals that are socially subordinate. Becker (1975b) and Marino (1975) also describe how the acclimation of animals to unfamiliar cages in itself constitutes a stress situation. It generally takes about 2 weeks before the animals have become so accustomed to their surrounding that the new cage situation is no longer stressful to them.

A fundamental problem in field experiments, but one which finally has been solved, is that of the control situation.

Without doubt, a well-shielded Faraday cage represents the definitive control condition with respect to any electric field situation. In many experiments it is unclear whether the experimental cage with the electric field conditions was also located within a Faraday cage, so that the electric environment of the laboratory was excluded in both cases. In biologic terms, of course, the absence of a field also constitutes a change from the familiar environment and will also evoke a stress response. However, normal laboratory conditions are basically unsuitable as control conditions because of the presence of atmospherics. Moreover, the electric power supply, air-conditioning system and diverse measuring instruments create an artificial electric environment which is by no means constant. Based on general experience, therefore, the recommended course is to employ Faraday conditions as the control situation, while allowing the animals sufficient time (as described above) to become accustomed to these shielded conditions. This does not solve the problem of the necessary identity of the initial biologic and physical conditions, of course, but it does provide an exact and defined procedure for undertaking such experiments. Largely the same considerations apply to experiments in the area of human physiology.

Biologic Sensitivity to Fields

With regard to the sensitivity of the biologic organism, it has already been stated that, according to Verheijen (1958), a physiologically meaningful perception of environmental factors can be expected and sought only in the intensity range that corresponds to the naturally occurring range of this factor in the undisturbed environment. Wiltschko and Fleissner (1976) made an interesting discovery in this regard when they observed that robins, which are capable of natural geomagnetic orientation, found their destination in artificial magnetic fields only if these artificial conditions were no more than 25% below or 50% above the natural field intensities.

A comparison of the free-choice experiments of Altmann and Lang (1973) with those of Brinkmann (1976) also shows how small mammals naturally react to the influence of biologically adequate fields by remaining in an artificial 10-Hz field during active periods, and sleeping under shielded conditions, which correspond to their underground refuges in the wild.

There is some evidence to indicate that the sensitivity to such environmental factors can also be acquired. If mice are bred for 12 generations in a Faraday cage that screens out all electric fields up to 10 kHz by more than 99%, the subsequent generation is no longer capable of distinguishing between 10-Hz square-pulse field conditions and Faraday conditions, as normal mice are. The control animals, which had also been raised in a Faraday environment for the same number of generations but were exposed concurrently to a square-pulse field at 10 Hz and 100 V/m, showed a clear preference for the field conditions familiar to them, regardless of their activity phase (Lang et al., 1977). The question previously addressed by Persinger (1969) on the extent to which these abilities are acquired in certain "imprinting" phases was investigated further by Hartmann (1978). According to his findings, there are two types of mice with respect to nesting and sleeping behavior: those that prefer a stimulus-poor environment for these functions, and those that prefer a more stimulus-rich environment (called "Faraday types" and "field types," respectively). Apparently there was no adaptation to atmospheric electric fields during rearing. However, an abrupt change of the atmospheric electric milieu (from Faraday to field conditions, or vice versa) in newborn mice effected a change of "type" in at least some of the animals tested. Older animals appear to be immune to this effect.

Fole (1972) and Fole and Dutrus (1974) observed a marked response to strong electric fields only in animals that had been previously exposed to a weaker field of 7 kV/m. These results suggest that the initial exposure to electric field conditions may produce a sensitization. While this reaction might be characterized more as a learned response from the standpoint of behavioral physiology, one is tempted to compare it with the immune responses of the organism. The findings of Persinger (1969, 1974) on the behavior of rats exposed prenatally or during early life to magnetic fields, as well as experimental results in man, could also be interpreted along these lines. As in antibody formation, the true response of the organism does not become manifest until the next encounter with the environmental parameter in question.

At the same time, a comparison of available results on the physiologic effects of electric, magnetic and electromagnetic alternating fields in the range of the Schumann resonances shows that the form of the oscillations and the associated responsivity of biologic tissues to these oscillations also play a decisive role. Direct and reproducible biologic effects could be achieved even at low field intensities with strongly harmonic, square-pulse electric fields at 10 Hz; with 10-Hz modulated electrical oscillations with a carrier frequency of 461 MHz; with 10-Hz modulated electric fields with a carrier frequency of 10 kHz; and with harmonic magnetic field oscillations with a fundamental frequency of 10 Hz. By contrast, sinusoidal electric field oscillations at this frequency usually produced no effects, while sinusoidal magnetic field oscillations at this frequency produced only partly corresponding effects (Lang, 1976; Altmann et al., 1976a; Anselm et al., 1976; Fischer, 1976a,b; Jacobi and Krüskemper, 1975; Bawin et al., 1973; Ludwig et al., 1976; Maxey, 1976; Wiltschko and Fleissner, 1976; König and Ankermüller, 1960).

Earlier authors had already found a differentiated response to electric environmental factors in the broadest sense for different constitutional types (Ranscht-Froemsdorff, 1966a, and others). H. Schulz (1970) was the first to identify psychic features as determinants of the behavior of subjects with respect to electric fields. If these findings are compared with the results of Lang (1975), Altmann et al. (1976a) and Anselm et al. (1976), the correlation found in these studies between the responses of experimental animals and human subjects to electric fields and the biotropy of the corresponding meteorologic situation (cf. Sönning, 1976; Faust, 1977) can be interpreted only to mean that the organism under stress responds more strongly than that under no stress. Apparently it does not matter in this case whether the stress results from exogenous factors, such as weather conditions or other environmental elements, or if it is already present by virtue of certain endogenous peculiarities (hypotonics/hypertonics, vagotonics/sympathotonics, mental stability/lability, extroverts/introverts, etc.). A brief anamnesis with no regard for fundamental psychic personality structures on the one hand, or for meteorologic environmental stresses on the other, is clearly inadequate as a means of selecting comparable experimental subjects. A fatal error in this regard is made by experimenters who, concerned with obtaining the greatest possible homogeneity, select their subjects on the basis of good health and willingness to cooperate (these are mostly mentally stable persons!). For reasons mentioned earlier, however, it is in just such persons that one would expect to find a low sensitivity to electrical environmental factors.

Discussion

In light of the arguments presented, it is apparent that the oft-cited "contradictory results" of biometeorologic research on the effects of electric, magnetic and electromagnetic fields on organisms are actually comparable in many respects and, with careful evaluation, can be interpreted as stress responses, field-specific adaptive responses and frequency-specific responses.

The stress or stimulative response of the organism to such fields has been described most clearly by Becker (1975b), Marino (1975), Lang (1975) and Fischer (1976a,b), and was found to be evoked by 50-Hz fields as well as by static and 10-Hz fields of high intensity. Data on threshold values that characterize the boundary between stress and simple stimulation or no effect can scarcely be found in the literature. Field intensity data alone, with no regard for the temporal action of the field, are insufficient. It should be remembered in this context that if the stressing effect of such environmental parameters is confirmed for relatively strong or even weaker fields, a specific type of environmental stressor must be established. Unlike all other stress-producing agents in our modern environment, man-made fields are more or less all-pervading. Atmospheric emissions, water pollution, food contamination, etc., are still relatively local in their scope. By contrast, it is no longer possible, at least in living areas, to find a space that is both "natural" and field-free.

During the course of the general adaptation syndrome in response to stress, the secretion of glucocorticoids marks the onset of an adaptive response by which the organism adjusts to the environmental influence or attempts to regenerate itself. The field-specific adaptive response, as it is called here, differs from the stereotypic adaptive response of the stress syndrome in certain critical hormonal regulatory processes and characterizes, rather, the efforts of the organism to adjust specifically to changes in its electric environment. Like the stress response, this reaction appears to be very old phylogenetically, for the same impairments found in relatively highly developed small mammals (mice and rats) have also been observed in much more primitive organisms such as frogs (Altmann et al., 1977).

The frequency-specific response of the organism to electric, magnetic and electromagnetic fields is certainly the most interesting of the biologic responses. It is doubtful that this response is limited just to the ELF spectrum in the Schumann resonance range, on which most research has been done. In any event, the findings of Bawin et al. (1973), Maxey (1976), Ludwig et al. (1976) and Jacobi and Krüskemper (1975) not only provide an interesting confirmation of earlier findings by König and Ankermüller (1960) and Hamer (1965), but (disregarding the still-unresolved question of the necessary carrier frequency and nature of the acting oscillation) also demonstrate a direct coupling of central nervous mechanisms to exogenous electric, magnetic and electromagnetic factors.

Ludwig (1971) calculated the physical circumstances that enable weak exogenous electric and magnetic fields in the ELF and VLF ranges to act on molecular processes at the cell membrane. He was able to show that a hydrated calcium ion at a cell membrane is moved away from a membrane pore by more than one pore diameter (several angstroms, or 10^{-10} m) during one-half period by an alternating field at 0.5 Hz with an electric intensity of 10^{-5} V/m and magnetic induction of 0.05 mT. If this calcium ion was previously blocking the pore of a synaptic membrane, its displacement will open the pore, and an action potential can be produced. According to Adey (1975), the confirmation of these calculations by Bawin et al. (1973) shows a possible connection between field-excitable calcium ions at the membranes of brain neurons and "biologic switching molecules" such as certain prostaglandins. These compounds are active only in the presence of calcium ions and thus may represent a link between electric stimulation and metabolic processes in the cell.

Conclusions

Electric, magnetic and electromagnetic fields present a stress, but have therapeutic applications as well. Scientific results currently available on this applied aspect of biometeorology indicate the following:

1. The danger of electric stress occur-

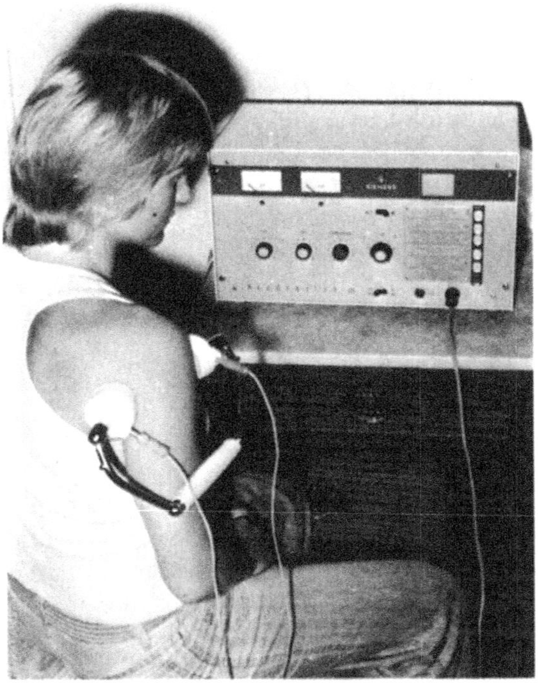

Figure 11-7. The use of electric currents and electromagnetic fields in medicine is on the increase. Here, direct currents and specific low-frequency alternating currents are employed for physiotherapeutic purposes.

ring in healthy persons exposed temporarily to high-voltage transmission lines of the usual dimensions is probably relatively slight, but cannot be ruled out with absolute certainty.

2. The effects of long-term exposure to strong 50-Hz fields should be investigated further in human subjects by neutral, experienced institutions under strictly-defined conditions and compared with results published by Russian investigators. Only in this way can final conclusions be drawn.

3. In persons with a preexisting physical or mental stress, particular care is warranted in judging the possible additional stresses associated with energy fields. Cumulative effects in organisms exposed to various stimulators and stressors are very likely to have adverse consequences in a particular case, described as "electrical stress" by Becker (1975b) and Marino (1975).

4. The frequency-specific response and field-specific adaptive response described above appear to be therapeutically useful in every case if field conditions are suitably optimized. The question of whether and to what extent these effects may be of general benefit to nonselected patients or subject groups cannot yet be resolved conclusively in accordance with the strictest scientific criteria.

Fields and Currents in Medicine*

We shall not present a comprehensive survey of this topic, for the application of electric currents in such areas of medicine as physical therapy (Fig. 11-7) and electronic diagnostics (ECG, EEG, etc.) is nothing new. However, it would appear that medicine is not exhausting all the possibilities in this area. Some potential is also offered by certain "lay methods," which might be useful in some areas and should be subjected to further research. We shall discuss

*Contribution of Dr. H.L. König

some possible applications of electromagnetic fields and electric currents that may not be generally known.

The medical significance of pulsating magnetic fields was investigated by Evertz and König (1977), with special consideration given to the questions involved in the physiotherapeutic application of such fields. The use of pulsed magnetic fields in medicine is the consequence of a development which has culminated in the discovery of biochemical and magnetic relationships in modern cancer research. The authors attempt to interpret the biologic and medical effects of these fields on biologic systems and discuss promising new therapeutic possibilities.

For example, magnetic fields are now being used for energy transmission, as means of generating electric potentials at the site of bone fractures—a technique which has produced amazing cures in the most stubborn cases. The technique developed by Kraus and Lechner (1973) involves the use of a coil arrangement (Figs. 11-8 and 11-9) in which a pulsed magnetic field of about 2.4 kA/m (30 Oe) is produced by means of an electric function generator. An implanted transducer (usually mounted inside the fracture pin) converts the magnetic energy into an alternating voltage of 0.3–0.5 V, which maintains a current of 1–2 $\mu A/mm^2$ at the electrode surface. The formation of new blood vessels, the most important prerequisite for successful bone healing, was observed in animal experiments. A survey of the technical aspects of this technique is given by Kraus (1978), and clinical data are reported by Lechner and Ascherl (1978). Excerpts from a complete photographic documentation of the course of congenital tibial pseudarthrosis in an 8-year-old female patient are shown in Fig. 11-10.

A similar application of magnetic fields for curative purposes is envisioned by Gleichmann (1974), based on the technique of Koeppen (1966). In contrast to

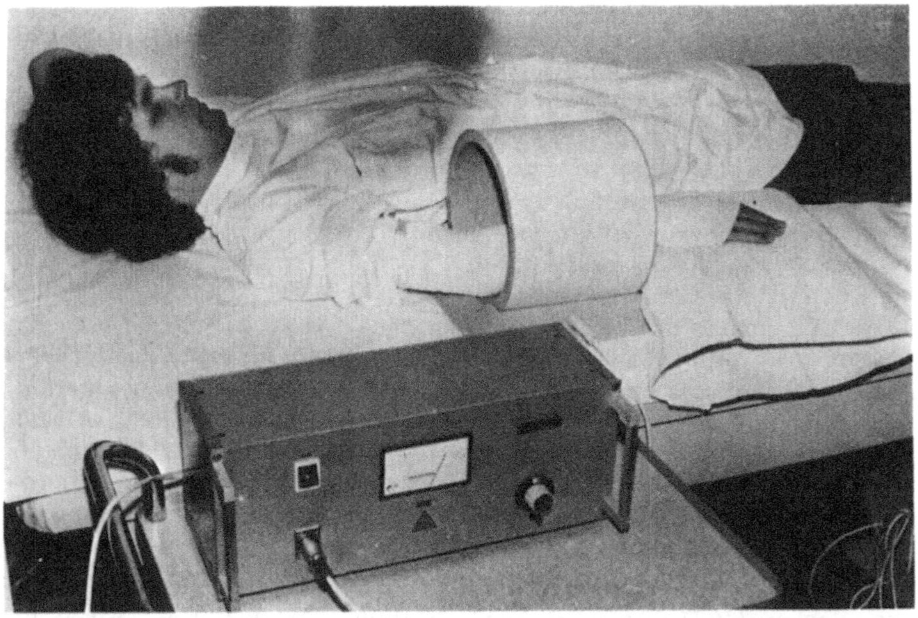

Figure 11-8. The magnetic field produced by a coil is used to generate electric currents directly at the site of bone fractures. The pins used to fix the fracture contain the apparatus necessary for converting the magnetic energy into the desired currents. Both the applied currents and the magnetic field promote the healing process.

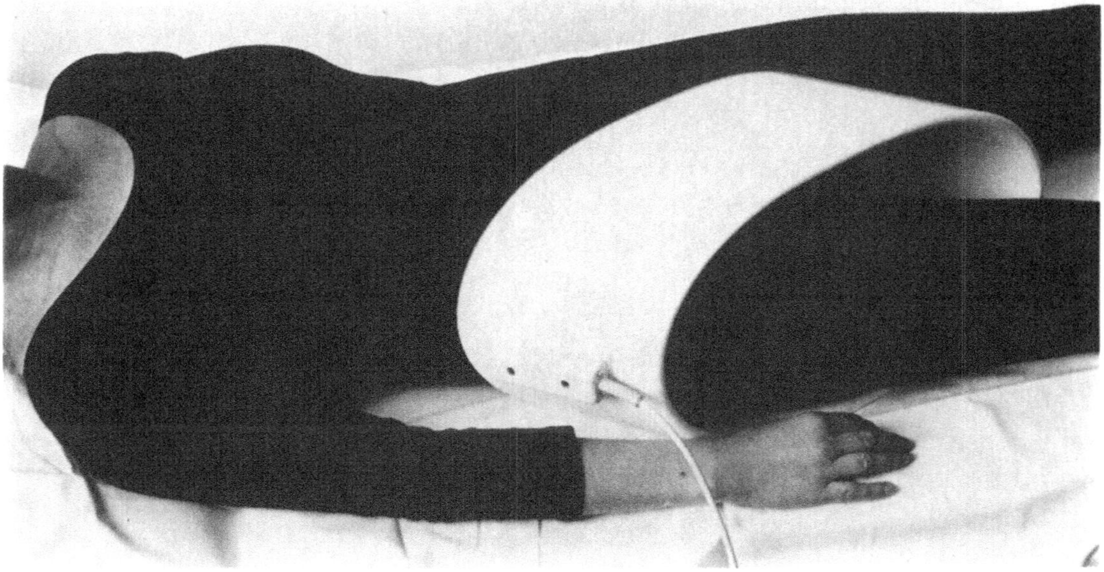

Figure 11-9. Magnetic field coil designed especially for the treatment of hip-joint arthrosis.

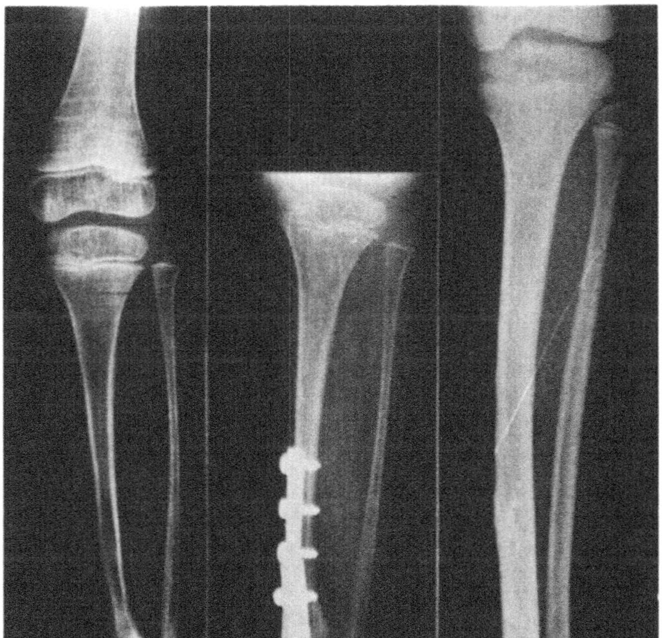

Figure 11-10. Typical course of congenital tibial pseudarthrosis in a patient born in 1969. Field therapy was begun in August, 1973 (29.8.73), and continued over an 8-month period for a total treatment time of about 500 hr. The photos above are an excerpt from a complete photographic record made throughout the period of observation.

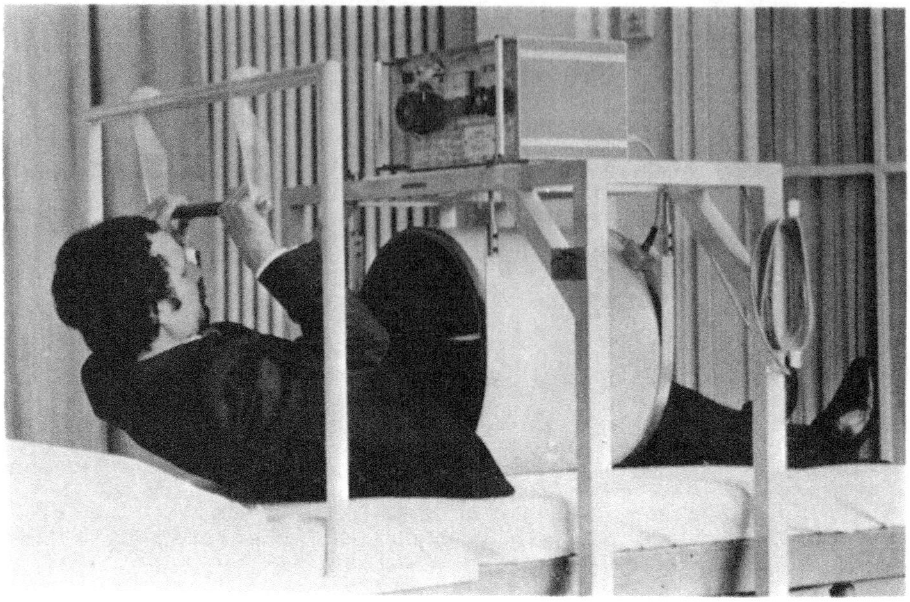

Figure 11-11. Large coil for administering magnetic fields at 1–100 Hz to the trunk region.

Kraus and Lechner (1973), he uses much larger coil arrangements (diameter 0.45 m), which can accommodate not only limbs, but the entire body. By means of a maximum effective current of 6 A, a magnetic field of 0.01 T (100 G) can be produced at the center of the coil whose time course corresponds to a pulsating DC field. This was accomplished by subjecting the 50-Hz alternating coil current to half-wave rectification, a technique also employed for currents in physical therapy (Fig. 11-11). It is claimed that this arrangement produced remarkable results in the treatment of cardiovascular complaints, renal insufficiency, changes in the vertebral bodies (intervertebral disks), hip-joint arthrosis and muscular atrophy.

Further experience and clinical tests are needed before conclusions can be drawn. In any case, the question arises whether the successes achieved by Kraus and Lechner through the direct application of current with electrode implants could not also be achieved without the use of implants. In other words, what part of the therapeutic effect is attributable to the current, and what part to the magnetic field?

The promising results achieved in fracture healing with magnetic fields led Birnberger (1973) to experiment with rats in order to study the effect of the field on severed nerve pathways. No curative results were obtained, but other marked effects were noted. The skin lesions (pressure sores) that developed on the part of the denervated leg which was in contact with the ground were noticeably less severe in the animals treated with the field. Moreover, a marked and unexplained muscular hypertrophy (increase in muscle volume) was observed in these animals.

Besides the magnetic field devices already discussed, smaller pocket units have also been placed on the market, as reported by Evertz and Ludwig (1977). These pocket devices have a variable frequency range, through which specific therapeutic effects can reportedly be achieved: 1–3 Hz for treating inflammatory processes; 4–6 Hz for sleep disturbances, nervousness and hypertension; 7–25 Hz for headaches and hypotension. Magnetic field-generating devices in the form of a rod are available for localized use. Ehrmann et al. (1976) report on clinical tests of such an electromedical pocket unit, which indicate that ELF fields have a stabilizing effect on cell membranes, while Ludwig (1977) discusses the biologic tolerance of such fields. The question of the possible physical and physiologic mechanisms of action of magnetic fields is addressed by Warnke (1978). He shows how changes in the membrane potential, with the attendant consequences, can be produced by means of induction, the Hall effect, the Lorentz force and the resulting physico-chemical effects.

"Electro-sleep therapy" has also been much discussed in recent times. Wageneder et al. (1973) state that this term is rejected by professionals, since the word sleep implies that the patient must fall asleep during the procedure, which involves the passage of an electric current through the head for therapeutic purposes. However, sleep is unnecessary for the treatment to succeed. The term electrotherapy or electroanesthesia is preferred, therefore.

Because the mechanism of action of the electric current passed through the brain is still unknown, the type and form of the current are selected on a purely empirical basis. Most commonly used are:

1. Different alternating currents with equal amplitudes but somewhat different frequencies, applied to the patient via two channels. Their interference yields a fluctuating alternating current;
2. Rectangular pulses superimposed on a smaller direct current; pulse frequency 100 Hz, pulse width 1 ms;
3. Sinusoidal alternating current, used primarily in veterinary medicine; frequency about 700 Hz;
4. A mixture of all frequencies between 10 Hz and 20 kHz (noise signal) with a limited amplitude;
5. A double-triangular pulse with a pulse frequency of about 60 Hz and a total width of 3 ms for both triangles;
6. A pulsed high-frequency current at 100 kHz with a pulse repetition frequency of 100 Hz and pulse length of 1 ms.

For cerebral electrotherapy, the current intensities are about 0.2–1.5 mA; for electroanesthesia, up to 50 mA. In insomnia patients, treatment with suitable currents on a short- or long-term basis has been successful in 85% of cases.

Busch and Busch (1966) made an interesting contribution to the use of physical methods of measurement in the study of human basal regulation. They report on the application of two known methods of measurement for such investigations, one electric and one thermal. Electric values were measured by means of an RC bridge circuit that operated with a current of 30–50 μA and a voltage of about 1 V at a frequency of 9 kHz. The impedance was measured by an active silver electrode with a diameter of 3 mm and a spring pressure of 100 g-force, as well as a broad, indifferent metal electrode that was held in the subject's hand. By applying the silver electrode to various points on the body, the impedance between these points and the indifferent electrode could be determined. The data obtained from such measurements, as well as from parallel radiation temperature measurements (by means of an IR-radiation measuring device), presented in overlay form, can give the physician important information on the patient's state of health based on deviations from normal values and symmetry characteristics. Apparently this technique makes it possible to study the function of the complex human regulatory system by physical methods of measurement, for mechanical and thermal stimulations of the human body surface evoke marked changes in conductance and radiation temperature, which can be detected by two entirely different physical measuring techniques and which follow parallel courses.

The technique of using an electric current to stimulate nerves and muscles, which is already in general use for the diagnostic and therapeutic treatment of neurologic problems, should also be mentioned in this context. For example, physiotherapeutic measures are employed in fracture cases with concomitant nerve injury. In order to preserve the affected muscle and protect it from atrophy, it is stimulated and thus activated by a direct current (up to about 60 mA) or alternating current (sawtooth wave with about 0.5 s rise time and abrupt fall, with pause equal to twice the rise time) as a substitute for the currents no longer delivered by the damaged nerves. Such currents are also suitable for diagnostic purposes. The functional performance of a nerve can be checked by employing a sawtooth current with various rise times, because only the healthy nerve is capable of responding to relatively short rise times (20–50 ms) during the transmission of artificial control impulses to the muscle. In the case of damaged nerves, this time is 2–5 s. The technique can also be applied for the purpose of muscular training, or for the treatment of pain attacks accompanying degenerative joint diseases (cervical vertebral complaints, back pain), painful shoulder stiffness, contusions, etc. Direct and/or alternating current (50 Hz or 100 Hz) in the range of 10–50 mA is best for this purpose.

By applying the appropriate salve to the electrode tip, current can also be used for the local introduction of medication. Sometimes a water bath is used instead of metal electrodes. Besides these techniques, which employ direct contact to produce a closed electric circuit, there is also shortwave and microwave therapy (diathermy), used for the localized heating of body tissues.

Apparently it is also possible to accurately locate the specific body points known from Chinese acupuncture by physical methods of measurement, for these points are distinguished from their surroundings by marked changes in electric resistance (Fig. 11-12). Bergsmann and Woolley-Hart (1973) conducted an exhaustive study of the differences in electric skin conductivity between acupuncture points and adjacent skin areas:

> All voltage-current measurements show a point of inflection where nonlinearity occurs. The current value of this inflection is termed the "threshold current." It is shown that the threshold current over the point Lung 9 on the cavity side and Liver 9 on the right have significantly higher values. This is

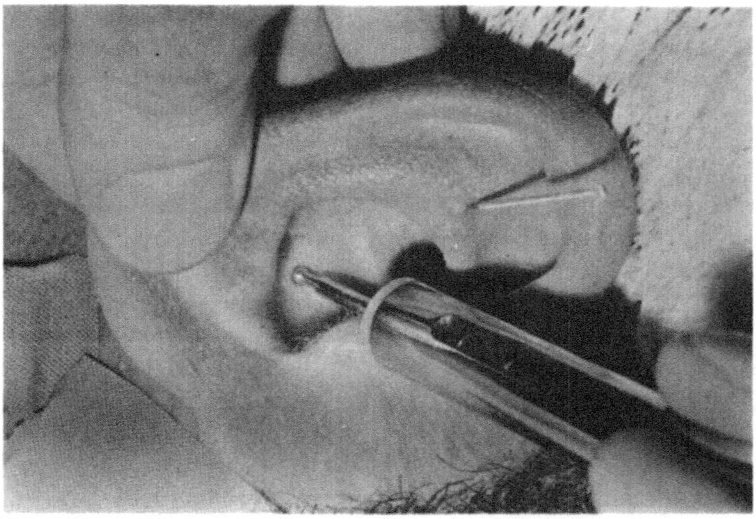

Figure 11-12. Special electrodes with a constant contact pressure make it possible to accurately locate acupuncture points by measuring electric resistance.

considered to be an indication of altered permeability, and recent mathematical reassessment of curves suggests semiconductor characteristics, thus reflecting a change in membrane characteristics. It is felt that these studies demonstrate a functional change in the skin of acupuncture points relevant to diseased organs.

Voll (1966) in particular has made many valuable contributions in the area of "electroacupuncture." According to Voll, the deviation of measured resistance values from a certain empirical norm serves as a diagnostic index. Moreover, by introducing electric currents at 0.9–10 Hz through the acupuncture points or through hand-held electrodes, the abnormal resistance values can be restored to their original value, presumably with a simultaneous therapeutic effect. Because carcinoma tissue is reportedly inactive bioelectrically—even external, healthy tissue is bioelectrically dead in the case of stomach cancer, for example—it would seem possible to treat such a disease with currents introduced from the outside. The anticipated therapeutic effects must first be confirmed by experimental results, however.

Although electroacupuncture in particular has found acceptance in classical medicine for some years (several thousand heart operations have been performed in West Germany alone using this technique), both acupuncture and electroacupuncture remain a subject of controversy in the medical community. Competent experts expressed their views on the subject in the medical forum of *Selecta* (1976). The problems involved in both forms of acupuncture were discussed in detail, supplemented by the observation that electroacupuncture must not be confused with the electric stimulation of individual points, which is particularly true for acupuncture analgesia.

The German Cardiac Center in Munich, Germany, has also performed over 1000 open-heart operations using electroacupuncture instead of anesthesia. Conventional pain-relieving agents can be dispensed with even postoperatively, for the effect of electroacupuncture persists for 12 hr on the average. A pulsed electric current is delivered to the patient simultaneously through electrode pairs applied to the ears, throat and forearms (Fig. 11-13). The time course of the current (Fig. 11-14) includes

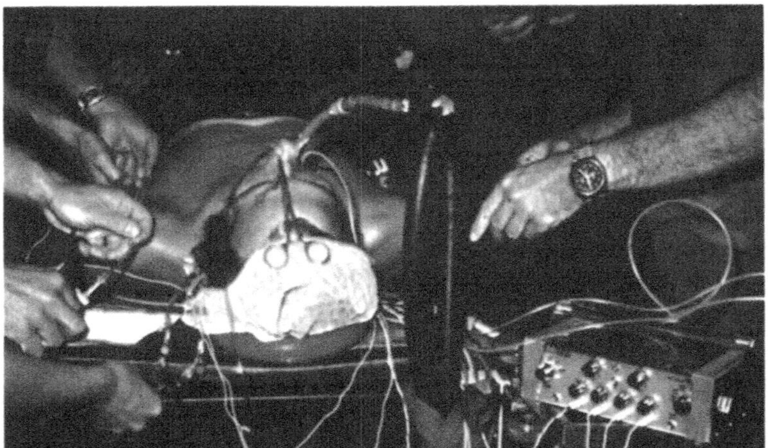

Figure 11-13. Open-heart surgery is made practically painless by electroacupuncture analgesia. The figure shows a pulsed-current device with four terminals for the electrode pairs, manufactured in the People's Republic of China.

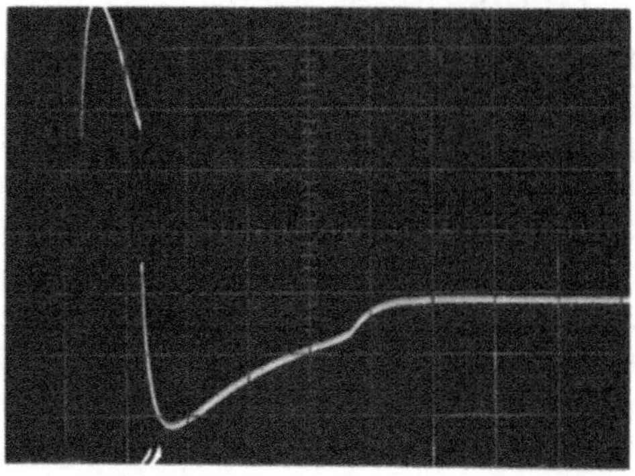

Figure 11-14. Time course of the current (~70 mA, ~ 13 Hz) employed in electroacupuncture analgesia.

the following specifics: voltage at the ear electrode, $80V_{pp}$; peak value of current, 60–80 mA; pulse repetition frequency, 12–15 Hz; pulse duration, positive phase, 0.5 ms, negative phase, 1.7–2 ms; negative amplitude about $^1/_3$ of positive peak value. The I/U ratio is exponential, and the analgesia is dependent on the pulse shape.

The mechanism of action of electroacupuncture is still largely unknown. Presumably a link is established with the body interior and the thalamus via sympathetic pathways. The externally applied current coud act via sympathetic nerve fibers to cause a blockage (pain barrier) of the slower pain signals at synapses in fast-conducting pathways.

All attempts to identify acupuncture points by purely anatomic means have failed. No large neural, vascular or lymphatic channels, nor any anatomic structural changes such as cell-lumen changes, are detectable in vivo or in vitro. Schuldt (1974a) suggests that the points may have to do with certain concentrations of force field lines, which are not connected with conductive pathways, but are determined by overall body morphology.

Vill and Jahnke (1975) employed a more technically refined method to investigate the diagnostic problem in particular. They recorded a pulsed current with a 10-Hz repetition frequency which was interrupted every 4 s by a direct current measuring a maximum of 10 μA. Eight such current–pulse intervals, alternatley negatively and positively poled, were passed successively through various parts of the body. This gave rise to a measurable return current in the body, which must be regarded as a type of discharge process like that occurring in storage batteries (as opposed to a capacitor discharge). This technique is capable of revealing pathologic changes in man, including, (1) meningeal and brain damage, (2) chronic foci in the head and neck region, (3) chronic disturbances of the intestine–lymph system, (4) chronic disorders of individual organs, (5) chronic erosional processes and intoxications (poisonings), (6) acute inflammatory organic disturbances, (7) acute severe shock phenomena in the organs, such as myocardial infarction.

Schuldt (1974a) conducted corresponding long-term experiments with the device developed by Vill and Jahnke (1975) from the fundamental work of Voll (1966). With regard to the measurement of electric conductance in test subjects, he reports:

1. An individual mean was present, depending on the overall reactive state of the body. This mean fluctuated in both directions due to various factors, such as food intake, oxygen supply, general state of health, temperature, inflammation, diurnal variations, etc.

2. Previous observations in a group of 10–16 subjects showed that overall fluctuations of conductance (besides individual fluctuations) tend to be in the same direction on certain days (in certain meteorologic situations). A decrease in the conductance values of the group as a whole coincided with the occurrence of phantom pains in amputees and with the exacerbation of joint pains in rheumatics.

Based on these measurements, Schuldt (1974b) proposed his simultaneous multiparametric analysis technique, in which several parallel series of measurements are made in a sufficiently large subject group over long periods of time. The results allow the necessary inferences to be drawn on a comparative basis with regard to the functional relationship between the external stimulus and biologic activity. This technique may well make a significant contribution toward the recognition of cause-and-effect relationships on a statistical basis.

Lampert (1968) reports on another interesting application of electric current in connection with the prevention of embolism and the mobilization of thrombi. Most experience in this area has been with phlebitis, venous thrombosis and thrombophlebitis; the results in relieving the obstruction were very good in 85% of cases, good in 12%, and poor in only 3%. A secondary indication was found to be the rapid absorption of massive blood effusions, which also represent subcutaneous thrombi. Coronary thrombosis could also be treated by this method without the need for coronary angiography. The current source was a power pack adjustable to any setting from 0 to 30 V, which delivered a current of about 20–30 mA through electrodes to the endangered zones. Starting at zero, this current must be increased very gradually (over a period of several minutes) to its final value, care being taken that the polarity of the direct current is unchanged when treatment is repeated. Lampert (1968) attributes the success of the treatment to the detachment of the thrombus on one side only by the galvanic current. The concentration of OH ions on the anodic side leads to the dissolution of fibrin and blood corpuscles, thereby reopening the vessel.

Further important results are expected from research into electric brain activity. For now we shall mention in passing that Dewan (1964) was able to show that communication can be accomplished by the voluntary, conscious control of EEG currents. Apparently the alpha rhythm of the brain could be voluntarily controlled sufficiently so that the EEG could be modulated with Morse code signals. According to the experimental results, any message desired can be transmitted to a recording instrument in this manner. It is still an open question whether information flow can be effected in the opposite direction, although new prospects are opened up by the utlilzation of electromagnetic fields.

Research into the production of subjective light patterns, or "phosphenes," by means of galvanic currents is only indirectly related to the biologic activity of electromagnetic fields, but it is nevertheless interesting enough to be discussed in closing. Knoll et al. (1962) mentioned that such phenomena were described as early as 1816 by Volta and 1819 by Purkinje. They produced "galvanic light patterns" (arcs and bands) with a 20-V Volta cell.

To produce this phenomenon, Knoll et al. (1962) attached two silver electrodes wetted with Ringer's solution to the foreheads of test subjects. To these electrodes was applied a rectangular voltage (pulsed direct current, duty factor of 1:4 or 1:1) of 0.5–3.5 V. The subjects—20 psychiatric patients and 10 polytechnical students —were then able to see patterns that appeared as a function of frequency. The patterns described could be produced in a third of the persons tested without any operative intervention. The following findings were reported:

1. Some subjective patterns occasionally changed their form during constant stimulation. The mean stimulation frequency for many patterns also varied with time.
2. Many of the patterns described could also be produced by nonelectrical stimulation, such as a mechanical, acoustic or chemical shock.
3. All of the patterns sketched by the subjects were abstract in nature (mostly geometric shapes).
4. The number of patterns that could be produced per subject was greater among the psychiatric patients than in the healthy group.

Phenomena similar to the phosphenes produced by pulsed electric stimulation (head electrodes) can also be triggered by

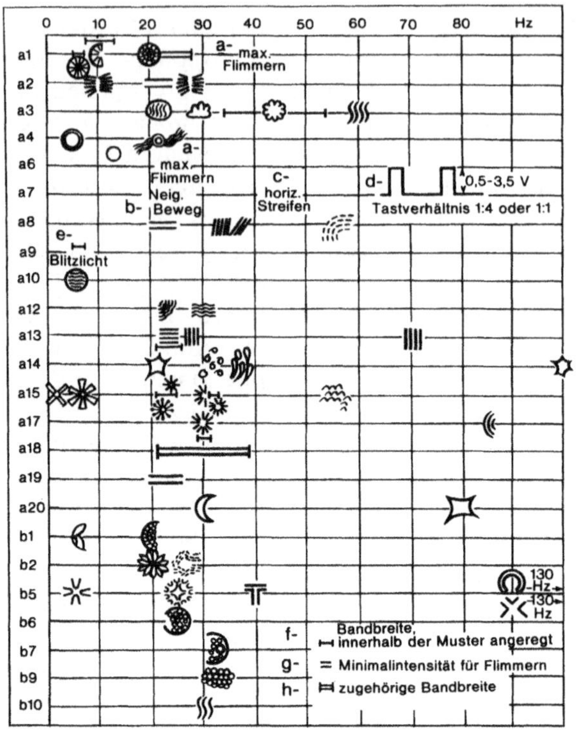

Figure 11-15. Spectrum of individual patterns clearly seen by various subjects when weak electric currents at the indicated frequencies (0–130 Hz) were passed through their heads. Subjects a1 through a20 were psychiatric patients, subjects b1 through b10 were students. After Knoll and Kugler (1959).
Key: a-Max. flicker; b-Tilting movement; c-Horiz. bands; d-Duty factor 1:4 or 1:1; e-Light flash; f-Bandwidth within which patterns are excited; g-Minimum intensity for flicker; h-Associated bandwidth

magnetic fields. Seidel et al. (1968) report that a stimulator coil placed near the head produces phosphenes by magnetic induction; the necessary induction is approximately 0.02–0.1 T (200–1000 G) in the frequency range of 2–50 Hz. All of the patterns found in the experiments are identical to the geometric phosphenes that were produced in the electric stimulation experiments described above (Fig. 11-15). However, their percentage frequency differs from that of the electric patterns. A consideration of the conductivity distribution in the cranial structure suggests that an increase in the density of induced stimulating currents near electrically conductive media (such as the eyeballs and thus the retina) may be responsible for the magnetic excitation of phosphenes; and this theory agrees with similar considerations of Schuldt (1974b) mentioned earlier.

12

Electromagnetic Energies in Man's Immediate Environment

Electric Power Supply Facilities

The increased utilization of electric energy in man's immediate environment raises the urgent question of whether such energy might not cause changes, perhaps even radical ones, in human living conditions. High-voltage transmission lines in particular may seem menacing to the layman and generate fears that their presence, especially near homes, may have adverse effects on health. But health questions are raised not only by high-voltage lines for 50- or 60-Hz power current and 16 $^{2}/_{3}$ Hz railway current (in central Europe), but also by electric railways themselves (15,000 V, 16 $^{2}/_{3}$ Hz), for very strong currents arise in the overhead contact lines at irregular intervals, combined with a return current which may flow over paths other than the rails. Because these rails are not specifically insulated from the gound, substantial stray currents may arise at points many kilometers distant from the rails (Fig. 12-1). Household power current, to which modern man must live in close proximity, is yet another factor, even though the voltages in this case are substantially lower (110 V, 220 V or 380 V). Other sources of electromagnetic fields, such as telephones, radios, television sets, etc., will be disregarded for the time being.

Concerning the effects of high-voltage transmission lines, it must be understood that there are various gradations ranging from harmless, biologically active and health-damaging to lethal. Differences in the initial state of each individual, other environmental influences, and the location factor make each case unique. General statements can be made only on a statistical basis, in terms of probabilities (see also Chapter 7).

Without special large-scale investigations, therefore, it is impossible to accurately assess the health hazard posed by high-voltage transmission lines. At a time when man is gradually beginning to develop an environmental awareness, the problem of possible biologic effects of high-voltage lines and related facilities in man's immediate environment can no longer be ignored.

High-voltage lines pose another, by no means trivial problem from the standpoint of environmental quality, for they represent both an eyesore and an acoustic annoyance to persons living nearby. Under certain wind conditions the humming and whistling noise produced by transmission lines is not

Figure 12-1. All electrically powered systems emit a more or less intense electric and magnetic radiation, or field. High-voltage transmission systems and electrically powered railway systems are key examples. But streetcar and subway systems in man's immediate environment also alter natural electromagnetic conditions without our being aware of it. The biologic activity of such fields is an established fact. Their harmfulness remains to be determined.

only annoying to noise-sensitive persons, it subjects them to nervous stress as well. Even more annoying are the corona discharges that occur on high-voltage lines under certain weather conditions (icing, fog, high humidity), because they create a loud, continuous, monotonous hum that is particularly irritating in contrast with the calm which otherwise prevails under such conditions.

High-voltage lines can also attract lightning discharges. That this property is an advantage (according to some experts) because it enables such lines to divert lightning from adjacent buildings is questionable, because the number of lightning strokes is simultaneously increased, with unpleasant consequences for persons in the vicinity.

We shall focus our attention, however, on the electromagnetic aspects of high-voltage transmission lines.

Physical Factors

Power-line fields. By virtue of its operating voltage (conductor–conductor, conductor–ground), a high-voltage transmission line generates an electric field about it whose intensity depends on the operating voltage of the line as well as on the overall geometry of the line (spacing of the wires and the environment in which the electric field is of interest) and of the observation site itself.

As long as current is flowing through the line, a magnetic field also exists whose intensity is determined mainly by the magnitude of the flowing current, the arrangement of the conductors and the properties of the environment. The technical aspects of this problem were investigated extensively by Schneider et al. (1974). Since the voltage and current in the conductors have a frequency of 50 or 60 Hz (or 16 $^2/_3$ Hz for railway current), this is also true for the fields produced by them. In addition to this fundamental frequency, however, higher-frequency signal components are also present in high-voltage transmission lines. The amplitudes of these harmonics are generally different for voltage and current.

Due to insulation conditions, it is to be expected that certain stray currents will flow in the ground in the vicinity of high-voltage transmission lines or between them. In addition, inductive or capacitive coupling between the wires and ground can lead to earth currents whose frequency spectrum generally differs from that of the line currents (or voltages), and which result in the formation of electric and magnetic fields that are essentially independent of the conduction system. Corona discharges, moreover, give rise to signals whose spectral components are essentially independent of the frequency of the alternating line current and normally lie in higher frequency ranges.

Finally, if high-voltage wires are moved by wind forces, changes occur in the overall line geometry which produce corresponding changes in electromagnetic conditions. This means that the electric and magnetic fields associated with the wires undergo an additional amplitude modulation in rhythm with the mechanical motion of the wires.

Another factor to be considered is that the voltage and current in high-voltage lines do not follow a strictly sinusoidal curve with respect to time. Transitory voltage or current surges (or peaks) occur, which set up an additional pulse-like field.

High-voltage lines are also capable of transmitting over great distances certain signals they have picked up by capacitive, inductive or galvanic means (from lightning strokes, radio transmitters, television transmitters, etc.). Effects from structural members, especially metal ones (poles, cables, etc.) on the field distribution of high-frequency fields (reflections, standing waves, etc.) produced mainly by commercial transmitters (radio stations) and their consequences are mentioned in passing.

Natural fields. The natural electromagnetic climate is also significantly altered by such facilities, for all natural electric and magnetic fields are strongly disturbed in the vicinity of high-voltage transmission lines. This is especially true for static fields and fields up to the kilohertz range, which, due to their correspondingly long wavelengths, have a field distribution that is comparable to the of static fields. Such low-frequency natural electromagnetic fields and their biologic significance have already been discussed elsewhere in detail.

The disturbance of natural electromagnetic and static fields also results in a disturbance of natural air ionization. The shielding or diverting effect of the grounded metal towers and high-voltage wires in particular leads to substantial changes in the static electric field, which is important for air ionization. Because ionized air consists of electrically charged particles, upon which a force is exerted by the atmospheric electric field, a close connection exists between the state of air ionization and the technical facilities that disturb the natural atmospheric electric milieu. Even the presence of a sufficiently intense 50- or 16 $^{2}/_{3}$-Hz alternating field is enough to cause a significant shift in the natural air ion balance. As measurements have shown, the space affected by the electric field becomes deionized, that is, the air ions are drawn off by the field-generating electrodes regardless of their polarity. This could be explained by the fact that beyond a certain field intensity, and if the alternating frequency is not too high, the ions could acquire very large amplitudes of motion in synchrony with the alternating field. The ions could be carried to the electrodes resulting in an ion depletion of the surrounding air. Owing to the tendency of charges to distribute themselves uniformly in space, the deionized space near the electrodes is continuously replenished by new ions from outside, which in turn are diverted by the electrodes. As a result of this process, a space filled with a relatively intense ($\geq$ 5 kV/m) alternating electric field (60 or 50 Hz) could become deionized with time. Other possible explanations for this effect are an increased recombination rate or the formation of large ions through enhanced aggregation caused by the alternating field.

The biologic importance of air ionization has already been discussed. Thus, the possibility of such a significant change in natural air-ionization conditions must not be ignored when the biologic effects of such facilities are investigated.

Field intensity values. The biologic significance of a given stray field depends in large measure on the intensity of the field as compared with intensity values that are known to be biologically active. On-site measurement is the most reliable method of determining the magnitude of such fields. The field intensity in the vicinity of a high-voltage transmission line apparently depends on several factors. The electric field intensity is determined primarily by the operating voltage of the line, which may exceed 100,000 V. The magnetic field intensity depends mainly on the current flowing through the conductors, or the operating current. Unlike the operating voltage, which should remain constant, the operating current—and, with it, the magnetic field—is subject to sizable fluctuations during the day. The field intensity is also strongly influenced by the overall geometry of the transmission line, as mentioned earlier (Schneider et al., 1974). But the properties of the space between the lines and the point at which the field intensity is of interest are also important. Both electric and magnetic fields can be screened out or greatly attenuated by certain materials and their arrangement (e.g., large window surfaces, steel-reinforced concrete, sheet-metal roofing). The distance from the high-voltage line is the most important determinant of field intensity, however, because the intensity of both electric and magnetic fields falls off sharply with distance, even under undisturbed conditions. Despite these complicated interrelationships, it is still possible to predict the field intensity values near high-voltage transmission lines with a fair degree of accuracy for certain conditions, even though on-site measurements should always be made in specific cases. The near-field situation of the undisturbed electric field is shown for various operating voltages and tower configurations in Fig. 12-2. The field intensity values at a distance of 10–20 m and a height of only 0.5 m above the ground are always in excess of 100 V/m, in the case of a 110-kV line. This intensity is several orders of magnitude greater than that which is now generally known to be biologically active.

The displacement current caused by persons standing upright in an electric field is approximately 15 μA per kV/m, and thus has a strength of 1.5 μA at a field intensity of 100 V/m (corresponding to 0.5 m above the ground).

The critical distance is somewhat smaller in the case of the magnetic field, although actual conditions may deviate considerably from estimates if earth currents are present. As mentioned earlier, the stray currents from electric railways play a special role in this regard. The following important points can be derived in connection with the electric facilities discussed thus far:

1. Natural electric and magnetic conditions are strongly disturbed in the vicinity of such facilities.

2. The principal source of artificial fields is high-voltage overhead transmission lines and indoor (household) wiring. Depending on the environment and other factors, their intensity ranges from extremely weak and therefore negligible values, to values corresponding to natural environmental conditions, to values several orders of magnitude greater than the intensity of these natural fields.

The "final report" of Bridges (1975) on the biologic effects of high-voltage fields, 315 pages long, indicates the importance of this question on an international level. The report contains a comprehensive bibliography and a survey of ongoing work, with emphasis on the following areas: (1) biologic effects of AC power-line fields between 45 and 75 Hz, (2) biologic effects of DC electrostatic fields, (3) effects of AC power-line fields and radiofrequency waves on patients with implanted cardiac pacemakers.

The material presented by Bridges is intended to assist biologists and engineers in planning with regard to the biologic effects of fields from overhead high-voltage power transmission lines and to guide the Electric Power Research Institute* staff and their counterparts elsewhere in the planning and conduct of biologic research. We shall not take the time to cite specific quotes from the report, as its principal contributions on the bioactivity of fields have already been treated elsewhere.

Biologic Factors

What indications are there that such alternating fields are biologically active or even harmful? Based on the current status of scientific research, it is useful to treat this question in two parts: (1) investigations dealing with the biologic effects of 50- and 60-Hz fields on animals as well as man; (2) investigations on the biologic effects of electromagnetic fields in adjacent frequency ranges, which allow corresponding inferences to be drawn.

Animal experiments. In a series of experiments in which microbes (*Listeria*) were injected into mice and guinea pigs, Odintsov (1965) investigated the immunobiologic effects of a magnetic field. The animals were exposed either once for 6.5 hr or continuously on 15 consecutive days to a 50-Hz magnetic field at 16 kA/m. A single exposure to the field influenced neither the lethal microbe dose, its distribution in the organism, the number of leukocytes or their phagocytic activity. Multiple exposure, however, caused a marked reduction in the natural resistance of the animals to the microbes; the number and phagocytic activity of the leukocytes were also reduced.

In similar experiments with 50-Hz fields, Lantsman (1965) observed phagocytic activity in mice under comparable conditions. A single field exposure had the effect of stimulating phagocytosis, while multiple exposure retarded it. Should it be shown that these results are applicable to humans, there would be far-reaching consequences concerning increased susceptibility to disease caused by exposure to excessive stray-field intensities.

The drinking behavior of rats in a 50-Hz high-voltage field was investigated by Spittka et al. (1969). They proceeded from the experiments done by Knickerbocker et al. (1967) with mice in a 60-Hz high-voltage field with an intensity of 160 kV/m. Although the animals were exposed to the field over a period of 1500 hr, they did not differ significantly from a control group in terms of weight or other physiologic data. Only the male offspring of field-exposed fathers showed a somewhat smaller weight increase than the sons of the controls. Spittka et al. (1969) therefore studied the drinking behavior of rats in alternating fields in an effort to detect possible field effects. The rats were placed into an apparatus containing a lever that was electrically linked to a drinking vessel in a manner such that a drop of water was released whenever the lever was pressed. The number of presses per unit time (15 s) was recorded as a function of the field situation (on or off). The main finding was a significant decrease in the lever-pressing rate when the field was switched on (for the first 15 s), as well as individual-specific changes in the pressing sequence. Spittka et al. (1969) states that his results conflict with the findings of Knickerbocker et al. (1967), even though the latter investigators used a higher field intensity (160 kV/m) than that employed in the present case (about 60 kV/m). This is attributed to the unfavorable experimental conditions of Knickerbocker et al. (1967): currents arising in the drinking vessels as well as slight discharges during drinking,

*3412 Hillview Avenue, Palo Alto, California 94304.

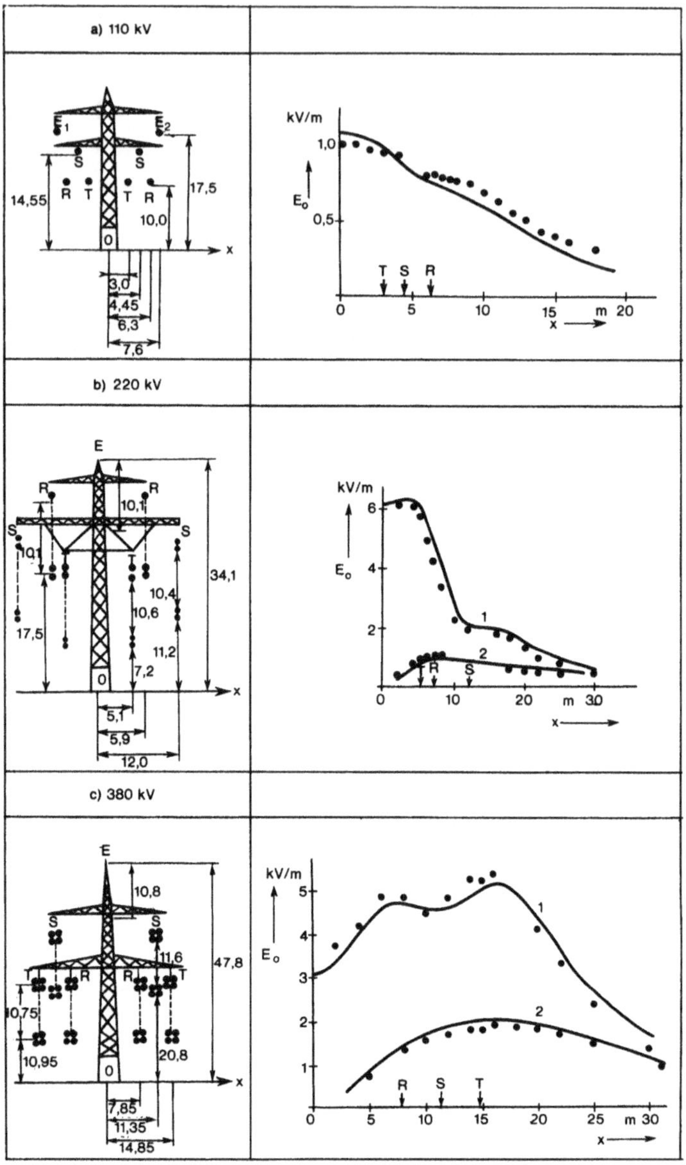

Figure 12-2. Electric field intensity E_0(——— calculated, · · · · measured) for various types of high-voltage transmission tower; field values 0.5 m above the ground, distance *x* from center-line, all measurements in meters; RST = circuit phases; E = ground wire; double-circuit system is employed in all cases; ***e*** asymmetrical design, all other cases are symmetrical; ***d*** influence of a ground elevation; ***f*** influence of vegetation, (1) calculated for 0.5 m above the ground, (2) calculated for 1 m above the ground; ***c*** (1) between towers, (2) at the tower. After Schneider et al. (1974).

which Spittka et al. (1969) took care to avoid. Based on his results, he was able to prove statistically that the operant drinking behavior of the rats was altered by the field. Both general trends and marked individual differences were observed (e.g., decline in lever-pressing rate or change in pressing sequence). Of particular interest are the changes in the lever-pressing sequence in the field. The question arises whether the increasing variance of the pressing rate (scatter) in the field is correlated with the decline of the pressing rate. In general terms, this would mean that behavior is dis-

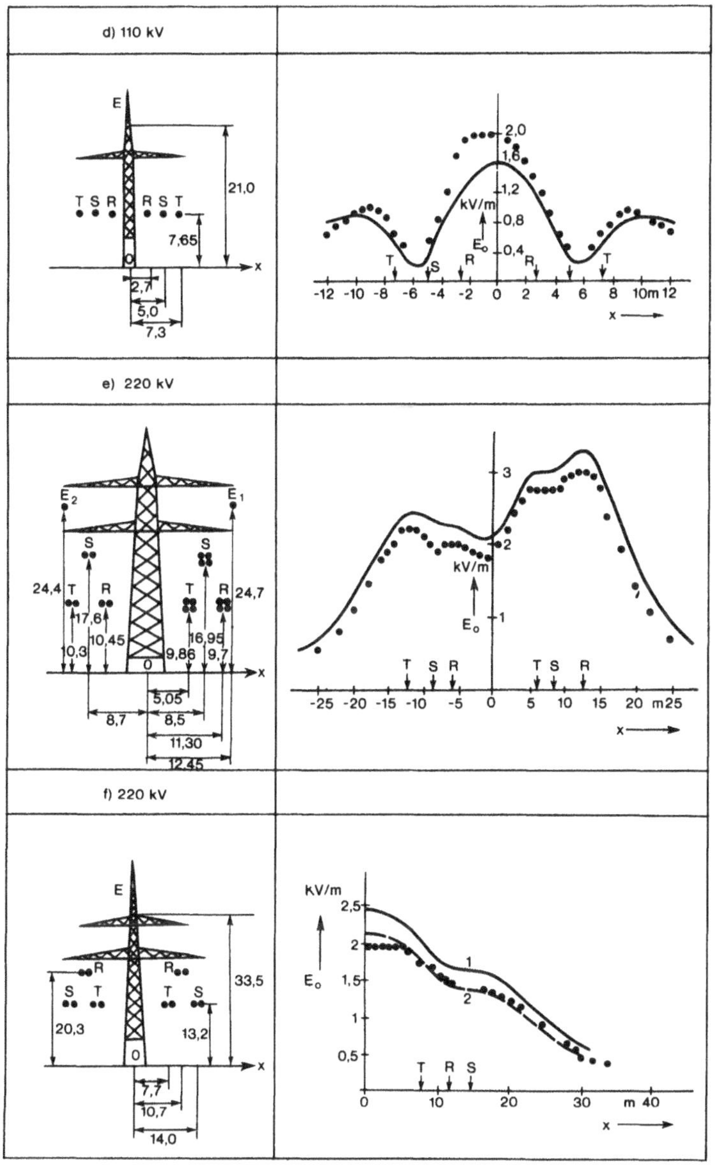

turbed by field exposure, resulting in a slowing and loss of precision in action sequences. Finally, the field was found to produce probably short- and long-term after-effects.

The effect of an electric field generated by a 110-V, 60-Hz household power current on the development of mice was investigated by Moos et al. (1967). They recorded the weight of the mice over a period of 1 month and compared it with corresponding controls. Mice exposed continuously to the field at about 1 kV/m gained weight more rapidly than the controls. The difference was slight but statistically significant. In further experiments the animals were first exposed to X-rays before entering the electric field. These animals died earlier than those which were only irradiated. In 3 of 26 cases, however, the electric field had a life-prolonging effect, while in 4 other cases no difference was apparent. An increased weight loss was noted in those irradiated animals which were also exposed to the artificially produced 60-Hz field.

Recent investigations on the biologic effects of 50-Hz electric fields were conducted by Fischer et al. (1974). They exposed mice to a field with a maximum intensity of 15 kV/m and monitored their O_2 consumption (respiration). A 10%–15% increase in oxygen consumption ($p < 0.01$) was found relative to the controls. The hepatic uptake of O_2 in these animals was significantly reduced ($p = 0.01$), however. No significant change in rectal temperature was noted in the experimental animals after 7 days' exposure. Additional experiments with rats produced no changes in the ECG, but led to a significant rise in the pulse rate from about 340 to 400 ($p < 0.01$).

Wellenstein (1973) dealt with the question of whether the behavior of bees is influenced by nearby high-voltage power transmission lines. Because the lanes cleared for high-voltage lines often make excellent beekeeping grounds, this problem has become increasingly important in recent times. One question, of course, is the effect of electromagnetic fields on the honeybee; but another pertains to the influence of air ionization, which is still poorly understood. Wellenstein (1973) assumes that the air beneath high-voltage transmission lines contains a higher-than-normal concentration of negatively charged ions. However, this conflicts with investigations mentioned above, which clearly showed a rapid ion depletion when the field intensity exceeds a certain value (about 5 kV/m).

While Wellenstein's investigations are not yet finalized, they have yielded some valuable results. Bees in experimental fields directly below long-distance lines (100 or 200 kV) showed a marked increase in activity and excitability in comparison with control colonies located at a distance of 600–800 m. Under good weather and foraging conditions, the experimental colonies gathered twice as much honey as the controls, but also exhibited a loss of substance (this corresponds to the findings of Altmann, 1962, 1968, 1969b). On cool and rainy days, however, the bees located below the 200-kV line were unusually excited and eager to swarm. On the basis of this experience, it is recommended that beehives not be set up directly below high-voltage transmission lines, but rather 50–100 m away.

Investigations by Altmann et al. (1974) with bees under the influence of a 50-Hz high-voltage field (maximum 6 kV/m, rms value) substantiated the findings of Wellenstein (1973). For this purpose, several bee colonies were placed into an experimental apparatus in which the high-voltage field could be switched on or off as desired. Altmann et al. made a motion picture record of the behavior of the bees under various experimental conditions. When the field was on, the colony (in the beehive and especially before the hive entrance) appeared highly agitated. The bees were seen running about, spreading their wings and making abrupt movements. Their aggressiveness was increased and was even directed against colony members, some of which were stung to death. Communication appeared to be impaired. The brood nest, which contained numerous flaws when constructed at low field intensities, was destroyed by the bees at intensities above 4 kV/m. Honey was no longer stored, and existing stores were consumed due to the high metabolic activity. Newly transferred colonies consistently abandoned the hives after 3 days under the influence of the field. If the colonies had already occupied the hives for some time before the experiment, exposure to the field led to vigorous sealing activity, especially at the hive entrance. Within 5 days the entrance was reduced to a tiny hole and was finally sealed completely. Because there was no longer any communication with the outside, the colonies perished from a lack of oxygen. This process was interpreted as a defensive measure by the bees against disturbing external factors.

The sensitive reaction of bees to electric

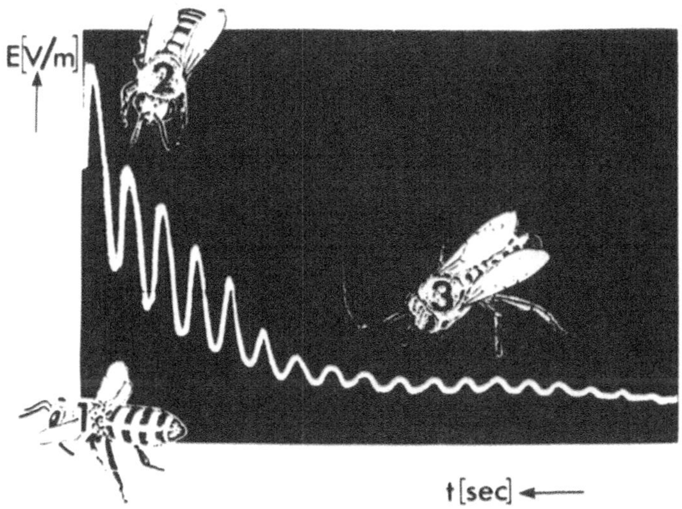

Figure 12-3. Oscillogram of the electric field of a bee flying past (1). The field intensity increases as the bee approaches the receiver (2) and decreases as it departs (3). After Warnke (1975a).

fields of the type present near high-voltage transmission lines becomes more understandable if one considers the endogenous electric field possessed by bees, as described by Warnke (1975a) (Fig. 12-3).

In his experiments with mice, Solov'ev (1963) showed that 50-Hz alternating fields can be lethal at sufficiently high intensities. Field intensities of 650 kV/m at 50–500 Hz produced a mortality of 70%–90% in the mice following an exposure of 60–120 min. In another experiment with a frequency of 50 Hz only, the mortality was 50% after an exposure of 270 min. In neither experiment could any field-related temperature increases be detected in the bodies of the animals.

Varga (1972a) reports on the effects of electric fields on the incubation of chicks. The field was generated by capacitor plates measuring 30 × 30 cm which were separated by a distance of 6 cm. A direct voltage of 4–50 V was applied to the plates in some experiments, and a 50-Hz alternating voltage in others. The effects were compared with a control incubator arrangement with no field. Over 160 measured values were obtained, indicating the following:

1. Of the eggs located in the electric field, only about 72% were hatched in comparison with the controls.
2. Of the chicks hatched in the fields, about one-third had visibly degenerate extremities (one or both feet not fully developed). Half these chicks died in 5–10 days.
3. The surviving chicks had a lower average body weight than the untreated animals.
4. The effect of the static fields was essentially the same as that of the 50-Hz alternating fields.

Reiter (1960) reports on the effect of metric-wave radiation with 50-Hz modulation (diathermy device) on the conjunctival pH of anesthetized guinea pigs, employing various electrode arrangements (Fig. 12-4). The immediate capacitor field, with its relatively high energy, produced a strong shift of the tissue pH toward more alkaline values. Slight pH shifts could still be produced at a distance of 4 m between the animal and a single transmitter electrode. Precise measurements of body temperature during the experiments again indicated a typically athermal effect, for no changes in body temperature could be detected during the experiments in the short-wave field.

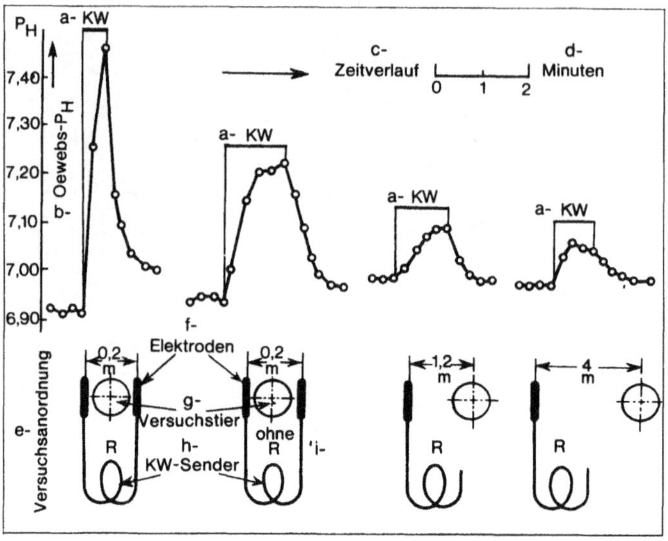

Figure 12-4. Conjunctival pH of the anesthetized guinea pig under the influence of metric-wave radiation with 50-Hz modulation for various electrode arrangements. R = resonance tuning, thus high energy emission. The animal's body is indicated by a circle. After Reiter (1960).
Key: a-SW; b-Tissue pH; c-Time; d-Min; e-Experimental arrangement; f-Electrodes; g-Experimental animal; h-SW transmitter; i-Without R

Distinct changes in tissue pH were also observed in experiments with low-frequency alternating fields in the range of 2–20 Hz. Unlike the experiments in the modulated short-wave field, however, the shifts were toward both more basic and more acidic values. The mean amplitude of these shifts was 0.1 pH.

The use of rectangular oscillations to generate a field with a pulse frequency between 50 Hz and 100 kHz and intensity of 10–1000 V/m also resulted in a characteristic shift of the tissue pH by 0.02 to 0.1 units from the resting value if the pH measurements were made on tissue located within the alternating field.

Human Experiments

The data obtained from experiments with animals (as well as with plants and bacteria, as reported by Varga, 1973) are supplemented by a substantial body of information acquired through human experimentation.

As early as 1902, Beer reported on his observations of the biologic effects of low-frequency electromagnetic fields. Thompson (1910) and Fleischmann (1922) also reported on this subject. The latter author recorded sensations of light flashes as well as headaches in persons who were exposed to a strong, 50-Hz alternating magnetic field close to their heads.

Most far-reaching, perhaps, in its biologic implications is the publication by Korobkova (1971, 1972) dealing with special occupational safety rules in the USSR for extremely high voltage (EHV) transmission systems. He reports on periodic medical examinations in various laboratories because of two adverse effects which have been observed on prolonged exposure to the electric fields of EHV facilities and in workers who service overhead EHV transmission lines:

1. The electric field has an adverse effect on the nervous and circulatory systems. This effect depends on the field intensity and the amount of time spent in the field.

2. Touching the parts of a system which has a potential difference with respect to man is also injurious. The effect of the

discharges associated with such a potential difference depends on the duration and amplitude of the discharge current and the frequency of the discharges. Such discharges may cause phase changes in electric brain rhythms as well as disturbances of respiration and the cardiac rhythm.

To prevent such discharges, Korobkova (1971, 1972) reports that the USSR has implemented special safety regulations which also specify the maximum allowable duration of exposure to strong electric fields (Table 12-1). These regulations apply as long as no other safety measures prevail. If the electric field intensity is not measured, the regulations call for a minimum distance of 20 m from EHV facilities with a 400- and 500-kV operating voltage, and 30 m from 750-kV facilities. As special safety measures, the regulations call for the use of equipment in the form of fixed or portable screens as well as special protective clothing of conductive fabric.

It should also be noted that according to the information of Hauf (1976b), the threshold values—i.e., the maximum allowable electric field intensities for prolonged exposure to artificial 50-Hz fields—have been defined as follows in Soviet territories: 12 kV/m in densely populated areas, 15 kV/m in sparsely populated areas, and 20 kV/m in uninhabited areas.

Hartmann (1976) investigated the electric skin resistance of test subjects under the influence of 50-Hz fields and observed corresponding reactions.

Tromp (1963) reported on the effects of 50-Hz electric and magnetic fields, which produced visual sensations and headaches, according to his observations.

In experiments with subjects in alternating fields between 10 Hz and 5 kHz (and thus also at 50 Hz), Reiter (1960) found, among other things, a change in the clotting ability of the blood. The prothrombin index was shifted by 4%–20% toward an increase in blood clotting ability after the experiment.

Hauf and Wiesinger (1973) and Hauf (1974, 1976a,b) investigated the biologic effects primarily of man-made electric and electromagnetic fields. They state that not only naturally occurring fields, but also man-made fields are of biologic significance in the frequency range below 100 Hz (e.g., 50 Hz, 60 Hz, 16 $^{2}/_{3}$ Hz). This raises the question, then, of where and to what extent the health and lives of persons exposed occupationally or otherwise to such electric fields are jeopardized. To explore this question, Hauf and Wiesinger set up high-voltage electrodes in a completely isolated climatic chamber (Figs. 12-5 and 12-6). With this arrangement, an undisturbed electric field with an intensity up to 25 kV/m and a frequency of 0–100 Hz could be produced in the environment of the test subjects. A constant chamber temperature of 22° C and a relative humidity of 50% were maintained to ensure uniform test conditions. The subjects did not know when the high-voltage field was switched on. A 50-Hz alternating field at 1 kV/m and 15 kV/m was employed in the first experiment. No significant changes were noted in the ECG, EEG, pulse or blood pressure under these conditions. Effects from the 1- and 15-kV/m fields were found only with regard to reaction time, which was always measured in the afternoon. Reaction time increased after the field was switched on and decreased somewhat after the field was switched off, without returning to the original value, however.

Ross Adey (1974) showed that important physiologic processes and biologic cycles are influenced by very weak electric fields

Table 12-1. *Allowable exposure to the 50-Hz electric field of high-voltage transmission systems in the USSR. After Korobkova (1971, 1972).*

Electric field intensity, kV/m	Maximum allowable daily exposure, min.
5	unlimited
10	180
15	90
20	10
25	5

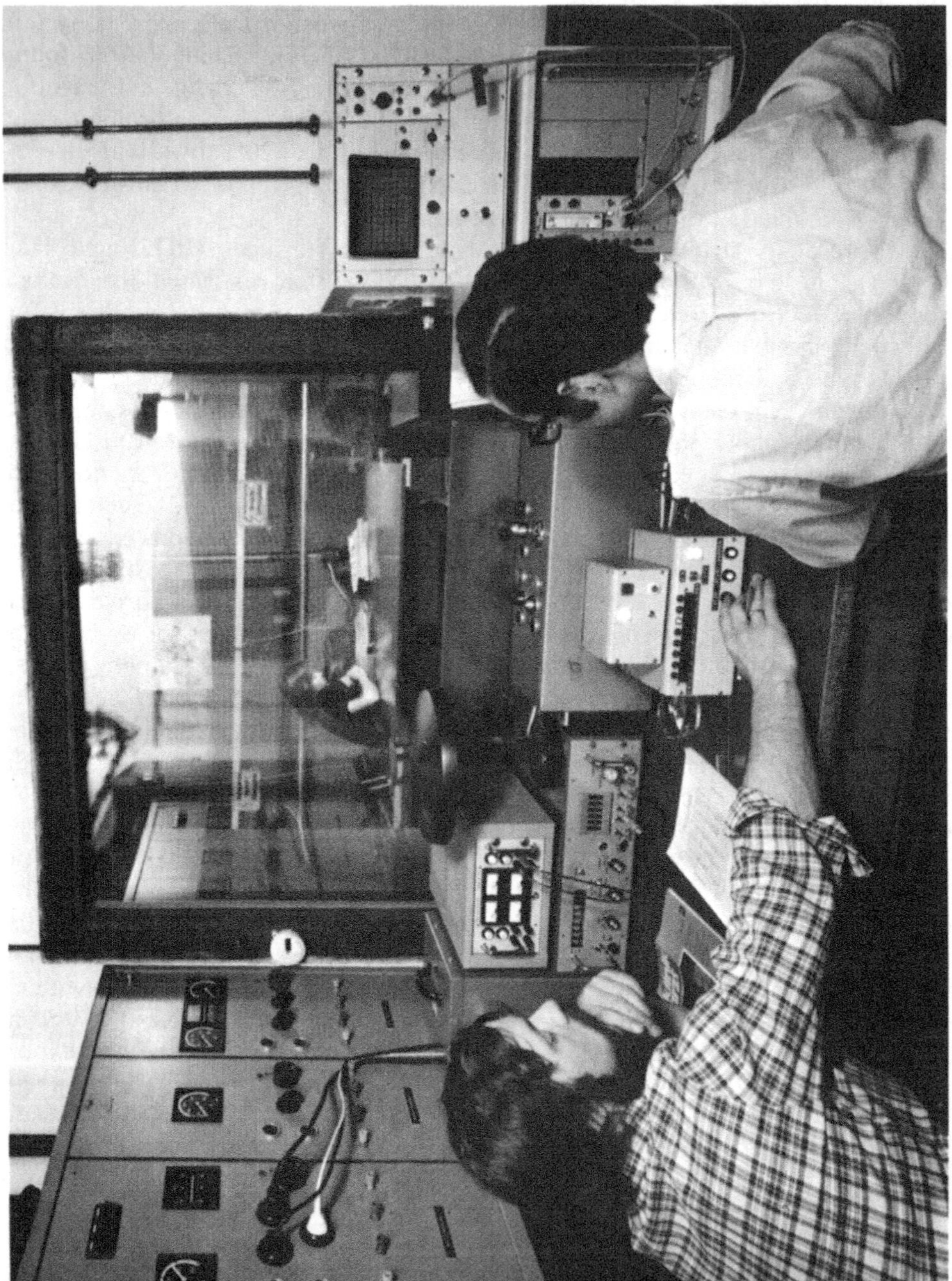

Figure 12-5. Reaction-time measurements under the influence of 50-Hz high-voltage fields were carried out at the High Voltage Institute of the Technical University of Munich in order to gather information of the biologic activity of such power fields. The photograph shows a view of the test subject from the control room.

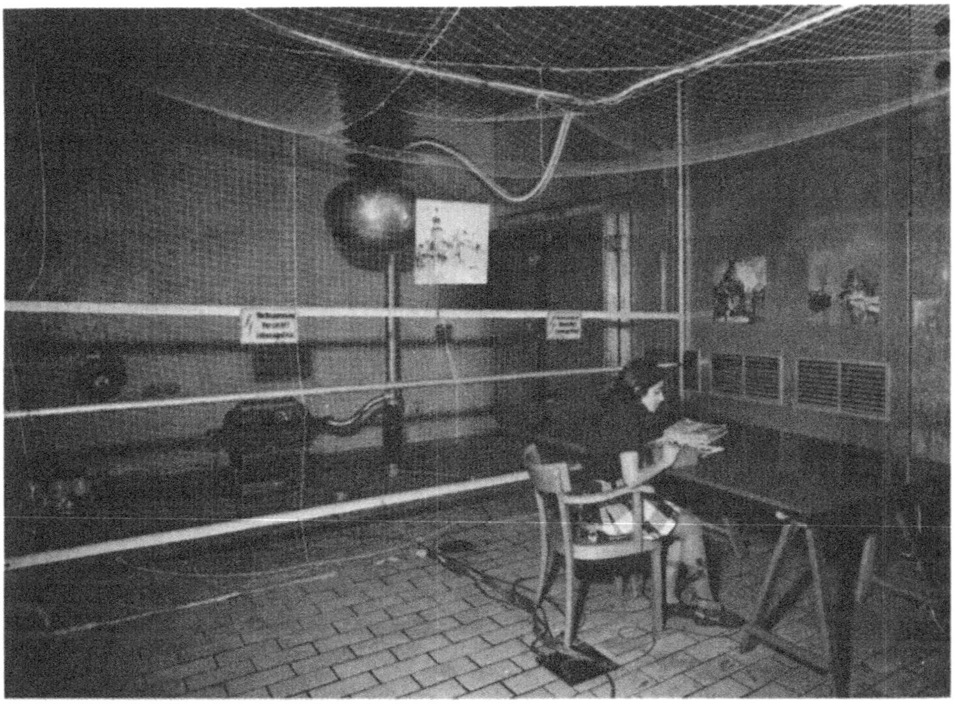

Figure 12-6. Test subject in the high-voltage experimental field. The necessary electric field is set up between the net-like ceiling electrode and the floor.

both in animals and, according to most recent research, in humans. Fields at low and very high frequencies, such as those in the radar band, were employed in these experiments. It was found that such fields are capable of influencing rhythmic activities of the brain, such as those characteristic of sleep, sleeplessness or quick reaction. This principle can perhaps be utilized to develop new methods of treating insomnia and chronic pain and of regulating biologic rhythms (e.g., to improve responsiveness or to "adjust" highly variable emotional states). Interestingly, it was also discovered that the frequencies customarily used in power current had similar effects when relatively high field intensities were employed.

In contrast to the effects of EHF fields, which stem mainly from the thermal action of the field, it was found that extremely weak fields with certain very high frequencies can support specific rhythms of brain activity, making it possible to alter behavioral patterns and hormonal cycles with no accompanying thermal injury. (Supplementary information on this subject is presented in Chapter 11.)

A topic of much interest in recent times is the cause of sudden infant death (SID), which was reviewed by Valdes-Dapena (1967). According to a definition adopted at an international congress in Seattle in 1969, SID is the "sudden and unexpected death of an infant, which is not explained by postmortem examination." In a survey report, Janssen (1978) examines the status of knowledge in this area based on papers presented at two medical conferences. According to Jansen (1978), numerous explanations are offered for this syndrome, including the influence of electromagnetic fields. Eckert (1976) explores this possibility in greater detail. According to his studies in Philadelphia, U.S.A. and in Hamburg, Germany, a high incidence of SID is found in the following locations:

1. in the vicinity of electric railways and their stations;

2. between railways and water;
3. between the railways of different systems (e.g., between train and streetcar tracks);
4. at the end of railway lines;
5. in the vicinity of magnetically operated water supply plants;
6. within the bends of electric railway lines.
7. Local concentrations were also found in the vicinity of wireless transmitters, airports and high-voltage transmission lines.

The hypothesis is advanced that the prolonged influence of certain magnetic and/or electric fields on the infant who is sensitive to them or made sensitive by such fields, effects an irreversible and therefore dangerous change in regulatory systems. Schaefer and Silny (1977) attack Eckert's working hypothesis, stating that it is an obvious and clearly demonstrable error. Eckert's reply to this criticism by no means resolves all doubts in the matter, but his views are supported by Janssen's report (1978), which states that the medical community is in agreement that infants succumbing to SID are often only apparently healthy. This raises the question of whether prolonged exposure to the aforementioned fields might not be responsible for this preexisting stress. (Evidence of the stress-producing effect of electromagnetic fields is presented in Chapter 11.) Eckert's hypothesis could thus be refined as follows: Man-made electromagnetic fields could shift the threshold in all disease pictures involved in SID to a critical point at which the slightest (apparently harmless) additional stress could have a lethal effect. This may account for a large number of seemingly inexplicable infant deaths.

Adjacent Frequency Ranges

Besides these reports, which deal mainly with the biologic effects of electromagnetic fields at the frequency of 50 Hz, we must also consider research results for the frequency ranges directly adjacent to the power frequencies. This includes the range down to a few hertz and up to higher frequencies of several kilohertz. The biologic activity of electromagnetic fields over this entire frequency range was demonstrated earlier in a comprehensive survey of the literature. Field effects were found even at intensities known only for natural fields. It can be inferred from this that if fields in a frequency range directly adjacent to the power frequencies are found to be biologically active, then this also applies to fields in the range of the power frequencies. It cannot be supposed that the power frequencies are exceptional with regard to their effects on biologic systems, as has been proved by the results of direct 50-Hz experiments. Regarding the "optimal" intensities necessary for biologic effects to occur, it is certain that there are zones somewhere in the vicinity of intense field generators that satisfy this requirement. In any case, the logical consequence of such considerations must be to incorporate all studies, experiments and research results for at least the ELF and VLF ranges (see Chapters 4–6) into the material presented here for 50 Hz, in order to supplement and lend further credence to experimental evidence on the biologic activity of 50-Hz fields.

Conclusions

Based on the current status of scientific research, it must be granted that electromagnetic fields of technical origin are at least biologically active, and possibly harmful. Every effort must be made, therefore, to prevent a worsening of conditions in this regard. This pertains in particular to the routing of overhead high-voltage transmission lines through populated areas; this is no longer necessary and should therefore be rejected, especially if populated areas can be circumvented by deviating somewhat from the intended route during the planning of such lines. Forestry and agri-

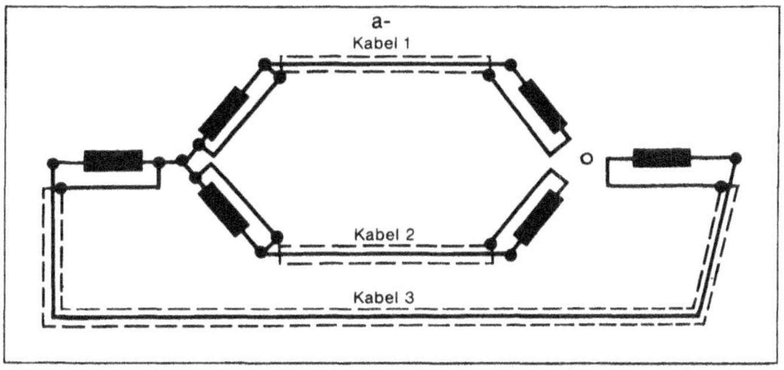

Figure 12-13. Shielded-cable design for a 3-phase circuit using coaxial cables to connect two transformers.
Key: a-Cable 1 etc.

cultural interests must come second to basic human needs. Another solution is the use of underground cables, which are now suitable for operating voltages up to 100 kV, all counter-arguments concerning finances and reliability to the contrary (see also Fig. 12-13). Coaxial cables have practically no external fields and, with proper circuiting, can be used for three-phase operation (one cable per phase). Many environmental problems would be solved by the use of such cabling, which should also be considered during the planning of transformer stations.

In a time when environmental protection is on everyone's lips, sufficient fiscal appropriations should be available for such measures, so that the environment will not be despoiled by the installation of additional power-supply facilities. These words of warning are not without opposition, of course. Highly qualified technicians, often without sufficient expertise in the field of biology, are attempting even now to show theoretically that electromagnetic fields, and especially power-line fields, cannot possibly exert biologic effects. Conversely, objections from the medical community often lack an adequate technical understanding of the problems involved. It is important to realize that this area of research is extremely far-reaching and complex. It is for this reason that all concerned disciplines should work together to clarify and explore the many questions at hand. It is also important that such a team possess a broad background of experience in the treatment of such problems. Aside from objective contributions, any kind of counter-argument must be received extremely critically from the outset, except when it comes from sources that can be considered truly qualified.

We shall return to the very important contributions of Becker (1975b) and Marino (1975) to the problems that arose in connection with hearings by the New York Public Service Commission on the health and safety aspects of 760-kV transmission lines. Regarding the question of whether the somewhat contradictory reports published on the harmful effects of energy fields implied that such fields are in fact harmless, it was determined that the inconsistencies in no way justify such a conclusion. They indicate only that such effects are not present in every case. The reason for this lies first in the fact that experimental conditions are not always easily reproducible, and second in the varying sensitivities of different experimental subjects. The question of the maximum allowable field intensity is also discussed, and reference is made to the safety factor of 1:100 employed in food chemistry in connection with toxic substances. This safety factor is based on the dose of a toxic substance for which a toxic effect can no longer be detected; this ensures that such substances will be harmless even in special cases. In the case of energy fields, the experimental results discussed so far would indicate a field intensity of about 5 kV/m as the critical value. Applying the 1:100 safety factor in this case, we find that prolonged exposure to areas with an electric field intensity greater than 50 V/m is inadvisable.

König (1976) summarized these recent findings as follows. High-voltage transmission lines generate energy fields about them, the nature of which depends primarily on the operating voltage (electric field) and the operating current (magnetic field) as well as the distance between the observation point and the line. The geometric arrangement of the line and trees, buildings, etc., in the space in question also plays a role. The time course of the field intensity (50 or 60 Hz, $16^2/_3$ Hz, etc.) follows that of the voltage or current producing the field. The fundamental question of whether or not high-voltage lines have harmful effects and what the worst-case consequences are, as well as the complex problem of the biologic activity of electric, magnetic and electromagnetic fields in general, cannot be answered in simple terms. However, several points should be taken into consideration in any theoretical, experimental or applied treatment of these questions:

1. Possible effects depend in every case on the field intensity at the observation site, which must first be determined if possible.
2. Most results that indicate adverse field effects have been obtained through animal experimentation. The attempt to induce adverse effects in humans does not appear to be opportune.
3. Where is the dividing line between harmfulness and an acceptable encroachment upon the quality of life?
4. There is a dearth of long-term experiments dealing specifically with the effects of relatively low-intensity fields on persons over at least a 1-year period, corresponding to the conditions encountered at a somewhat greater distance from energy facilities.
5. Clarifying studies should be conducted primarily by independent research groups with special experience in the area of interest.
6. Man, as a biologic system, is highly diverse. Consequently, the problem of field effects must be approached mainly from a statistical standpoint, taking into account such factors as age and state of health.
7. What safety margin should be required for maximum allowable field intensity, especially where long-term exposure is concerned (minimum distance of residences from each side of the centerline of high-voltage transmission lines)?
8. What are the acceptable limits for the financial burden imposed by the construction of less environmentally harmful power transmission facilities (cabling) or the rerouting of power lines?

Even this incomplete list gives an indication of the many spheres of interest involved in such problems. Although professional opinion on the adverse effects of high-voltage transmission lines is still divided due to discrepancies in research results, some general conclusions can be drawn:

1. The biologic activity of electric and magnetic fields must be considered a proven fact.
2. The biologic effects of electric and magnetic fields are generally manifested in the form of:
   a. frequency-specific (time course) effects of biologically adequate field conditions,
   b. a stress effect under biologically inadequate field conditions (power-line fields, for example),
   c. a field-specific adaptive response.
3. Even if it cannot now be proved with absolute certainty that power-line fields are generally harmful, neither can it be proved conclusively that they are harmless.
4. It must be added that a harmful effect

Table 12-2. Minimum safe distances from the center line of high-voltage transmission lines.

	Distance from the right-of-way		
Operating voltage	Field intensity <50 V/m (undoubtedly safe)	Field intensity <150 V/m (probably safe)	Field intensity <5 kV/m (probably dangerous)
380 kV	180–250 m	100–140 m	15–20 m (tower height $>$ 50 m)
220 kV	140–180 m	75–90 m	6–10 m (tower height $>$ 30 m)
100 kV	80–120 m	45–60 m	—
50 kV	50–70 m	34–45 m	—

cannot be expected in every case, although there is a possibility of it, especially in persons with impaired health. But there is still no experimental evidence which would prove that power-line fields are generally harmful in a medical sense to healthy persons.

5. Field effects doubtlessly represent second-order effects in a biometeorologic sense (as opposed to first-order effects from temperature, humidity, etc).
6. Employing a safety factor of about 1:100, as described earlier, we obtain a critical field intensity of approximately 50 V/m for prolonged exposure to a 50- or 60-Hz field. The minimum safe distances from the path of three-phase high-voltage transmission lines are given in Table 12-2 for a freely propagating electric field in the vicinity of such lines.

Marino and Becker (1977) discuss the situation in the United States in a survey article:

> High-voltage transmission lines cause ELF electric and magnetic fields. ELF electric and magnetic fields cause biological effects. Presently, in the United States, high-voltage transmission lines are designed and built with no regard or weight given to the biological effects produced by their electric and magnetic fields. The next generation of high-voltage transmission lines is now materializing. It would appear appropriate to determine the spectrum of biological effects and to adequately assess the hazards to man of exposure to such fields. In view of the large population at risk, all of the health hazards associated with the electric and magnetic fields of high-voltage transmission lines should be carefully examined in the context of the specific design for each facility, prior to construction.

Apparently the problems addressed here are also recognized by the competent authorities. In connection with the aforementioned hearings, to which Becker (1975b) and Marino (1975) contributed, the Public Service Commission of the State of New York (1978) issued its opinion #78-13, the conclusion of which is reprinted in Appendix F.

In summary, the following can be said with regard to the problem of biologic effects of electromagnetic fields in the low range of power frequencies. It is certain that the quality of life of many persons is not noticeably impaired by the presence of such fields. But this circumstance does not mean that such fields are entirely harmless. There are enough studies on an international basis that indicate that these fields are at least biologically active. These results can no longer be ignored and can no longer be called into question by dubious counter-arguments. Previous findings should provide a sufficient stimulus for all competent institutions to investigate the problem on an even broader scientific, and

Figures 12-7 and 12-8. The differences between a steel-reinforced concrete building and a traditional wooden house are more than visual. The natural electromagnetic climate is greatly altered in steel and concrete structures. Architectural biology is concerned with such questions. Recent scientific findings show an urgent need to determine the effect of such differences on human health.

thus statistical, basis; for this is the only way in which the issue can be clarified in its entirety. The public is entitled to know the risk posed to the individual by the presence of such fields in his environment. Aside from this, however, environmental considerations alone dictate that such practices as routing high-voltage transmission lines through densely populated areas should be avoided wherever possible. The present discussions are in no way intended to advocate restricting the further expansion of power supply facilities, which is unquestionably necessary, but are meant only to assert the priority of human interests.

Architectural Biology

As previous chapters have shown, biometeorologic and geobiologic problems are closely linked to the field situation and other electrophysical parameters (air ionization, for example). It follows that these new findings also have a bearing on questions pertaining to man's living climate. We are concerned here both with problems of large-scale city planning and with the living climate of individual rooms, for it is apparent that the electrophysical factors that surround man can be directly or indirectly influenced by both.

The extent to which the subject at hand may be related to the effects of pyramid-shaped structures—the preservation of mummies and other phenomena, such as the retardation of bacterial growth (Jaekkel, 1955)—is unclear. In any case, the existence of certain biologically unfavorable regions or zones, as described in geobiology, should be taken into account in the future, especially if their existence is confirmed. With a far-sighted system of city planning, it should not be difficult to avoid such areas in the construction of dwellings.

Figure 12-8.

Moreover, the different effects of various building materials (wood, brick, concrete, etc.) and of various types of construction (classical, steel-reinforced concrete, etc.) should also be taken into account with a view to achieving an optimal electromagnetic climate indoors (Figs. 12-7 and 12-8). Even now, certain measures should be taken to screen out potentially harmful alternating fields of the conventionally used power frequencies. In the bedroom, too, it is important that the bed be placed in a "safe" spot as a preventive measure against long-term exposure to biologically unfavorable physical energies, as minute as they may be.

Once one has acknowledged the biologic activity of electric, magnetic and electromagnetic fields, ionization conditions, etc., and realizes that they can have a serious biologic impact, it is evident that this knowledge should also be applied in the construction of dwellings.

The term architectural biology has been coined to describe this branch of science. Although this field is still in its early stages, ample evidence has been gathered to show that too much is at stake, for the health of the general public, simply to dismiss the problem as irrelevant.

We cannot discuss specific details within the limited scope of this chapter, only general relationships; however, this is a subject with which Kaufmann (1968) has dealt for some time. Palm (1968), on the other hand, addresses himself more to problems of structural engineering. But while many important points have been brought out by these studies, they lack the exact, scientific method of proof which is indispensible, even here. This is not meant to be a value judgment of work already done, rather it is meant to be a statement on the current situation in architectural biology. There is still much that must be raised from the status of assumption to that of precise knowledge.

Lang (1973a) discusses the problems and potentials of architectural biology in connection with the "electroclimate." Atmospheric electric oscillations (atmospherics) are a physiologically essential parameter of the ecosystem. Experiments have demonstrated their influence on behavioral psychology (circadian rhythm, motor activity, reaction time). The effects on chronobiologic processes are an important index of the significance of the electric climate in terms of the physiology of work. For example, highly significant correlations have been found between atmospherics and the responsiveness and working capacity of test subjects and students. However, the intensity of atmospherics is strongly reduced by certain weather conditions, as well as by the screening action of building materials. It is recommended, therefore, that "electroclimate control" be employed indoors to recreate optimal natural conditions. According to Lang, such measures are absolutely necessary in living and working areas that are climatically unfavorable.

In a joint project, Lang et al. (1976), Ludwig (1976a) and Usemann and Vogel (1976) constructed an experimental device to measure the ability of atmospheric electric fields to penetrate certain building materials. It consisted of two semiconductor-type electrometer probes with a frequency analyzer (Bach and Lang, 1976; Lang et al., 1976). The probes were installed on the inside and outside of various buildings at equal heights and at a predefined distance from the wall material. The intensity of incident atmospherics was measured in the frequency range around 0.3, 1.0, 1.5, 2, 3, 5, 7, 10, 12, 15, 20 and 30 Hz. At each measuring site, ten measurements were made in this frequency range on various days with favorable atmospheric radiation and high barometric pressure. The difference between the intensities measured inside and those measured outside (taken as 100%) was calculated to give the permeability in percent. Measurements were made in buildings of brick, prefabricated sections, wood, quarry stone, hollow pumice blocks and steel-reinforced concrete. No one was allowed in the rooms during measurements because of the alternating electric fields set up by electrostatic charges. The power supply was also shut off. A further series of measurements was made in a dense spruce forest and in an automobile for purposes of comparison. The results of the measurements are shown in Fig. 12-9. It is seen that the mean permeability over the entire frequency range was 46% for the spruce forest and 10% for the automobile. In the buildings, it decreased from 57% in brick to 3% in steel-reinforced concrete.

Ludwig (1976a) and Lenke and Bonze (1975) recorded the static atmospheric electric field inside and outside enclosed rooms. They found that while no natural static field component could be detected in the interior of the rooms, the normal use of these rooms gave rise to periodic fluctuating potentials in the ULF or lower ELF range due to electrostatic charges on clothing, movements of people, and the use of electric switches. These artificial ELF fields were superimposed onto the natural atmospherics measured within the rooms as soon as these rooms were used normally. However, the interpretation of these results to mean that differences in the shielding properties of different building materials are irrelevant because the normal use of rooms creates an artificial electric climate much like the natural one appears questionable. The periods of measurement were

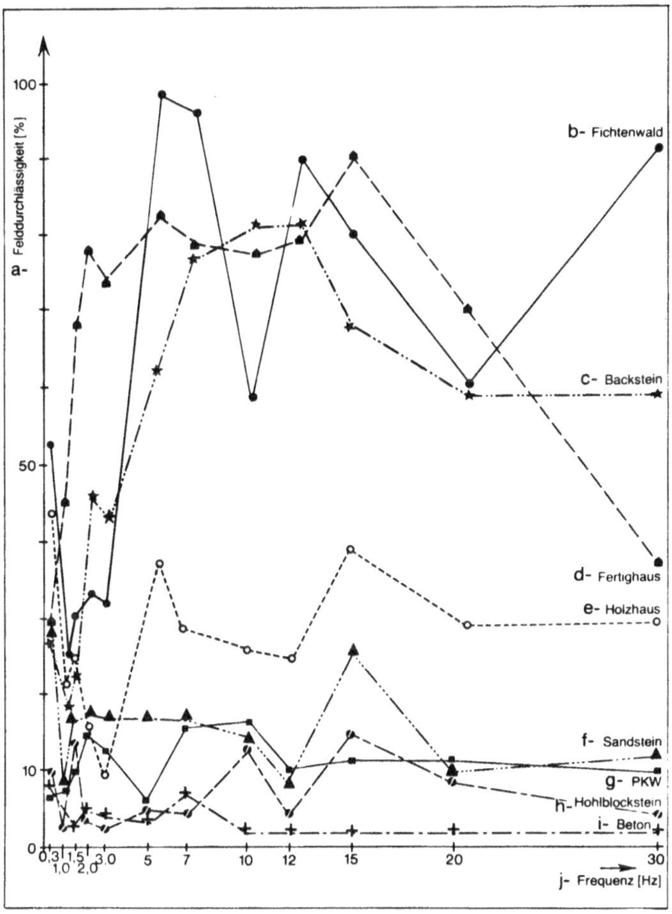

Figure 12-9. Ability of various materials to transmit atmospheric electric fields in the range from 0.3 to 30 Hz as a function of frequency.
Key: a- Field transmission (%); b- Spruce forest; c- Brick; d- Prefab house; e- Wooden house; f- Sandstone; g- Automobile; h- Hollow blocks; i- Reinforced concrete; j- Frequency (Hz)

only about 30 min, insufficient to reveal any daily periodic processes that would have shown differences between the artificial and natural electroclimate. Moreover, the actual time course of the ELF pattern of the artificial electroclimate cannot be compared with that of natural atmospherics in this frequency range. Besides, the measurements did not show which components of the artificial electroclimate were attributable to electrostatic charges on moving surfaces and which to switching processes in electric systems.

In other experiments, Lang et al. (1976) limited their study of the electric environment in working spaces to the measurement of 50-Hz fields, small ion concentration and electrostatic charge effects, in order to develop a standard measuring technique based on these important parameters. Nine measuring sites were selected in the classrooms of a school, and the potentials of the 50-Hz field were measured with the lights on and off, at heights of 60, 90, 120, 150, 180, 210 and 240 cm above the floor. The situation in one of the classrooms, whose rear wall was constantly damp due to a leak in the roof of the building, is shown in Fig. 12-10. Compared with the other rooms, in which the spacial distribution of the existing 50-Hz field was relatively symmetrical, this rear wall produced

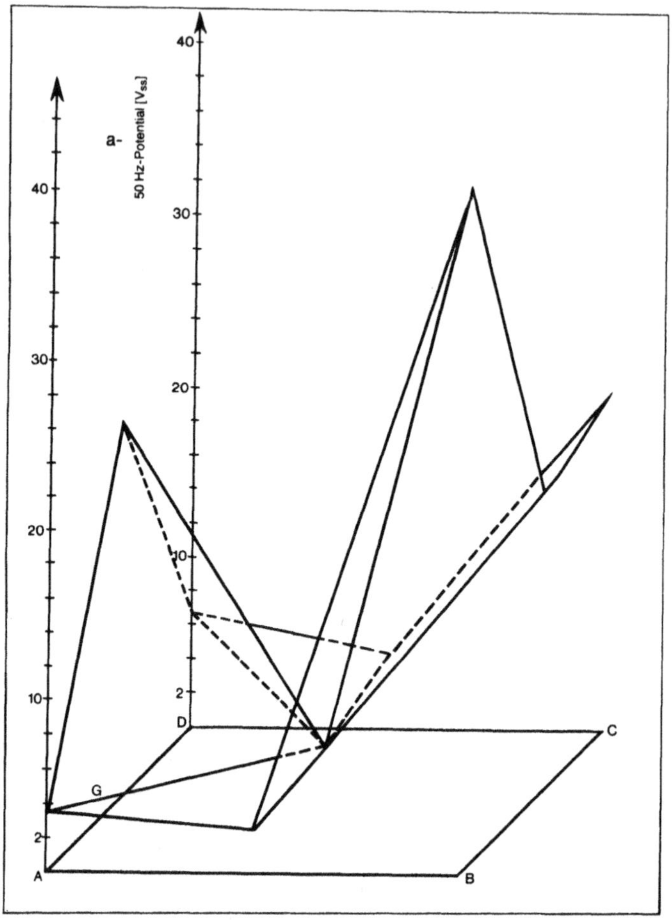

Figure 12-10. Spatial potential distribution of the 50-Hz field in a school classroom with area ABCD (6.4 × 9.2 m), measured at a height of 1.80 m.
Key: a-50-Hz potential (V_{pp})

a potential 10 times higher. No flaws could be detected in the grounding contact or grounding system of the wires and sockets installed in the wall. When the ceiling lights were switched on, the potential of the 50-Hz field increased with height above the floor. An ungrounded lamp at the front of the room produced potentials up to 56 V_{pp} with respect to ground. Overall, it was found that 50-Hz potentials were normally on the order of 2–20 V_{pp} near the wall of a room, but could be increased up to tenfold by a defective wiring system or structural flaws in the building.

The concentration of small ions in the same rooms showed the profile illustrated in Fig. 12-11. Electrostatic charges caused by the movement of people could hardly be detected in the classrooms. Near the PVC floor covering, a relatively low average positive field intensity of about 100 V/m was measured. The plastic-coated table and chair surfaces carried a charge during normal classroom use that was manifested in an electric field of about 800 V/m in the immediate vicinity of these objects.

The technique described here offers a standarized approach to investigating enclosed spaces with regard to (1) their shielding properties with respect to atmospheric electric fields; (2) the spectrum of artificial electric fields in the ULF and ELF range; (3) electrostatic charging phenomena; and (4) small ion concentration.

Though this concept must still be supplemented and revised in many respects, it shows a standardized, practical means for the recording of such architectural climatic parameters.

The illumination of interiors with artificial light, and especially with light produced by fluorescent tubes, represents another interesting problem in architectural biology. Studies by Hollwich et al. (1977) indicate that besides the brightness of the light, its spectral composition plays a decisive role:

In contrast to natural light, artificial light has an unnatural and hence unphysiological stimulating effect; the greater the intensity of the light involved, the greater this effect. This is especially true for windowless working areas whose only source of light is fluorescent light in the shape of tubes. Unnecessarily high intensity of artificial light leads to stresslike metabolic and endocrine changes detectable in blood chemistry.

Several possible causes of the biologic effects of artificial lighting are

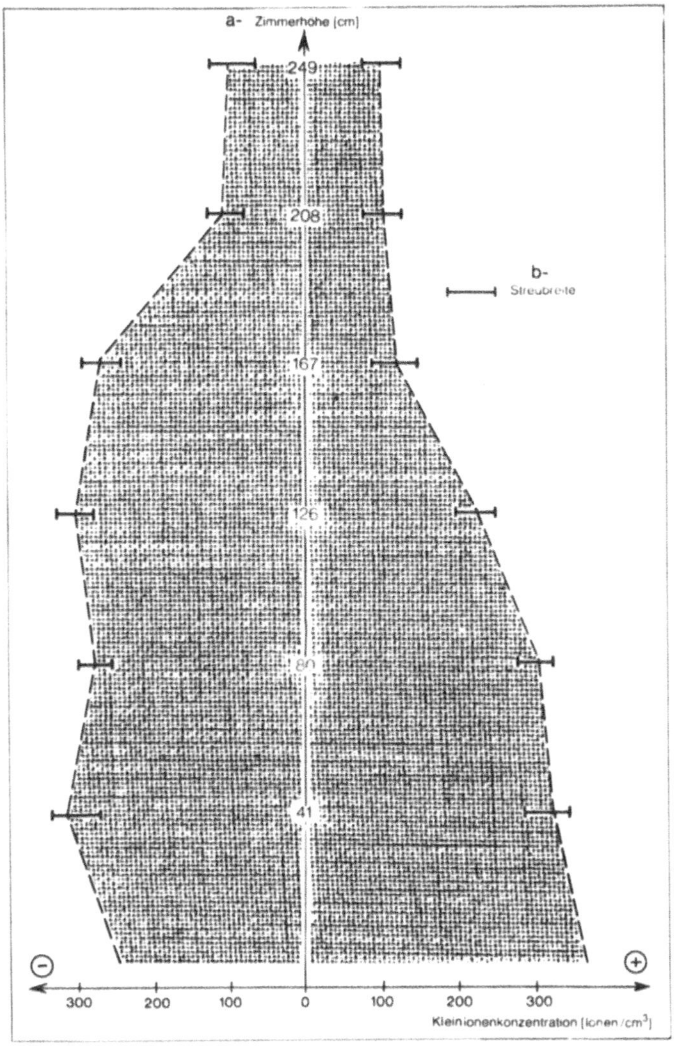

Figure 12-11. Vertical distribution of small ions in a school classroom with conventional climate control.
Key: a-Height above floor (cm); b-Scatter; c-Small ion concentration (ions/cm^3)

spectral composition (extra intense mercury lines), excessive brightness, monotony of illumination, flicker frequency, glare and the direction of light incidence.

If the importance of the problems discussed here is fully recognized, it must be acknowledged that a living climate corresponding to natural conditions is the most desirable state. Natural environmental conditions, including the special atmospheric electric and magnetic conditions discussed here, should be allowed access to living spaces, while unfavorable artificial electromagnetic energies should be avoided there. However, it is seldom possible for the modern, technically oriented building industry to meet such requirements in practice, primarily for engineering and financial reasons. The avoidance of geopathogenic zones is simple by comparison, but their existence is rarely acknowledged, or else there is a complete lack of awareness of such possibilities. If the use of neutralizing and similar devices is to be rejected in favor of avoidance for the fundamental reasons explained earlier, it is advisable at least to employ technical means to compensate for deficiencies of atmospheric electricity indoors. Such means are already available commercially in the form of air-ionization devices, for example. Systems are also available for generating electric fields; one such system designed for use in large office areas is illustrated in Fig. 12-12. (Details on such devices and their use in motor vehicles are given in Chapter 4.)

The current status of science with regard to the questions discussed here has been summarized. It is clear that many questions remain unanswered. It is not yet possible to draw final conclusions, and ultimately it must be left to the individual to judge the significance of previous findings.

From a scientific standpoint, the main task at present is to investigate the most fundamental aspects of architectural biology. Lotz (1972) describes the difficulties involved in such research very well when he points out that the parameters necessary for an accurate appraisal of building materials from a biologic standpoint are still largely unknown. In view of the consequences of an imprecise assessment due to a lack of scientifically valid data, some degree of caution is advised. Nevertheless, Lotz (1972) analyzes the problem in detail by examining the standards for appraising building materials and the method of their selection, and finally establishes ordering factors such as areas of application, properties and comparisons, characteristics of building materials, and problems with the radioactivity of building materials. He subdivides the latter subject into questions of indoor radiation sources, the shielding of external gamma rays by building materials, radiation from building materials, cosmic radiation and airborne radioactivity in houses, and compares indoor and outdoor radiation dose levels as well. But radioactivity is only one of the parameters to be considered in the biologic assessment of building materials. Others are electrostatic charges, thermal conductivity and heat capacity, insulating action against geopathogenic influences, capability of gaseous exchange (ability to "breathe"), hygroscopicity (regulation of air humidity), permeability to static and alternating electric fields, and shielding action against man-made radiations.

This list illustrates the vast range of questions in the area of architectural biology that need to be explored by basic research on an interdisciplinary basis. If new possibilities for improving the quality of human life are recognized on the basis of the discussions presented, or new impetus is given the search for answers to the many unanswered questions, one of our main goals will have been achieved.

Figure 12-12. Large office area with commercially installed ceiling electrodes for creating an artificial electrical climate. The successful use of such systems confirms their architectural-biologic importance.

A

Technical and Electrophysical Terms

In our discussion of the biologic activity of electromagnetic forces, it will be necessary to use technical and electrophysical terms with which every reader may not be familiar. We shall therefore attempt to explain the most important of these terms in a manner which is generally understandable and to illustrate them where necessary.

Aerosol: A colloidal system in which solid (smoke, dust) or liquid (mist) substances are dispersed in a gaseous medium. Only systems in which the suspended particles actually remain in suspension for a sufficient period of time can be called true aerosols, however. Any solid, liquid or gaseous substance that is suspended in a gas and has a diameter of 1 nm–10 μm as a physicochemical unit can be considered an aerosol.

Atmospherics: Electromagnetic signals of natural origin, chiefly in the form of waves with frequencies of 0.3 Hz–0.3 kHz (extremely low frequency atmospherics) and 0.3 kHz–0.3 MHz (very-low frequency atmospherics), which propagate mainly between the earth's surface and the ionosphere. Atmospherics are caused primarily by lightning discharges.

Current Density: That amount of electric current j passing at right angles through a standard cross-sectional area of 1 m^2 (unit of measure: A/m^2); j is proportional to the electric field intensity E and the specific conductivity σ, $\vec{j} = \sigma \cdot \vec{E}$.

Electric Charge: Synonymous with "quantity of electricity" (unit of measure: 1 coulomb (C) = 1 ampere (A) × 1 s).

A characteristic of electric charges is the force they exert upon one another. There are two kinds of charge: positive and negative. Like charges repel and unlike charges attract each other. If equal quantities of unlike charge types are combined, the charges cancel, and are said to be compensated (Fig. A-1). It has long been known that a positive charge develops on a glass rod that is rubbed with a wool cloth coated with tin amalgam, and a negative charge on sealing wax or sulfur rubbed with fur.

Electric charges are not infinitely divisible. The smallest possible charge, of which all charges are some integral multiple, is the

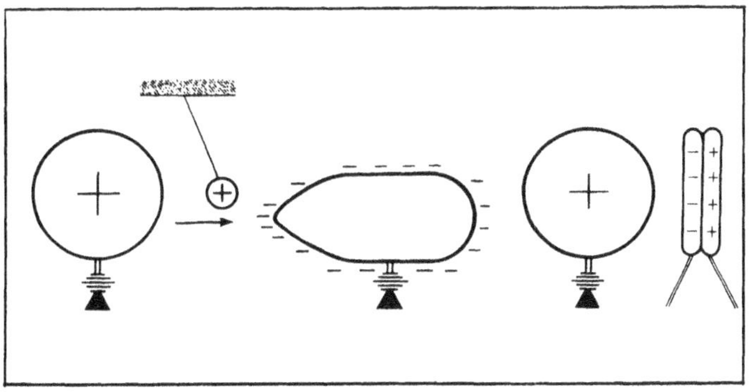

Figure A-1. Detection of electric field by its action on small test charges; charge distribution on a conductor; charge displacement on double metal plate by electrostatic induction (from left to right).

elementary charge of the electron, $e = 1.602 \times 10^{-19}$ C.

Electric Charge Transport: Synonymous with electric current (unit of measure: the ampere).

Electric Conductance: The reciprocal of the electric resistance R (unit of measure: the siemens (S) or mho). The electric conductance, $G = 1/R$, is proportional to the quantity of current I flowing between two points across which there is a voltage U, $I = G \times U$.

Electric Current—Ohm's Law: Electric current is a motion of electric charge. The current I (unit of measure: the ampere) flowing in a conductor is, according to Ohm's law, proportional to the voltage U that is applied to this conductor and inversely proportional to its electric resistance R, $I = U/R$.

Electric Field: Following the thinking of J. C. Maxwell, we can describe an electric field as the state that is produced in space by electric charges or by time-varying magnetic fields. Electric charges exert forces on one another. The "electrostatic force" between two point charges q_1 and q_2 separated by the distance r is given, according to Coulomb's law, by the formula, $F = kq_1q_2/r^2$ (where k is a proportionality constant). The electric field intensity, $\vec{E} = \vec{F}/q$, is defined as the force per unit charge, where the vectors of E and F have the same direction (see Fig. A-1 and A-2).

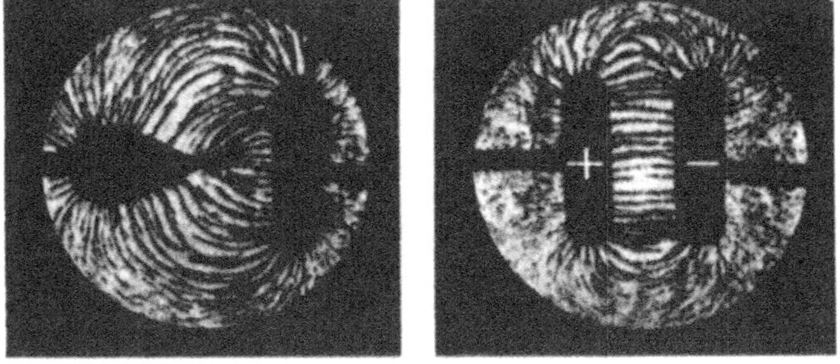

Figure A-2. Electric field patterns. *Left:* Between a point and a charged plate. *Right:* Between two flat, charged plates.

An electric field can also be produced in a magnetic field which varies with time, in accordance with the law of induction. The electric field intensity is measured in V/m.

Electric Resistance: According to Ohm's law, electric resistance is a proportional measure of the voltage U which drops between the two end-points of an electrical resistance R through which a certain current I is flowing, $U = R \cdot I$. In a wire, resistance is computed from its specific resistance ϱ, length d and cross-section a, $R = \varrho \cdot d/a$ (unit of measure: the ohm).

Electromagnetic Waves: According to the law of magnetic flux, a magnetic field always forms around an electric current. Moreover, the law of induction states that every time-varying magnetic field produces an electromotive force which is capable of moving electric charges and thus driving an electric current (Fig. A-3). A changing magnetic field is also capable of setting up a similarly changing electric field in a vacuum that contains no charges.

The term displacement current has been introduced to describe these conditions in a vacuum, in an analogy of the relationship between an electric field and moving charge. This current has the same effect as moving electric charges, in that it also generates a magnetic field. Thus, a changing electric field sets up a magnetic field, which in turn sets up an electric field, and so on. Since this process propagates through space, it leads to a transport of energy in the form of electromagnetic waves. These waves were first described theoretically by J. C. Maxwell in 1864, and were demonstrated experimentally by H. Hertz in 1887. They consist of electric and magnetic fields that have the ability to propagate in space. To a fixed observer, a passing electromagnetic wave of a certain frequency appears as a sinusoidal variation of the electric and magnetic field intensity. Observers in different locations will measure the maximum values of such fields with a certain shift in time or phase. In free space, the components of the electric and magnetic fields are perpendicular to each other and to the direction of propagation, and are related by a mathematical ratio called the wave impedance, $W = E/H \approx 376\ \Omega$.

The velocity of propagation c of electromagnetic waves in free space and that of light, which also consists of such waves, is approximately 300,000 km/s. It was first measured experimentally by A. H. Fizeau in 1849.

During one period T, the wave propagates over a distance which is called the *wavelength* λ, where $\lambda = c \cdot T = c/f$. According to this equation, a fixed relation exists between the wavelength λ and frequency f of a wave, since the product of these quantities in a vacuum or in air is always equal to the velocity of light c (Fig. A-4).

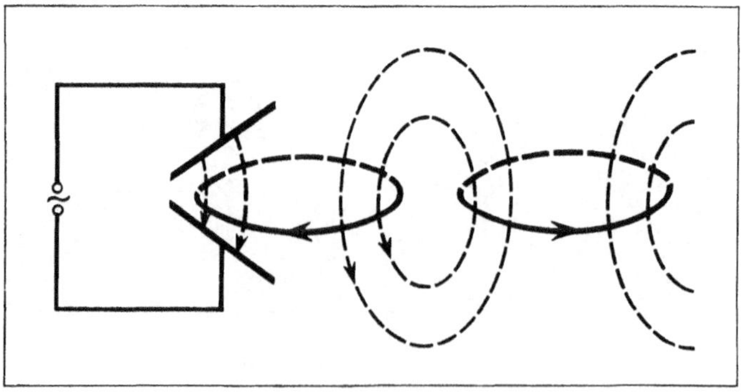

Figure A-3. Interlinking of electric and magnetic alternating fields, resulting in the propagation of electromagnetic waves.

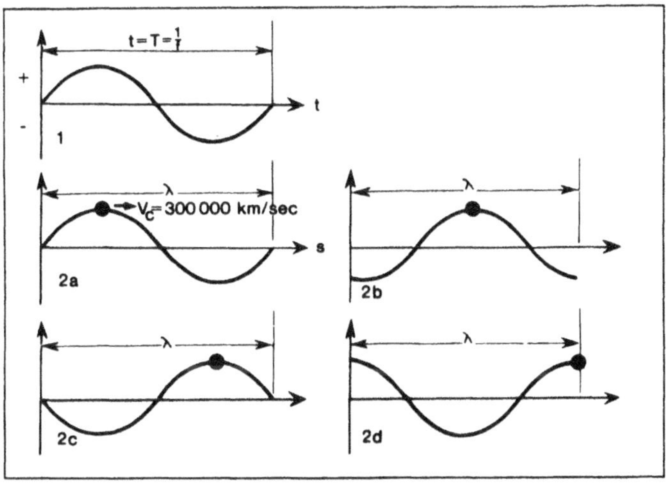

Figure A-4. *1* One oscillation is completed during the period T. The number of oscillations per second $1/T$ gives the frequency f of the process expressed in Hertz. *2* Propagation of a wave at the velocity of light v_e. During the period T, the wave covers the distance λ; this is the wavelength of the radiation.

The conditions discussed here exist in the far field (distance of at least several wavelengths from the point of origin) of an electromagnetic wave (Fig. A-5). The conditions in the near field, such as those occurring in the immediate vicinity of a transmitting antenna, may be quite different. It should be noted that the term antenna is used here in a general sense and includes everything which contributes to the formation of a rapidly changing electric or magnetic field (e.g., a dipole transmitting antenna; but a kilometer-long lightning discharge, fluctuations in atmospheric electricity associated with charged clouds, an AC-driven capacitor with an arbitrary plate spacing, a wire coil, etc., also constitute antennas in terms of wave emission). Until an electromagnetic wave of the type described above has formed at some distance from the antenna, complicated processes take place near a transmitting antenna which

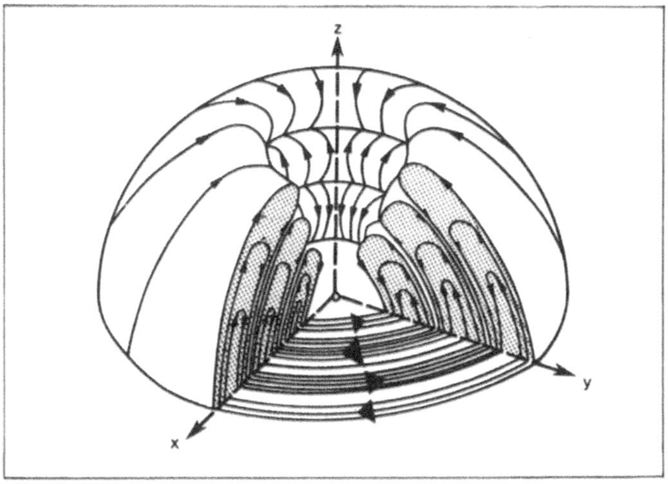

Figure A-5. Instantaneous diagram of the electric and magnetic distant field of a dipole.

cannot be fully discussed here. It is sufficient to point out that while changing electric fields always give rise to changing magnetic fields, and vice versa, the electric and magnetic field intensities near an antenna do not obey the relation $W = E/H$, as they do during propagation in free space. The near field is considerably smaller than would be the case in a regular electromagnetic wave. This effect is particularly marked when the dimensions of the antenna arrangement are considerably smaller than the wavelength corresponding to the transmission frequency (in a vacuum), which is especially true for relatively low frequencies. A strong wave can be emitted from an antenna system (without special technical tricks) only if the dimensions of the system enable it to set up electric and magnetic fields whose full dimensions are formed within a distance of about one wavelength from the antenna (e.g., one-half or one-quarter wavelength, as achieved by VHF dipole antennas).

The laws of induction and magnetic flux also state that the intensity of the electric field produced by a changing magnetic field, like that of the magnetic field produced by a changing electric field, is proportional to the time variation of the field: as the frequency decreases, the mutually generated fields become weaker.

In view of this, let us consider electric field fluctuations with a frequency of 10 Hz, as might be produced by an antenna arrangement several meters in length (laboratory capacitor array). Such an arrangement could never generate a magnetic field with a local or remote intensity that could be detected with the best measuring instruments or could be regarded as biologically active in comparison with the ever-present background level of other fields. An electromagnetic field with a frequency of 10 Hz has a wavelength of about 30,000 km in free space. The artificially produced electric field is many orders of magnitude smaller by comparison. Of course the capacitor also generates a stray field which can theoretically be infinite in extent, but at a distance of 30,000 km the intensity of this field is entirely negligible, as is the magnetic field this weak electric field produces.

It can be said, in summary, that time-varying electric fields generate magnetic fields, and time-varying magnetic fields generate electric fields. However, at low frequencies for which the dimensions of the antenna system are considerably smaller that the corresponding wavelength in free space, the intensity of these mutually generated fields is negligible.

Energy: The ability to do work. If a system contains energy, it can release this energy to the outside and so is capable of doing work. Energy can exist in various forms: potential energy (the energy of position), kinetic energy (the energy of motion), electric energy, magnetic energy, thermal energy (heat), chemical and atomic bonding energy, and mass energy (described by the relativity theory as $E = mc^2$). The theorem of the conservation of energy in a closed system, which asserts that the amount of energy contained in the universe is constant, is a basic axiom of physics. Energy in one form can be changed to energy in another form, but some transformations are irreversible: for example, any form of energy can be completely transformed into thermal energy, but thermal energy can never again be completely converted to any other form.

Force: A physical quantity which possesses the characteristics of a vector and can accelerate a material body or, when combined with other forces, deform it. Acceleration is not a prerequisite for the existence of a force, since forces acting on bodies at rest or in uniform straight-line motion may be in equilibrium and so exert a deforming, rather than an accelerating, effect. Three variables are needed to define a force uniquely: its magnitude, its direction, and the position of its line of action (Fig. A-6).

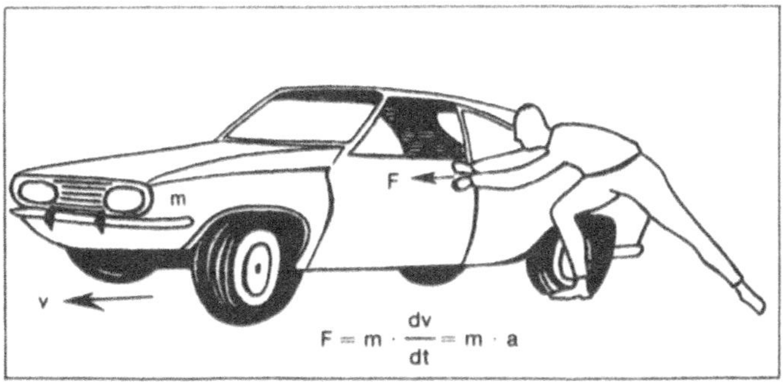

Figure A-6. Schematic diagram of the relationships between force F, mass m, velocity v and acceleration a.

Frequency: The number of cycles completed by any periodic process in a unit of time. If the unit of time is 1 s, the unit of frequency is the hertz (Hz). Frequency is a measure only of the number of cycles completed, without reference to the type of process involved, which may be an alternating current, an electric signal, an alternating field, a vibrating tuning fork, etc.

Gradient: A term from vector analysis. If u is a constant scalar function of spacial coordinates, as is true in physics for temperature, density or electric potential, then the spatial variation of u depends on the direction of progression in space. The gradient is a vector whose direction coincides with the direction of the greatest increase in the function u, and whose value is equal to this greatest change in the scalar position function u per unit distance.

Ion: An atom or molecule that has acquired a positive or negative electric charge by losing or gaining one or more electrons (see "Ionization"). Radical ions are formed when molecules break up into two electrically charged components, neither of which is a single electron (dissociation). Large ions are formed by the aggregation of many molecules, some of which are charged (cluster ions). Ions can thus exist in the form of atoms, radicals, molecules and clusters. A certain amount of energy is required for ionization or dissociation to occur: the "ionization energy" for the former, the "dissociation energy" for the latter.

Ionization: The process whereby an electron is detached from an atom or a molecule by the input of its ionization energy, forming a positive ion. But ionization can also mean the production of ion pairs, generally by some form of radiation (cosmic rays). This process also forms positive ions initially, but the detached electrons attach themselves to neutral molecules, forming one negative ion for each positive ion (Fig. A-7). The four most important agents of ionization are high temperature, electron impact, the collision of moving ions, and wave radiation (which forms the ionosphere, for example).

Lines of Force in an Electric and Magnetic Field: If the electric or magnetic field intensity is defined as a function of location, then every point in the field can be assigned a certain direction. The curves whose tangents coincide with the direction of the field at every point are called lines of force or lines of flux, for they indicate the direction in which the field force is acting. Because the field direction is uniquely defined at every point at a given time, one and only one line of force passes through each regular point in the field (exception: singular points in the field, such as charge locations or neutral points in an electric field). According to Faraday, the field intensity of an area is proportional to the

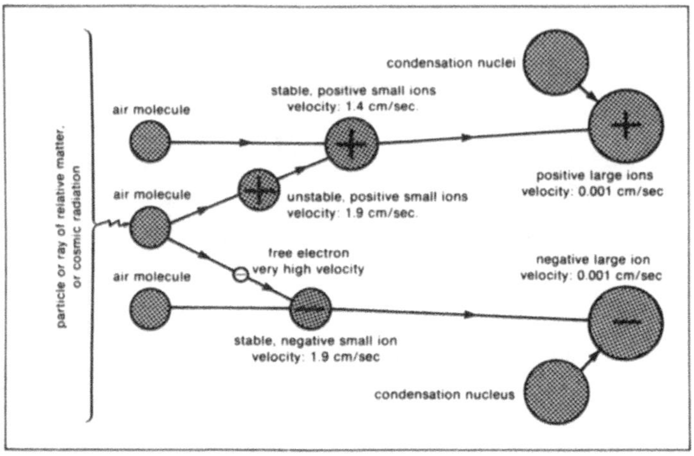

Figure A-7. The production of various types of air ions. After Wait (1934).

number of lines of force that intersect at right angles the standard cross-sectional area of 1 m^2 (Fig. A-5).

Magnetism

Magnetic Field: Magnetic fields are produced by charges in motion, that is, by currents and by rapidly changing electric fields (Fig. A-8). In turn, a magnetic field exerts a force, called the Lorentz force, on moving electric charges, and thus on current-carrying conductors. The magnetic field of permanent magnets is the result of tiny interatomic or intermolecular current loops. Like all force fields, the magnetic field is a vector field, meaning that it has a specific magnitude and a specific direction at every point. The magnetic field is characterized by the magnetic field intensity $\vec{H}$ (measured in A/m) or by the magnetic flux density (mag-

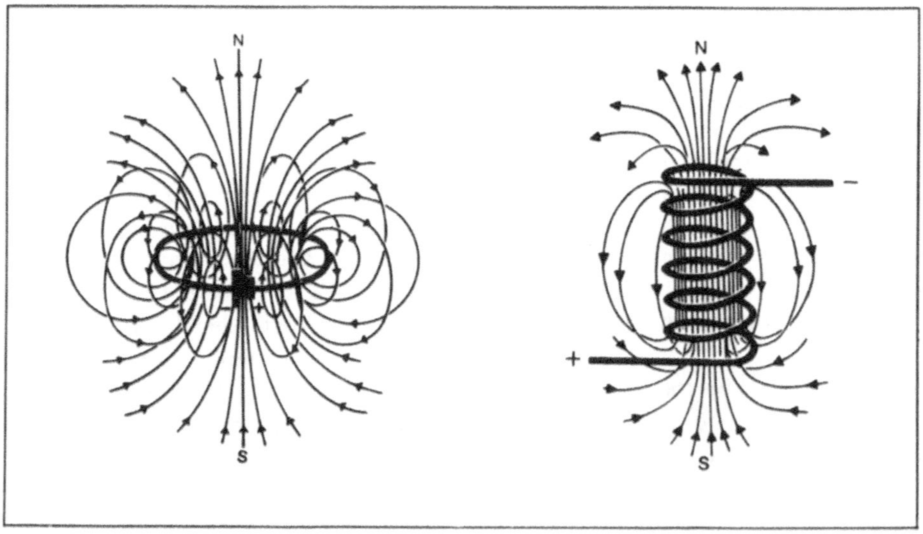

Figure A-8. Electric current produces a magnetic field. From left to right: Magnetic field pattern about a straight, current-carrying conductor; magnetic field pattern about an annular current; the magnetic field of a coil.

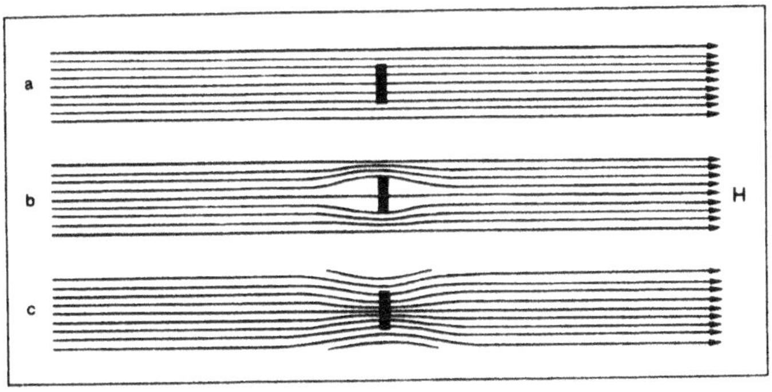

Figure A-9a–c. Bodies with different magnetic properties. ***a*** Magnetically neutral body with magnetic properties like those encountered in a vacuum, relative permeability constant $\mu_r = 1$; the path of the magnetic flux lines is not influenced. ***b*** Diamagnetic body, $\mu_r < 1$; the flux-line density is reduced. ***c*** Paramagnetic body with $\mu_r > 1$, or ferromagnetic body with $\mu_r \gg 1$; the flux-line density is slightly or greatly increased, respectively.

netic induction) $\vec{B} = \mu \vec{H}$ (measured in teslas (T), 1 T = 1 Vs/m^2), where μ is the permeability (q.v.).

Diamagnetism: When introduced into a current-carrying coil, diamagnetic materials (bismuth, mercury, sulfur) cause a decrease in the field set up by the current in the windings (Fig. A-9). Thus, the relative permeability μ_r of such materials is less than 1.

Paramagnetism, Ferromagnetism: Paramagnetic materials (platinum, oxygen) cause an increase in the field generated by a current-carrying coil (Fig. A-9). The relative permeability μ_r of such materials is only slightly greater than 1. Ferromagnetic materials (iron, cobalt, nickel) have the same effect, but to a greater degree, $\mu_r >> 1$.

Noise: Even if no external voltage is applied to a conductor, a small, rapidly fluctuating alternating voltage can be measured in the conductor. The free electrons in the conductor undergo irregular thermal motions, and occasionally many electrons will move in a common direction. This results in a short-lived voltage peak between the ends of the conductor. Because the spectrum of every voltage peak has a very broad, continuous frequency band which extends down into the audible range, the irregular voltage fluctuations in it are similar in spectral composition to noise: when amplified through a loudspeaker, they create the impression of noise. A theoretically important form of noise is "white noise," in which the amplitudes of the signals of all frequency components lying within a given bandwidth are equal.

Nonlinearity: The term nonlinearity is used in connection with electric processes when, for example, the proportionality of Ohm's law is no longer guaranteed; in other words, when the current through a certain load is no longer in direct proportion to the applied voltage. A typical example of a load with a nonlinear resistance is the incandescent light bulb. When the filament is heated, its resistance increases. If an increasing amount of current is passed through an incandescent bulb, its resistance does not remain constant, but increases. The necessary voltage increases faster than the current, and a nonlinear relationship exists between current and voltage.

Biologic cell systems also possess a nonlinear electric resistance. If a sinusoidal alternating voltage is applied to such a system, for example, the resulting current is not proportional to the voltage, because the electric resistance of cell systems varies with the polarity of the voltage. Thus, the positive and negative half-waves of the resulting current are unequal in size, and so the quantity of charge that is transported in one direction during the positive half-wave is not as large as that which flows in the opposite direction during the negative half-wave. The net result is a unidirectional transport of charge through the cell over a period of time, or a "rectifying effect." In alternating current, by contrast, there is no net transport of charge over a period of time.

Period: The period T of an oscillating process is the time required for the process to undergo one complete oscillation (Fig. A-3). In the case of an alternating current, this could be the interval of time T between two successive positive current maxima. If we then compute the number of oscillations that occur per second, this gives, by definition, the frequency f. Thus, $f = 1/T$.

Permeability: A magnetic field can be described in terms of the vectors of the magnetic induction $\vec{B}$ (flux density) and magnetic field intensity $\vec{H}$, $\vec{B} = \mu\vec{H}$. The proportionality factor between the two is the permeability μ. The absolute permeability μ_0 describes the relationship in a vacuum, while the relative permeability μ_r describes the permeability relative to a vacuum, $\mu_r = \mu/\mu_0$, for $\mu_0 = 4\pi \times 10^{-7}$ Ωs/m (Fig. A-9).

Potential: In an electrostatic field $\vec{E}$, the force, $\vec{F} = q \cdot \vec{E}$, acts on an electric charge q. If the charge is moved so that its path is always perpendicular to the force, and thus perpendicular to the field intensity, no work is done. This is possible on the "equipotential surfaces" of the electric field. The potential difference between two equipotential surfaces (1 and 2) is the voltage;

$$U_{12} = \psi_1 - \psi_2 = \int_1^2 E \cos \alpha \, ds$$

where E is the electric field intensity, ds is an element of the distance from point 1 on the equipotential surface with potential ψ_1 to point 2 on the surface with ψ_2 and α is the angle between the direction of ds and E. If a charge q is moved from 1 to 2, the work $W_{12} = qU_{12} = q(\psi_1 - \psi_2)$ is done. Like the voltage, the potential is measured in volts.

Power: Power is equal to the work done (W) divided by the time ($\pm$) required to do it, thus $P = W/t$, or more generally, $P = dW/dt$. The unit of power is the watt, 1 W = 1 J/s = 1 Nm/s = 1 m^2kg/s^3.

Often the following conversions must be made:

1 hp = 75 mkp/s = 735.6 W; 1 mkp/s = 9.81 W; 1 kW = 1.36 hp = 860 kcal/h; 1 W = 0.239 cal/s.

Pulse Duty Factor: If periodic pulses are formed, as by switching a current on and off at regular intervals, the ratio of the interval in which a signal is present (on-time) to the interval in which no signal is present (off-time) is called the pulse duty factor. The sum of the on-time and off-time is equal to the period of the process.

Pulse Repetition Frequency: If any kind of signal is repeated periodically, such as a current switched on and off at regular intervals, the rate of recurrence of the resulting pulses is called the pulse repetition frequency, which is simply the number of signals that occur per second. The pulse repetition frequency f (in Hz) is also defined as the reciprocal of the time interval T (in s) between two successive pulses ($f = 1/T$).

Radiation: Radiation occurs in two basic forms: electromagnetic wave radiation (e.g., X-rays, light, infrared radiation), and corpuscular radiation (e.g., electron beams, neutron beams). Various types of radiation are recognized (dtv, 1970).

Characteristic Radiation: In X-radiation, a distinction is made between the nonspecific bremsstrahlung (braking radiation), which shows a continuous spectrum, and the characteristic radiation, which exhibits a specific emission and absorption in the atom.

Extraterrestrial Radiation: The radiation, mostly solar, which strikes the outer boundary of the earth's atmosphere, where atmospheric absorption is not yet a factor. The nature of the extraterrestrial spectrum, and thus the intensity of the extraterrestrial radiation (solar constant), can be determined only indirectly by intensity measurements at various altitudes and extrapolation to the atmospheric boundary.

Cosmic Rays: Cosmic rays were discovered in 1913 by Hess and Kolhörster, who independently observed that ionization in the atmosphere increases with altitude. It is now known that radiation from outer space triggers highly complex secondary and tertiary processes within the earth's atmosphere. It is a common practice to refer to this radiation together with its products as cosmic radiation or cosmic rays, neglecting the fact that the products are entirely of terrestrial origin. In a stricter sense, cosmic rays include only the primary radiation itself, which consists of nucleons.

Owing to their extremely high energies, the particles in the primary radiation shatter any atoms and atomic nuclei which they encounter. This gives rise to new elementary particles such as mesons, pi mesons and, if these are electrically charged, electrons, neutrinos and protons, to name only a few of the most important components.

Natural Radiation: Planck's hypothesis of natural radiation asserts that the phases and amplitudes of the infinitely numerous, strictly monochromatic wave trains of which even the narrowest spectral line is composed (according to the Fourier theorem) are entirely independent of one another (incoherent radiation).

Corpuscular Radiation: All types of radiation which consist of moving material corpuscles: atomic and molecular beams, electron beams, meson beams, ion beams, neutron beams, etc. The "opposite" of corpuscular radiation is wave radiation, although the distinction is blurred by the duality of waves and particles, for wave radiations also exhibit corpuscular properties. The only significant difference between the two is that the particles in corpuscular radiation have a resting mass different from zero and therefore travel at sublight velocities.

Rectification: See "Nonlinearity."

Spectrum: The distribution of the components of any physical process over another physical quantity. The most familiar is the frequency spectrum, such as the multichromatic spectrum of sunlight, which shows light intensity as a function of frequency. In the frequency spectrum of an electric signal, the signal intensity is shown as a function of frequency.

Vector: A physical quantity which possesses both magnitude and direction. A vector is drawn as an arrow in space. The length of the arrow represents the magnitude of the quantity as well as its position in space; the direction of the arrow indicates the path along which the physical quantity acts. Forces, field intensities, moments, etc., can be represented as vectors. The symbol for a vector quantity is $\rightharpoonup$ or $\rightarrow$ (e.g., $\vec{E}$ or $\vec{F}$).

Voltage: A potential difference (see "Potential"), or an ability to do work per charge unit (unit of measure: the volt). In electrical units of measure, the work W is given by the product of the voltage U, current I and time t ($W = U \cdot I \cdot t$, in volt–ampere–seconds or joules). Since the unit of charge is 1 As (ampere–second), the voltage U is the amount of work that can be done per charge unit.

Work: "Work equals force times distance" is applicable in the special case where the direction of the force and the direction of the displacement coincide, and the force is constant. The following serves as a more general definition of mechanical work. If ds is an element of the distance s by which the point of application of a variable force F is displaced, and α is the angle between the direction of the force and the direction of ds, then the elementary work is given by the

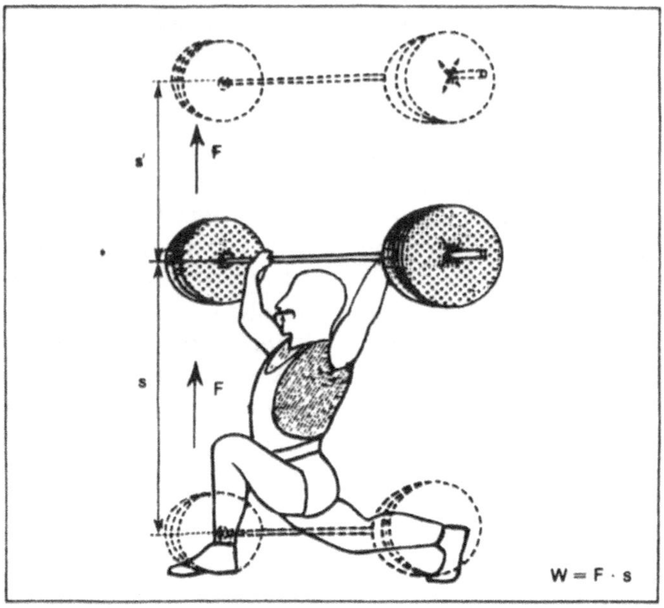

Figure A-10. Schematic diagram of the relationships between the work W to be done and the force F required to do the work over the distance s.

equation $dW = F \cdot \cos\alpha \cdot ds$, and the total work done in displacing the point by the distance s is given by $W = \int F(s) \cdot \cos\alpha(s) \cdot ds$.

Work can be added to or taken from a system by mechanical, thermodynamic, electric and other physical processes. Work added to the system increases the amount of energy in the system, while work taken from the system reduces it. Energy can be stored in various forms (see "Energy"). Let us assume, for example, that a certain force must be exerted over a distance h in order to lift a weight to a certain height above the floor (Fig. A-10). The work performed in doing this is stored in the form of potential energy. If the weight is allowed to fall, the potential energy is changed to kinetic energy (velocity of weight) during the fall. When the weight strikes the floor, this energy is transformed partly into thermal energy and partly into work deforming the weight or flooring.

Wavelength: See "Electromagnetic Waves."

B
Parapsychological Observations

According to Presman (1970), one cannot discuss the transfer of information by organisms with the aid of electromagnetic forces without also touching upon the problem of parapsychology. The observation of parapsychological phenomena began with the founding of the Society for Psychical Research in London in 1882. This branch of psychology deals with certain abilities that are apparently derived from psychic powers and are independent of the known sense organs, including the ability to receive and transmit information (telepathy), to determine the nature and location of objects which are imperceptible by the normal senses (telesthesia), to move objects by the force of the mind (psychokinesis), to perceive past occurrences (retrocognition) and to foretell the future (precognition). The techniques employed in the parapsychological research of telepathy and telesthesia are essentially as follows:

1. The observation and analysis of cases of "spontaneous telepathy," in which persons experience a sudden feeling of anxiety with regard to a close friend or relative, with no apparent cause. This feeling may take the form of a vague apprehension or a more specific feeling that someone has fallen ill, had an accident or died.

2. Experiments with a "sensitive" person who attempts, with or without personal contact (hand holding), to receive information from another person.

3. Experiments with cards (5 cards with 5 different geometric symbols, called Zener cards), in which one person tries to impart his knowledge of a random series of cards to another person by "thought transference."

4. Experiments in which one person tries to guess an object viewed by another person. Such investigations are the subject of various publications, which describe cases of telepathy and telesthesia with "mediums," or persons credited with special powers in this regard. In some experiments it is reported that communication was effected over distances of several thousand kilometers, even when one of the participants was contained in a shielded room. Sensational experiments in psychokinesis involving the bending of eating utensils received considerable publicity some time ago. They prompted more rigorous experiments by the members of an official scientific institution. In almost every person tested, it was found ultimately that some sort of trick had been resorted to. There were only two cases in which no evidence

of tricks or deception could be found, despite the most stringent investigative methods and the most modern measuring apparatus. One case involved a 12-year-old girl who was reportedly able to cause spoons or forks to bend 90° simply by touching them with her index finger. Ordinary metal paper clips placed in a closed cardboard box were noticeably bent when the girl shook the box gently. In the second case, a man in a restaurant attracted attention by breaking spoons in two with a touch of his index finger, and then rejoining the pieces with the handle of the spoon reversed. Under the strictest precautionary measures, this man was subjected to a test which was relatively simple but provided objective, recordable results. The man was given an ordinary 68-ohm electrical resistor to hold between his thumb and forefinger and was to attempt to change the resistance value, shown on a digital display. The progress of the experiment was recorded automatically. About 10 min after the start of the experiment, the record showed sudden, completely unmotivated jumps in resistance ranging from the design value of 68 Ω up to about 250 Ω—more than a threefold increase. Analysis of the resistors after the experiment showed no peculiarities. No explanation could be found for these phenomena.

The results of all such investigations prompted the parapsychologists to declare that telepathic communication, at least, appears to be independent of distance and material barriers. To explain the various phenomena, Wassermann (1956) postulates the existence of energy forms such as "psi fields" and "biologic quanta." Deep-penetrating neutrinos and the gravitational field are also mentioned. But the possibility that telepathic communication may be an electromagnetic phenomenon is also suggested, perhaps based on the utilization of signals in the 10^{19}–10^{29} Hz range. Signals of a lower frequency range, extending down to the extremely long radio waves, are also mentioned in connection with telepathic communication.

The Italian researcher Cazzamali (1941) even reports on the emission of electromagnetic waves from the brain. No reports are yet available which confirm Cazzamali's measurements. However, his other experiments in which hallucinations were induced in persons exposed to electric fields were successfully reproduced by Jaski (1960).

The poor reproducibility of previous experimental results has raised justifiable doubts as to the accuracy of reported results. Hansel (1966), in particular, comes to the conclusion that absolutely all results of telepathic investigations are unreliable, the product of faulty experimental procedure, or stem from an incorrect interpretation of statistical data. Despite his attitude, Hansel (1966) does not rule out the possibility of telepathic phenomena, but does demand that evidence be based on properly conducted and reproducible experiments.

Presman (1970) agrees with Hansel in principle, but states that the problem is not only faulty technique, but also the selection of humans as experimental subjects. This conclusion can be reached from the standpoint of evolutionary theory. The ability of animals or organisms to exchange information at a distance without the aid of the known senses can be viewed in evolutionary terms as a means of increasing the individual's chances of successfully competing in the struggle for survival and thus promoting the preservation of the species. As man evolved, he continuously improved his means of artificial communication, so that his ability to transmit and receive such signals gradually diminished and was finally lost. Thus, the man of today possesses these abilities only in extremely rare cases and is the least suitable subject for investigations on the transmission of bioinformation. Animals should be used for this purpose, and the results can then be applied to man.

The ability of man to transmit bioinformation should also be manifested in emotional responses. From this standpoint, the traditional methods of investigating telepathic communication are unsatisfactory. It is unlikely that experiments with symbols printed on cards would elicit any mood responses in the medium who is to receive the telepathic information, for such responses are inconceivable based on such a primitive information content. In any case, previous experimental results make it clear that such information transfer in animals is associated with the production and reception of electromagnetic energies by methods of an as-yet unknown physical nature.

Techniques are known for the direct reception of signal transmissions between animals. At present, however, there is no theoretical or experimental indication that man possesses either telepathic or psychokinetic abilities. It is also the international consensus that the results of parapsychological research in this area apparently cannot be regarded as sufficiently reliable. It is possible, of course, that many experimental results of responsible institutions are not made public simply for fear of being discredited in the sientific community.

C

Acceleration

"Acceleration" refers to the increased rate of growth and development which has been observed in children during the past 100–150 years. Presman (1970) discusses this phenomenon and points out the following quantitative and qualitative characteristics of acceleration:

1. Accelerated growth is noted even during the fetal stage of development. Newborns have increased an average of 5–6 cm in length and 3%–5% in weight during the past 100 years.

2. The acceleration of sexual maturity is expressed in an earlier onset of puberty—about 2 years earlier for boys and 3 years earlier for girls, compared with 100 years ago. In women, moreover, the onset of menopause has been delayed an average of 3 years during the last 50 years.

3. Acceleration is not proceeding at a uniform rate. From the period from about 1830 to the beginning of the 1930s, the average height of juveniles increased by about 0.5 cm per decade. During the last 40 to 45 years, however, juvenile growth has increased rapidly, by up to 5 cm per decade. These statistics suggest that the increase, like other biologic processes, may proceed exponentially.

4. Acceleration is a worldwide phenomenon.

5. Acceleration is most marked in large cities and is least evident in rural areas (Fig. C-1).

Presman (1970) discusses the various possible causes of acceleration. According to the solar hypothesis, it is due to the more frequent exposure of children to the sun. According to the most widely accepted hypothesis, it can be traced to improved nutrition. Other causes could be the increased vitamin intake of mothers and children (the use of vitamin B_6 in particular stimulates growth), the increased intellectual demands made on children, and progress in medicine. It is also suggested that acceleration has to do with increasing exposure to ionizing radiation. This is hardly likely, however, because the epochal shift was noted long before the widespread use of radiography and atomic testing. Each of these factors probably plays a certain role, but none of them can explain all the manifestations of acceleration: its global character, the duration of the periods of its occurrence, and its nonuniform rate of progression. Presman (1972) advances the hypothesis that the main cause of acceleration may be the grad-

Figure C-1. The hypothesis that growth is accelerated by an increase in the energy level of natural and artificial electromagnetic fields was advanced by Presman. One is struck, in any case, by the number of young people who are tall (Kaufinger and Neuhauser Strasse in Munich, Germany).

ual increase of natural and artificial electromagnetic field energies in the low- and radio-frequency range in the biosphere, and that acceleration is therefore a sign of man's adaptation to these environmental changes. There is certainly much to support this hypothesis, but strong objections must be raised if a gradual increase in the natural electromagnetic fields in the biosphere is submitted as a significant contributing factor. Aside from certain long-term fluctuations, which have probably existed since the earth was born, the concentration of tall buildings in large cities plays a special role, according to Presman (1972). The electromagnetic phenomena caused by discharges, due to thunderstorms or otherwise, are supposed to increase in such areas. However, Presman fails to take into account the shielding effect of modern buildings. Electric fields, which play a dominant role in the processes mentioned, are much more strongly attenuated by the field-distorting effect of buildings than in, say, a village where construction is far less dense, the houses are smaller, and wood is a principal building material. The objections become even stronger when one considers the time spent by persons in dwellings that are constructed mostly of steel-reinforced concrete. Natural electric fields are greatly weakened by such structures, which would lead one to expect a retardation of growth in cities rather than an acceleration.

It is probable, therefore, that man-made electromagnetic energy sources, which have steadily increased since the beginning of the 1930s, have become an important factor in the biosphere: power supply facilities, telephone systems, radio stations and television transmitters, to name only the most widespread. Surprisingly, Presman discusses only the radio waves emitted by radio and television stations in various frequency ranges, apparently overlooking the existence of artificial, low-frequency electromagnetic energies. Of particular importance in this regard is the power-supply network (50- or 60-Hz household power supply, 16 $^2/_3$ Hz railway power supply, local electric transportation systems and other DC-powered systems), which has by now penetrated into the farthermost reaches of human civilization. The biologic activity of such electromagnetic fields can no longer be disputed. But apparently Presman is not well informed on the current status of low-frequency field research, for while he discusses the pulsed form and higher intensity level of natural electromagnetic fields at night, he does not point out that this is true only for high-frequency radiation. In the range of the Schumann resonances (10 Hz), conditions are exactly reversed.

If electromagnetic energies are responsible for the phenomenon of acceleration, they are mainly the artificially produced high- and low-frequency energies that are present in every home (television, radio, telephone, lighting, refrigerator, cooking appliances, household appliances, etc). This does not take into account the effects of nearby high-voltage transmission lines, overhead railway lines, or AC- and DC-powered streetcars and subways.

Presman (1972) further points out that chronic exposure to electromagnetic fields of higher intensity than natural fields at corresponding frequencies can apparently lead to an impairment of reproductive ability and thus to developmental retardation as well. This change in environmental conditions has taken place in much too short a time, however, to permit an adaptation to the new situation at the genetic level. More probably, the adaptation is phenotypic in character. The phenomenon of acceleration—the prolonged reproductive period, earlier sexual maturity and increased growth rate—must be regarded, therefore, as an overcompensation for the increased level in the biosphere of electromagnetic forces that retard these processes.

D
Heredity

The studies of Gauquelin (1961) on "planetary heredity" give cause for speculation concerning the influence of electromagnetic radiations. Gauquelin reports that a study of more than 40,000 comparisons between the planetary positions occurring during the births of parents and their children showed that certain configurations can apparently be transmitted from parent to child. The rule can be stated as follows: Mars, Jupiter, Saturn, the Moon and Venus tend to occupy celestial positions during the birth of children which show similarities to the positions they occupied during the birth of the parents. For example, children are born more frequently after the ascension or culmination of one of these planets if the same planet was in the same region of the sky during the birth of the parents. The results depend on the circumstances of the birth. They are more apparent in the case of natural deliveries with no surgical intervention, while they are correspondingly adulterated in deliveries that are artificially influenced. There are significant differences among the five planets mentioned with regard to the degree of the hereditary effect. The statistical relationship decreases from the Moon to Saturn and is scarcely evident in the case of the latter planet (Fig. D-1). The question remains whether a scientific basis for astrology can be seen in this phenomenon.

A similar question is raised by the statistical studies of Kop and Heuts (1974) on connections between the month of birth and choice of partner in marriage. They found that men born during the second half of January show a statistical preference for marrying women born during the second half of October. The same applies to men born from April through mid-May, who marry women born during the same period. Men born from mid-June through July tend to marry women born from mid-July through mid-August. People most frequently marry a partner whose month of birth is 0–2 months earlier or later than their own. As for divorce, it was found that men born in March who marry women born in July are eight times more likely to divorce than in any other combination of months. A similar, though weaker tendency is found in men born in February whose wives were born in May. Men born during the second half of November are twice as likely to divorce as other men.

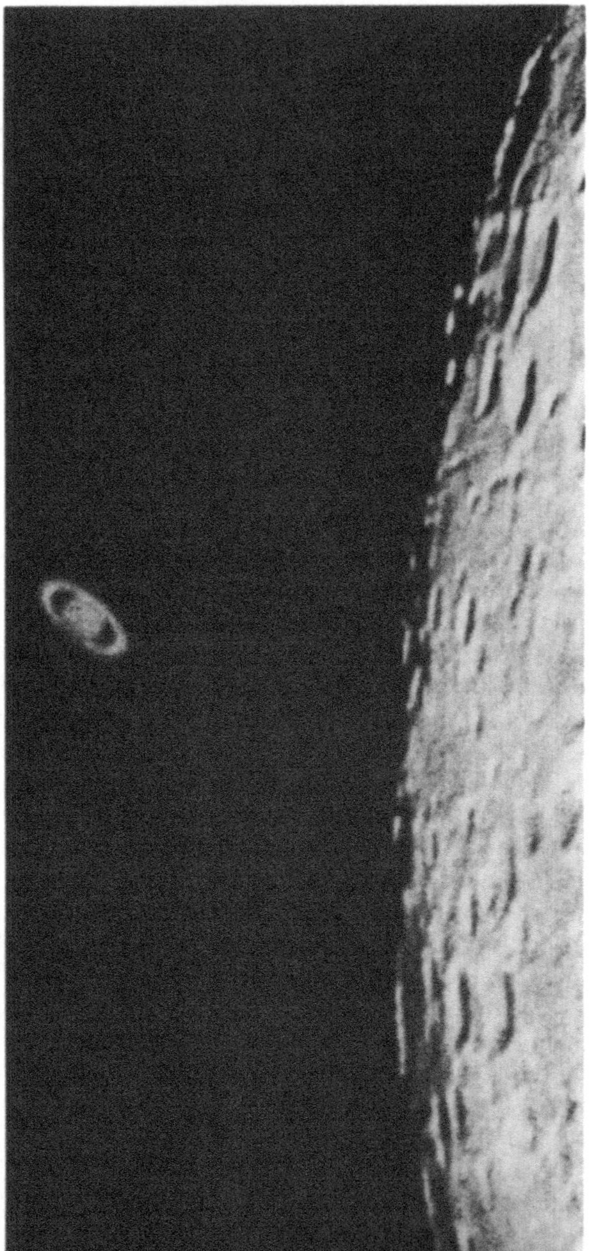

Figure D-1. No firm scientific basis has yet been found for the astrological significance of the planets. Nevertheless, statistical studies have revealed certain interesting aspects (photograph of Saturn with lunar craters in foreground, from the Bochum Observatory in Germany).

E

Heliobiology

It is certain that solar activity and everything connected with it plays a decisive role, directly or indirectly, in the course of various biologic processes (Fig. E-1). According to a report by Gnevyshev and Novekova (1972), this has led in the USSR to suggestions that a new branch of science be founded, "heliobiology." This move was motivated in part by the results of a study showing a statistical connection between heart attacks and days with high geomagnetic activity. An especially strong correlation was found for heart attack cases with fatal outcomes. A corresponding comparison of data on stroke cases yielded the same results. These studies are considered confirmation that geomagnetic factors associated with fluctuations in solar activity have an important direct or indirect influence on living organisms.

High sunspot activity is indicative of increased solar activity, which in turn is accompanied by an intensified solar wind. Whether this intensified solar activity (increased sunspots and chromospheric flares) will affect the earth depends on whether the flow of ejected particles envelopes the earth or bypasses it. This flow of particles (intensified solar wind) contains cosmic-ray protons and electrons whose velocity and energy depend on the magnitude of the chromospheric flare. These particles rarely reach the earth's surface, but this may occur after very powerful chromospheric flares once every several years. Usually the geomagnetic field prevents these particles from reaching the earth's surface.

Another solar phenomenon that is emitted toward the earth is wave radiation (electromagnetic waves, light, X-rays). The light and electromagnetic waves reach the earth's surface, while the gamma and X-rays are completely absorbed by the atmosphere. When the intensified solar wind envelopes the earth, it deforms the magnetosphere. Charged particles can penetrate through the neutral layers of the magnetosphere and the polar auroral zones, causing an ionization of the upper atmospheric layers. As a result of these deformations and the ionization of the upper air, geomagnetic storms are formed. Mikhov (1976) seconds the opinion that intensified solar activity influences the earth mainly indirectly through an increase of geomagnetic activity. Proceeding from this assumption, he investigated the correlation between geomagnetic activity and infant mortality, as well as deaths from cerebrovascular disease, ischemic heart disease and stom-

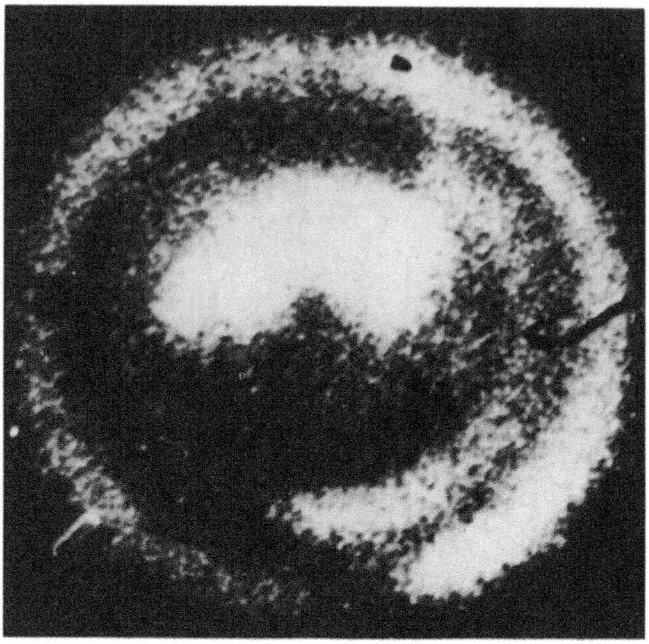

Figure E-1. The X-radiation of the sun, photographed 200 km over Mexico with a pinhole camera. Visible light was filtered out with a thin aluminized plastic film. Zones of intense radiation can be seen at the center and limb of the solar disk. The black spots are caused by dust particles.

ach cancer. Over a period of 6 years, 20,000 to 70,000 mortality cases were correlated. A correlation coefficient greater than 0.5 was found for cerebrovascular disease, ischemic heart disease and infant mortality, while no correlation was found between stomach cancer deaths and geomagnetic activity. In the opinion of Mikhov (1976), organisms respond nonspecifically to changes in the geomagnetic field. This nonspecific response, similar to the stress response, could lead to a fatal outcome under certain circumstances (impaired vital functions or diminished compensatory capacity).

Another example of solar influences is the effect of sunrise on the flocculation of human blood, described some years ago by Takata and Dohmoto (1936, 1937). The problems associated with this phenomenon are clarified to some degree by recent research results, especially in the area of ELF atmospherics—especially since Baudrexl (1952), on repeating Takata's experiments, found that the sunrise effect of Takata and Dohmoto (1936, 1937) is accompanied by a parallel change in the albumin/globulin ratio, as indicated by electrophoretic serum analysis. The sunrise effect, moreover, can be observed only if the blood sample is taken under insulated conditions, as pointed out by Takata. The presence of an artificial electric charge on test subjects was also sufficient to cause a change in the protein composition of the serum. The occurrence of special ELF atmospheric electrical processes at sunrise (see Fig. 3-12 and Fig. 5-2, type IV) may provide a bridge to understanding the nature of these electric phenomena.

The possibility of biologic effects from solar activity is a subject of particularly intensive research in the USSR. Gauquelin and Gauquelin (1975) present a review of recent Soviet studies in the area of plant growth, animal behavior, disease epidemics and circulatory disorders.

Both Soviet and British scientists (Brookes et al., 1976) have pointed out that the sun pulsates with a period of 2 hr,

40 $\pm$ 0.5 min. The surface of the sun reportedly moves inward and outward in the line of sight (i.e., radially) with a mean velocity of 2 m/s and amplitude of approximately 10 km. This not only poses certain theoretical problems for astronomers regarding the internal structure of the sun; it also suggests that there may be geophysical rhythms that have the same period. Certain connections with parallel biologic rhythms, in turn, cannot be rejected out of hand.

F

Conclusion of the Public Service Commission of the State of New York's Opinion #78-13 on 760-kV Transmission Lines

We find, on the basis of the record in the common record hearings and in Case 26529, that the operation of the facilities proposed by PASNY in Case 26529, the construction of which we have previously approved, is needed to serve the public interest, convenience and necessity, and will have, if conducted in accordance with the conditions described in this Opinion and specified in its ordering paragraphs, the minimum adverse environmental impact considering the state of available technology and the nature and economics of the various alternatives. We also find that the standards and conditions here adopted should also be applied to the transmission line proposed by Rochester Gas and Electric Corporation and Niagara Mohawk Power Corporation in Case 26559. We shall issue presently an order requiring Niagara Mohawk Power Corporation to show cause why these standards and conditions should not be applied to its Volney-Edic 765 kV transmission line, which we certificated in 1974 but which has not yet been built. It is also our intention to apply these standards and conditions, to the extent pertinent, to the operation at 345 kV of the transmission lines already certified in Cases 26462 and 26758 and Case 26717 and to all future proceedings pursuant to Article VII of the Public Service Law.

The Commission orders:

1. Subject to the conditions set forth in this Opinion and Order and in all previous applicable orders, the certificate of environmental compatibility and public need previously granted, in Case 26529, to the Power Authority of the State of New York is extended to authorize the operation of the transmission facilities to which it applies.

2. The operation of the 765 kV transmission lines here authorized is conditioned upon the following:

(a) PASNY is to acquire a right-of-way sufficient to exclude existing residences in an area extending 175 feet on each side of the centerline of the certified route.

(b) PASNY must acquire permanent rights to bar future residential development within a zone extending 125 feet on each side of the centerline of the certified route; it must also acquire rights to preclude, for a period not less than seven years, future residential development within a zone extending an additional 50 feet on each side of the centerline. The Commission reserves the right to require those additional rights to be made permanent or to permit earlier development if warranted by the result of the program of studies described in Ordering Paragraph 5.

(c) PASNY shall report to the Commission and attempt to resolve all complaints concerning audible noise produced by the lines. In the event such a complaint is made by the owner of a house located within a zone extending from the edge of the right-of-way to a point 600 feet from the centerline of the certified route and cannot be satisfactorily resolved by other means, the Commission may require PASNY to offer to purchase or move that house. This option shall exist for a period of 18 months from the date on which the 765 kV transmission line is made fully operational. The resale by PASNY of any such house shall be on notice to the buyer of the events that resulted in PASNY's having acquired it.

(d) PASNY shall contribute an amount, to be determined by the Commission but not to exceed 2% of the total cost of constructing the facilities here certified, toward the funding of the program of studies described in Ordering Paragraph 5.

(e) PASNY shall undertake a suitable program, consistent with this Opinion, for grounding and bonding fixed metal objects on the right-of-way and large movable metal objects likely to be brought on the right-of-way. It shall also undertake a suitable program for informing persons living near the right-of-way of the possibility of induced shocks from the lines and the best methods for avoiding them.

(f) PASNY shall serve a copy of this Opinion and Order, together with a cover letter to be composed by the Commission's staff, upon every manufacturer of cardiac pacemakers in the United States and upon every association of cardiologists in New York State.

(g) PASNY shall establish a procedure for receiving, responding to, and reporting to the Commission every complaint concerning the operation of the transmission lines here certified.

List of References

Adey WR (1975) Introduction: Effects of electromagnetic radiation on the nervous system. Ann NY Acad Sci 247:15–20

Al'pert Ya L, Fligel DS (1970) Propagation of ELF and VLF waves near the earth (Wait JR, ed and trans). Consultants Bureau, New York

Altmann G (1959) Der Einfluss statischer elektrischer Felder auf den Stoffwechsel der Insekter Z Bienenforsch 4:199–201

Altmann G (1962) Die physiologische Wirkung elektrischer Felder auf Tiere. Verh Dtsch Zool Ges Wien 11:360–366

Altmann G (1968) Weitere Untersuchungen der physiologischen Wirkung elektrischer Felder auf Tiere. 62. Jahresversammlung der Deutschen Zoologischen Gesellschaft in Innsbruck (preliminary notice)

Altmann G (1969a) Die physiologische Wirkung elektrischer Felder auf Organismen. Arch Meteorol Geophys Bioklimatol [B] 17:269–290

Altmann G (1969b) Untersuchung der physiologischen Wirkung elektrischer Felder auf Tiere. Umsch Wiss Tech 69:242–243

Altmann G (1976) Physiological-ethological aspects of the influences of electric fields on the organism. Arch Meteorol Geophys Bioklimatol [B] 24:109–126

Altmann G, Lang S (1973) Die Revieraufteilung bei weissen Mäusen unter natürlichen Bedingungen, im faradayschen Raum und in künstlichen luftelektrischen Feldern. Information from: Inst. f. Zoologie, University of Saarland Saarbrücken

Altmann G, Soltau G (1974) Einfluss luftelektrischer Felder auf das Blut von Meerschweinchen. Z Angew Bäder-Klimaheilkd 21:28–32

Altmann G, Warnke U (1971) Einfluss unipolar geladener Luftionen auf die motorische Aktivität der Honigbienen. Apidologie 2 (4):309–317

Altmann G, Warnke, U, Paul R (1974) Physiologisch-ethologische Aspekte der Einwirkung elektrischer Felder auf den Organismus. Rep 1st Bioklimatol Wirk Luftelektr Faktoren, Graz.

Altmann G, Lang, S, Andres, G, Lehmair, M, Stucky, G, Schmidt G. (1975) Leistungsphysiologische Untersuchungen über die Beeinflussung von Schülern durch die Biotropie des Wettergeschehens. Z Angew Bäder-Klimaheilkd 22(4):330–336

Altmann G, Lang S, Lehmair M (1976a) Psychotrope Wirkungen des Wettergeschehens und eines künstlichen elektrischen Rechteckimpulsfeldes der Frequenz 10 Hz. Z Angew Bäder- Klimaheilkd 23, (5):407–420

Altmann G, Lang, S, Reuss T (1976b) Analysis of the lipid metabolism of mice in shielded condition and in an electric field with rectangular impulses of 10 Hz. Arch Meteorol Geophys Biokimatol [B] 24:109–126

Altmann G, Lang S, Rothe H-J (1976c) Some

effects of electric fields on the metabolism of pathogen nutrified rats. Arch Meteorol Geophys Bioklimatol [B] 24:109–126

Altmann G, Lang S, Niklas G, Sievert S (1977) Einwirkungen luftelektrischer Bedingungen auf den Stoffwechsel von Rana esculenta (in preparation)

Amineev GA, Sitkin MI (1965) The effect of a low-frequency alternating magnetic field on the behavior of mice in a T-shaped maze. Questions of hematology radiobiology and the biological action of magnetic fields. Tomsk, p 372

Andersen I (1972) Effects of natural and artificially generated air ions on living organisms. Biometeorology 5 (Part II).

Anselm D, Danner M, Kirmaier N, König HL, Müller-Limmroth W, Reis A, Schauerte W (1976) Untersuchungen des Einflusses von luftelektrischen Impulsfeldern auf das Fahr- und Reaktionsverhalten von Probanden im Kraftfahrzeug-Fahrsimulator. Ber Allianz-Zentrum Tech, TU München and Inst. Biomed Tech, München

Arendse M C (1976) Magnetic orientation of tenebrior molitor. 2nd Kolloq Bioklimat Wirk Luftelektr Faktoren, Munich

Asanova T, Rakov A (1966a) The state of health of persons working in the electric field of outdoor 400 kV and 500 kV switchyards. Gig Tr Prof Zabol 10:550

Asanova T, Rakov A (1966b) Health condition of workers exposed to an electrical field of 400–500 kV open distributing installations. Labor Hyg Occup Dis (Engl Transl) 5:50–52

Aschoff D, cited from Hartmann (1967, 1976)

Aschoff D (1973) Kann die offizielle Wissenschaft die Theorie der Krebsentstehung auf Reizzonen heute noch ablehnen? Z Biol Heilmethoden Fachfortbild Nr 4

Assael M, Pfeifer Y, Sulman FG (1974) Influence of artificial air ionisation on the human electroencephalogram. Int J Biometeorol 18:306–312

Assmann D (1963) Die Wetterfühligkeit des Menschen. Fischer Verlag, Jena

Athenstaedt H (1972) Pyroelectric behavior of integument structures and of thermo-, photo- and mechanoreceptors. Z Anat Entwicklungs gesch 136:249–271

Athenstaedt H (1974) Pyroelectric and piezoelectric properties of vertebrates. Ann NY Acad Sci 238:68–94

Bach S (1961) Changes in macromolecules produced by alternating electric fields. Dig Int Conf. Med. Electron 21:1

Bach S et al. (1961a) Effects of R-F energy on human gamma globulin. J Med Electron Vol. 9.

Bach S et al. (1961b) Effect of radio-frequency energy on human gamma globulin. In: Barnothy MF (ed) Biological effects of microwave radiation: Vol. 1. Plenum Press, New York, p 117

Bach W, Lang S (1976) Messung niederfrequenter elektrischer Felder mit einer halbleiterbestückten Elektrometersonde. Biomed Tech 21 (6):185–188

Bachmann CH, Reichmanis M (1973) Some effects of high electrical fields on barley growth. Int J Biometeorol 17 (3):253–262

Bachmann CH, McDonald RD, Lorenz PJ (1965) Some physiological effects of measured air ions. Int J Biometeorol 9(2): 127–139

Bachman CH, McDonald RD, and Lorenz CJ (1966) Some effects of air ions on the activity of rats. Int J Biometeorol 10:39–46

Bachman CH, Hademenos DG, Underwood LS (1971) Ozone and air ions accompanying biological applications of electric fields. J Atmos Terr Phys 33:497–505

Baenninger R (1971) Lecture Series on Motivation. Temple University, Philadelphia, Pa., May, 1971

Balser M, Wagner CA (1960) Observation of earth-ionosphere cavity resonances. Nature 188:4751

Barnell FH, Brown FA, Jr (1961) Magnetic and photic responses in snails. Experientia 17:513

Barnothy MF (ed) (1964) Biological effects of magnetic fields, Plenum Press, New York

Barnothy MF (ed) (1969) Biological effects of magnetic fields. Plenum Press, New York

Barrett AH, Myers PC (1976) Wellen aus der Tiefe. Selecta 6:530; cited in Science 190:669 (1975)

Barron CI, Dreher JJ (1964) Effects of electric fields and negative ion concentrations on test pilots. Aerosp Med 35:20–23

Bartels J (1960) Geophysik. Fischer Bücherei, Frankfurt

Basset CAL, Pawluk RJ, Pilla AA (1974) Augmentation of bone repair by inductively coupled electromagnetic fields. Science 184:575

Baudrexl A (1952) Über den Nachweis des "Sonnenaufgangseffekts nach Takata" mit

Hilfe verschiedener Untersuchungsmethoden. Dissertation, Med Fak Ludwig-Maximilians-Universität, Munich

Bawin S M, Gavalas-Midici R, Adey WR (1973) Effects of electric fields on specific brain rhythms. Symposium and workshop on the effects of low frequency magnetic and electric fields on biological communication processes at the 6th annual meeting of the Neuroelectric Society, Vol 6

Beccaria G (1775) Della Elettricitá Terrestra Atmosferica a Cielo Sereno. Torino.

Becker F (1955) Ergebnisse des Königsteiner medizinmeteorologischen Arbeitskreises. Ergeb Phys.-Didt. Ther 5:216

Becker F (1956) Arbeitsgrundlage der medizinmeteorologischen Vorhersage im Königsteiner Arbeitskreis. Med Meteorol 11:21–25

Becker F (1962) Meteorologische Grundlagen der Klimaheilkunde. In: Handbuch der Bäder- und Klimaheilkunde. Schattauer Verlag, Stuttgart

Becker F, Ströder U (1962) Wirkungen kurzfristiger Änderungen der Biosphäre. In: Handbuch der Bäder- und Klimaheilkunde. Schattauer Verlag, Stuttgart

Becker G (1975a) Einfluss von magnetischen elektrischen und Schwere-Feldern auf den Galeriebau von Termiten. Umschau 6:183–185

Becker G (1976a) Influences of magnetic, electric and gravity fields on termite activity. Mater Org [Beih] 3:407–418

Becker G (1976b) Einflüsse elektrischer and magnetischer Felder auf Termiten. 2nd Kolloq Bioklimatol Wirk Luftelektr Faktoren, Munich

Becker G (1976c) Reaction of termites to weak alternating magnetic fields. Naturwissenschaften 63 (4):20

Becker G, Kraus H (1964) Über neue Möglichkeiten zur Raumentkeimung in der Lebensmittelhygiene. Arch Lebensmittelhyg 15 (10):1–4

Becker G, Speck U (1964) Untersuchungen über die Magnetfeldorientierung von Dipteren. Z Vergl Physiol 49:301–340

Becker R (1963a) The biological effects of magnetic fields. A survey. Med Electron Biol Eng 1:293

Becker RO (1963b) Relationship of geomagnetic environment to human biology. NY State J Med 63:2215–2219

Becker RO (1975b) Cases 26529 and 27559. Common record hearing on health and safety of 765 kV transmission lines. Public Service Commission, New York

Becker RO, Bachmann CH, Friedman H (1961) Relation between natural magnetic field intensity and the increase of psychiatric disturbances in the human population. Int Conf High Magn Fields.

Beckett JC, Krueger AP (1961) Experiments with air ions on the trachea. Trans Am Inst Electron Eng 79:695–699

Beer B (1902) Über das Auftreten einer subjektiven Lichtempfindung im magnetischen Felde. Wien Klin Wochenschr 15:108–109

Begu del Blanco J, Romero-Sierra C (1973) Microwave radiometry techniques: A means to explore the possibility of communication in biological systems. Symposium and workshop on the effects of low frequency magnetic and electric fields on biological communication processes at the 6th annual meeting of the Neuroelectric Society, Vol 6

Bein W (1976) Experimente zur Wirkung des elektrostatischen Gleichfeldes auf menschliche Aktivität and Leistung. Dissertation. Universität Graz

Beischer D E (1969) Contribution in Barnothy (1969)

Beischer DE, Miller FF, (1964) Cited by Beischer (1964)

Beischer DE, Miller FF, II, Knepton JC (1967) Exposure of man to low intensity magnetic fields in a coil system, NAMI-1018, NASA R-39 Naval Aerospace Medicine Institute, Pensacola, Florida

Beischer DE, Grissett JD, Mitchell RE (1973) Exposure of man to magnetic fields alternating at extremely low frequency, NAMRL-1180. Naval Aerospace Medical Research Laboratory, Pensacola, Florida

Beischer DF (1964) Biological effects of magnetic fields in their relation to space travel. In: Schaefer KE (ed) Bioastronautics. Macmillan, New York, pp 173–180

Belyanskil VB, Mikhailova GA (1961) Investigations of the properties of Aeronomiya 1 (3)

Bergsmann O, Woolley-Hart A (1973) Differences in electrical skin conductivity between acupuncture points and adjacent skin areas. Am J Acupunct 1:

Bernstein SL, Burrows ML, Evans JE, Griffith AS, Mcneill DA, Niessen CW, Willim DK (1974) Long-range communication at extremly low frequencies. Proc IEEE 62, (3):292–312

Bertholon, L'Abbé P (1983) De l'électricité des végétaux. Paris.
Bhartendu, also see Elkieg et al. (1977) Menon IA (1978) Effects of atmospheric small negative ions on the oxygen consumption of mouse liver cells. Int J Biometeorol 22 (1): 43–52
Birnberger K (1973) Neurologisches Institut der Technischen Universität, Munich (personal communication)
Biró G, Sváb F (1970) Effect of negative ions on micro-organisms. Medicor News (Budapest) 4:121
Biró G, Novak J, Sváb F (1969) Application of aeroionizator to prevent hospital contaminations. Medicor News (Budapest) 4
Bisa K (1954) Die Reagibilität des menschlichen Organismus auf unipolar aufgeladene Aerosole, dargestellt am Verhalten der Flimmerverschmelzungsfrequenz. Aerosol Forsch 3:39–51
Bisa K, Weidemann J (1955) Die Elektro-Aerosole. Z Aerosol-Forsch Ther 3:220–251
Blackman VH, Legg AT (1924) Pot culture experiments with an electric discharge. J Agric Sci 14:268
Blakemoree R (1975) Magnetostatic bacteria. Science 190:377–379
Blanchi D, Cedrini L, Ceria F, Meda E, Re G (1973) Exposure of mammalians to strong 50 Hz electric fields. Arch Fisiol 70:30
Blondlot R (1904) N-ray controversy. Rev Sci 74:73–79, 545–552, 620–625, 705–709, 718–722, 783–785
Blondlot R (1905) "N"-rays, a collection of papers (Garcin J, trans) Academy of Sciences, London.
Boenko ID, Shakhgeldyan FG (1968) On the role of reflexogenic vascular zones in changes of blood coagulation during the effect of sound frequency electromagnetic field. Sechenov Physiol J USSR 54:8
Bonka H (1975) Strahlenbelastung in der Bundesrepublik Deutschland. Das natürliche Risiko. Bild Wiss Nr 6
Bonnevie T (1966) Pap 4th Congr Soc Biometeorol
Bookman MA (1977) The sensitivity of the homing pigeon, *Columba livia,* to an earth strength magnetic field. Nature 267:340–342
Bortels H, Massfeller D, Wedler E (1970) Zeitliche und örtliche Variationen der Atmungs- und Gärungsintensität von Sacharomyces-Hefe. J Interdiscip Cycle Res 1 (1), 107
Bose JC (1913) Researches on irritability of plants. Longmans, Green, New York
Bossert WH, Wilson EO (1963) The analysis of olfactory communication among animals. J Theor Biol 5:443–469
Boulatov PC (1968) Traitement de l'asthme bronchique par l'aéroionisation négative. In: Gualtierotti R et al. (eds) Bioclimatology, biometeorology and aeroionotherapy. Carlo Erba Foundation, Milan
Boyco AD, Sverchkov AN, Yaroshenko VA, Voytsekhovskiy OA (1964) All Union conference in Leningrad on aeroionization in industrial hygiene, November
Brauner L, Bünning E (1930) Geoelektrischer Effekt und Elektrotropismus. Ber Dtsch Bot Ges 48:470–476
Brezowsky H (1962) Biosphäre, Wetter und Organismus. Arztl Sammelbl 51(8):265–276
Brezowsky H (1964) Morbidity and weather. In: Medical climatology. Physical Medicine Library, New Haven, Connecticut, Chapter XIII
Brezowsky H (1965) Meteorologische und biologische Analysen nach der Tölzer Arbeitsmethode. Meteorol Rundsch 18 (5):132–143
Brezowsky H, Grimmeisen H (1967) Pathologische Reaktionen auf biotrope Wetterphasen. Ther Ggw 106(3):341–359
Brezowsky H, Ranscht-Froemsdorff WR (1966) Herzinfarkt und Atmospherics. Z Angew Bäder-Klimaheilkd 13
Bridges JE (1975) Biological effects of high voltage electric fields: Bibliography and survey of ongoing work, 1975, Res Proj 381–1, Final rep. IIT Research Institute, Chicago
Brinkmann J (1976) Die Langzeitwirkung hoher elektrischer Wechselfelder auf Lebewesen am Beispiel frei beweglicher Ratten. Dissertation, Technische Universität, Hanover
Brookes JR et al. (1976) Sonnen-Puls schlägt Wellen. Selecta 11:1078–1080; cited in Nature 259:87 (1976)
Brown FA, Jr (1962) Responses of the planarian, dugesia and the protozoan, paramecium to very weak horizontal magnetic fields. Biol Bull 123:264–284
Brüche E (1954a) Bericht über Wünschelrute, geopathische Reize und Entstörungsgeräte. Naturwiss Rundsch 9:367–377
Brüche E (1954b) Naturwiss Rundsch 11:454–465
Brüche E (1960) Problematik der Wünschelrute. Firma JR, Geigy AG, Basel

Budko LN (1964) Change in blood carbohydrate content due to the action of electromagnetic vibrations of audio- and radio-frequency ranges on organisms. Some questions of physiology and biophysics. Voronezh p 73

Bürklin F (1965) Vortrag auf der Jahrestagung des Forschungskreises für Geobiologie, Eberbach, Neckar

Busby DE (1968) Space biomagnetics. Space Life Sci 1:23–63

Busch HJ (1961) Nachweis des Einflusses eines elektrostatischen Feldes auf lebende Zellen. Naturwissenschaften 48:654

Busch HJ (1963) Nachweis des Einflusses eines elektrostatischen Feldes auf lebende Zellen. Naturwissenschaften 50:474

Busch HJ, Busch L (1965) Entwicklung von Massnahmen zur Beseitigung schädlicher Einflüsse auf den menschlichen Organismus in Panzern, U-Booten, Unterkünften und Schutzräumen durch künstliche elektrostatische Felder. Gutachten BVM (unpublished)

Busch HJ, Busch L (1966) Beitrag zur Messbarkeit der menschlichen Basisregulation mit physikalischen Methoden. Phys Med Rehabil 7 (7)

Cabanes J (1976a) Action des champs électriques et magnétiques sur les organismes vivants et très particulièrement l'homme. Revue générale de la littérature. Rev Gen Electr spec No, July

Cabanes J (1976b) Etude bibliographique sur l'action des champs électriques et magnétiques sur les organismes vivants. Rev Gen Electr Spec No, July

Cabanes J (1976c) Action des champs électriques et magnétiques sur les organismes vivants et très particulièrement l'homme. Revue générale de la littérature. RGE-FRA, ISSN 0035-3116, 07 Spec No, 19–26

Callot FJ, Lecoeur J, Revolier J (1973) Etudes du Gradient de Champ Electrique dans ces Rapports Physio-Pathologiques chez des Sujets Hivernant à Kerguelen. Int J Biometeorol 17 (3):233–238

Caymaz G (1976) Effect of increased atmospheric electricity on the blood electrolytes of airplane crew. Agard, Feb. 1976. Done by Turkish Air Force, Ataturk Sanatorium, Ankara

Cazzamali F (1941) Il nuovo apparato radioelectrico rivelatore dei fenomeni electromagnetici radianti del cervello umano. Energ 18:28

Cerretelli P, Malaguti C (1976) Recherches effectuées en Italie par l'ENEL sur l'influence des champs électriques à haute tension. RGE-FRA, ISSN 0035–3116, 07. Spec. Nr, pp 65–74

Chalmers JA (1967) Atmospheric electricity, 2nd edn. Pergamon Press, Oxford

Chapman FH, Jones DL (1964) Earth-ionosphere cavity resonances and the propagation of extremely-low-frequency radio waves. Nature 202:654

Charry J (1976) Meteorology and behavior: The effects of positive air ions on human performance, physiology, and mood. Univeristy Microfilms International, Ann Arbor, Michigan

Chiles WD, Fox RE et al. (1962) Effects of ionized air on decision making and vigilance performance, Tech Doc Rep Nr. MRL-TD-R-62-51. Wright-Patterson Air Force Base, Ohio

Cholodny NG, Sankewitsch EC (1937) Influence of weak electric currents upon the growth of the coleoptile. Plant Physiol 12:385–408

Clayton MD, Polk C, Etzold H, Cooper WW (1973) Absolute calibration of antennas at extremly low frequencies. IEEE Trans Antennas Propag ap-21 (4):514–523

Coate WB, Reno FE (1970) Rat fertility studies. In: Coate WB et al. (eds) Project sanguine. Biological effects test program pilot studies, Final rep. Prepared for the Naval Electronic Systems Command Headquarters, Hazelton Labs, Inc, Falls Church, Virginia, Chapter C

Cohen D (1975) Developments in ways to measure the extremely weak magnetic fields emanating from organs. Phys Today pp 35–43

Cone J (1971) Unified theory on the basic mechanism of normal mitotic control and oncogenesis. J Theor Biol vol. 30

Conley CC (1969) Contribution in Barnothy (1969)

Cook ES (1969) contribution in Barnothy (1969)

Crombie DD (1958) Difference between the E-W and W-E propagation of VLF signals over long distances. J Atmos Terr Phys 12:110

Crombie DD (1963) Non reciprocity of propagations of VLF radio waves along the magnetic equator. Proc. IRE 51(4)

Curry M (1946) Bioklimatik. Riederau (Ammersee Bavaria, Germany)

d'Alibard TF (1752) Letter to Academy of Science

Damaschke K, Becker G (1964) Korrelation der Atmungsintensität von Termiten zu Än-

derungen der Impulsfolgefrequenz der Atmospherics. Z Naturforsch [B] 19(2): 157–160
Daniel KWO (1965) Neue Therapie-Erfolge unter elektrostatischer Feldwirkung. Erfahrungsheilkunde 14:119
Dardymov IA et al. (1965) The effect of water treated with a magnetic field on plant growth. Questions of hematology, radiobiology and the biological action of magnetic fields. Tomsk, p 325
Daubert K (1958) Spezifische Reizkomponenten des Wetters und ihre Beziehungen zum gesunden und kranken Organismus. Med Meteorol 13 63–76
David T, Derau JV, Kornblueh I, McGurk FC, Minehart JR (1962) Ionization of the air. The sedating effect of polarized air In: Tromp SW (ed) Biometeorology. Macmillan, New York, pp 484–489
Davidson I, Henry JB (eds) (1969) Todd-Sanford's Clinical Laboratory Methods, 14th edn. Saunders Philadelphia
de Lorge J (1972) Operant behavior of rhesus monkeys in the presence of extremely low frequency–low intensity magnetic and electric fields: Experiment 1, NAMRL–1155. Naval Aerospace Medical Research Laboratory, Pensacola, Florida
de Lorge J (1973a) Operant behavior of rhesus monkeys in the presence of extremely low frequency—low intensity magnetic and electric fields: Experiments 2, NAMRL–1196. Naval Aerospace Medical Research Laboratory, Pensacola, Florida
de Lorge J (1973b) Operant behavior of rhesus monkeys in the presence of extremely low frequency—Low intensity magnetic and electric fields: Experiment 3, NAMRL–1196 Naval Aero Space Medical Research Laboratory, Pensacola, Florida
de Lorge J (1975) Do extremely low frequency electromagnetic fields influence behaviour in monkey? Int J Biometeorol 19, (Suppl 6): 1
de Lorge JO, Grissett JD (1977) Behavioral effects in monkeys exposed to extremely low frequency electromagnetic fields. Int J Biometeorol 21 (4): 357–365
de Rudder B (1938) Grundriss einer Meteorobiologie des Menschen, 2nd edn.
de Rudder B (1952) Grundriss einer Meteorobiologie des Menschen, 3rd edn.
Dessauer F (1931) Zehn Jahre Forschung auf dem Physikalisch-Medizinischen Grenzgebiet. Thieme, Leipzig
Dewan EM (1964) Communication by voluntary control of the electroencephalogram, XV-7. Datascience Laboratory, Air Force Cambridge Research Laboratories, Bedford, Massachusetts
Dewey ER (1971) Cycle synchronies. J Interdiscip Cycle Res 2 (3).
Diehl JC, Tromp SW (1954) Probleme der geographischen und geologischen Häufigkeitsverteilung der Krebssterblichkeit. Haug Verlag, Ulm
Dorland J, Brinker N (1973) Fluctuations in human mood. J Interdiscip Cycle Res 4 (1):25–29
Dowse HB, Palmer JD (1969) Entrainment of Circadian activity rhythmus in mice by electrostatic fields. Nature 222:564–566
Drischel H (1975) Organismus und geophysikalische Umwelt. Sitzungsber Sächs Akad Wiss Leipzig [Math-Naturwiss, Kl] 111 (2)
D'Souza LA et al. (1969) Contribution in Barnothy (1969)
dtv-Lexikon der Physik. (1970) Deutscher Taschenbuchverlag
Duffee RA, Koontz RH (1965) Behavioral effects of ionized air on rats. Psychophysiology 1:347–359
Düll B (1941) Wetter und Gesundheit, Vol I. Dresden
Düll B, Düll T (1936) Medizin-Meteorolog Statistiken. Berlin
Düll B, Düll T (1938) Erd- und sonnenphysikalische Vorgänge in ihrer Bedeutung für Krankheits- und Todesauslösung. Nosokomeion 9:103
Düll T, Düll B (1935) Zusammenhänge zwischen Störungen des Erdmagnetismus und Häufungen von Todesfallen. Dtsch Med Wochenschr 61:95
Dumanskiy Yu D, Popovich VM, Rokhvatilo EV (1976) Hygienic effect of an electromagnetic field created by high-voltage lines of electro-transmission (in Russian, Eng. summ) Gig Sanit 8:19–23
Durfee WK, Chang P, Polk C, Smith L, Yates W, Plant P, Muthkrishnan S, Chen H (1975) Extremely low frequency electric and magnetic fields in domestic birds , Tech Rep, Phase 1. University of Rhode Island
Eccles Sir JC (1953) The neurophysiological basis of mind. The principles of neurophysiology. Oxford University Press (Clarendon), London and New York

Eccles Sir JC (1957) The physiology of nerve cells. Johns Hopkins Press, Baltimore, Maryland

Eckert EE (1971) Wetter Boden Mensch 11:637

Eckert EE (1976) Plötzlicher and unerwarteter Tod im Klein-Kindesalter und elektromagnetische Felder. Med Klin 71(37):1500–1505

Edmiston J (1972) The effect to the field of a permanent magnet on the germination and growth of white mustard seeds. Int J Biometeor. 16:13–24

Edwards DK (1960) Effects of experimentally altered unipolar air ion density upon the amount of activity of the blowfly Calliphora vicina. RD Can J Zool 38:1079–1091

Effenberger E, Jatho K (1973) Unterschätzte Umweltnoxe. Selecta 38:3414

Eggert D (1974) Eysenck-Persönlichkeit-Inventar E-P-I. Verlag für Psychologie, Göttingen

Ehrmann W, von Leitner H, Ludwig W, Persinger MA, Sodtke W, Thomas R (1976) Entwicklung eines elektromedizinischen Taschengerätes. Med-Markt/Acta Medicotech 24:282–285

Eichmeier J (1960) Investigation of the possible influence of atmospheric ions on human reaction time, Final Tech Rep, US Contract Nr DA-91-581-EUC-1950

Eichmeier J (1963) Über den bioklimatischen Einfluss künstlich erzeugter atmosphärischer Kleinionen auf die Atmungsfrequenz, die Pulsfrequenz und den kortikalen Alpharhythmus des Menschen. Elektromedizin 8 (1)

Eichmeier J (1964a) Eigenschaften und biologische Wirkungen atmosphärischer Kleinionen. Umschau 64:420–422

Eichmeier J (1964b) Wirkung atmosphärischer Ionen auf Bio-Rhythmen bei Menschen. Ergeb Phys-Diat Ther 5:1–5

Eichmeier J, Büger P (1969) Uber den Ein fluss elektromagnetischer Strahlung auf die Wismut-Clorid-Fällungsreaktion nach Piccardi. Int J Biometeorol 13 (3,4):239–256

Eisemann B (1975) Untersuchungen über die Langzeiteinwirkung kleiner Wechselströme 50 Hz auf den Menschen. Dissertation Med Fak Universität Uni Freiburg

El'darov AL, Kholodov Yu A (1964) Effect of a constant magnetic field on motor activity of birds (Engl summ). Zh Obshch Biol 25:224–229

Elkiey TM, Pelletier RL, Bhartendu, Barthakur, N (1977) Effects of small air ions on net blotch disease of barley. Int J Biometeorol 21 (1):1–6

Ellenby C, Smith L (1968) Observations on the piezoelectric properties of nematode cuticle. Comp Biochem Physiol 26:359–363

Elster J, Geitel H (1899) Über die Existenz elektrischer Ionen in der Atmosphäre. Terr Magn Atmos Elect 4:213–234

Endrös R (1969) Neue Erkenntnisse über die physikalische Wirkung unterirdischer Wasserführung. Wetter Boden Mensch 7:327–340

Erb HF (1969) Der Mensch im Nullfeld. Beton-Verlag GmbH, Düsseldorf

Evertz U, König HL (1977) Pulsierende magnetische Felder in ihrer Bedeutung für die Medizin. Hypokrates 48:16–37

Evertz U, Ludwig W (1977) Magnetfeldbehandlung. Grenzgeb Wiss II, 26:106–119

Eysenck H-J (1959) Das Maudsley personality inventory Hogrefe Göttingen

Fadini A (1971a) Zur statistischen Auswertung geopathogener Untersuchungen. Wetter Boden Mensch 11:609–617

Fadini A (1971b) Zur statistischen Auswertung geopathogener Untersuchungen. Wetter Boden Mensch 9:745–766

Fadini A (1972) Nachweis der Nichtzufälligkeit von physikalischen und bio-physikalischen Ortungen krankheitsverursachender geopathogener Zonen. Wetter Boden Mensch 15:945–951

Fanger PO (1970) Thermal comfort. Analysis and application in environmental engineering. Copenhagen

Faust V, Harlfinger O, Neuwirth R, Wedler E, Wehner WF (1977) Biometeorologie. Hippokrates-Verlag, Stuttgart

Fischer G (1973) Die bioklimatologische Bedeutung des elektrischen Gleichfeldes. [Orig B] 157:115–130 Zentralbl Bakteriol

Fischer G (1976a) Einige Untersuchungen zur bioklimatologischen Bedeutung des restwellenfreien elektrostatischen Feldes. 2nd Kolloq Bioklimatol Wirk Luftelektr Faktoren, Munich

Fischer G (1976b) Beeinflussungen physiologischer Parameter durch das 50 Hz-Wechselfeld. 2nd Kolloq Bioklimatol Wirk Luftelektr Faktoren, Munich

Fischer G, Richter TR (1976) The influence of the 50 Hz alternating field on the heart-frequency of rats. Arch Meteorol, Geophys Bioklimatol [B] 24:109–126

Fischer G, Richter T, Platzer P (1974) Bisherige Untersuchungen über die biologische Wirkung der 50 Hz-Felder. Rep 1st Kolloq Bioklimatol Wirk Luftelektr Faktoren, Graz

Fischer G, Strampfer H, Riedl H (1977) Wirkung eines künstlichen Elektroklimas auf physiologische und psychologische Messgrössen. Z Exp Angew Psychol, 24 [2]:193–208
Fischer WH (1966) The radio noise spectrum from E.L.F. to E.H.F. J Atmos Terr Phys 28:429–430
Fischer WH, Sturdy GE, Ryan ME, Pugh RN (1968) Laboratory studies on fluctuating phenomena. Int J Biometeorol 12 (1):15–19
Fisher R, Severs R, Hardy R (1977) Parallelism of atmospheric electrical potential and mortality in London air pollution episodes. Int J Biometeorol 21 (4):311–318
Fleischmann L (1922) Gesundheitsschädlichkeit der Magnet-Wechselfelder. Naturwissenschaften 10:434
Fleming J et al. (1961) Microwave radiation in relation to biological systems and neural activity. In: Barnothy MF (ed) Biological effects of microwave radiation: Vol 1. Plenum Press, New York p 239
Fole F (1972) Phänomen 'PAT' in den elektrischen Unterwerken. 2. Coll. Int. der Internationalen Vereinigung für soziale Sicherheit. Sektion für die Verhütung von Arbeitsunfällen und Berufskrankheiten durch Elektrizität, Cologne
Fole F, Dutrus E (1974) Nueva aportacion al estudio de los campos electromagneticos generados por muy altas tensiones. Med Segun Trab 22:87
Fournier MH (1960a) Description des installations d'une station d'enregistrement des variations très rapides du champ magnétique terrestre Extr C R Acad Sci 251:671–673
Fournier MH (1960b) Quelques aspects des premiers enregistrements magnéto-telluriques obtenus à Garchy dans la gamma des variations très rapides. Extr C R Acad Sci 251:962–964
Fox P (1978) Personal communication, 17 December
Franklin B (1752) (No title) Philos Trans R Soc Lond) 47:289
Frey AH (1961a) A rational framework for interpreting the behavioral effects of atmospheric ions. IRE Trans Bio-Med Electron 8:12–16
Frey AH (1961b) Human behavior and atmospheric ions. Psychol Rev 68:225–228
Frey AH (1965) Experimental controls and biological experiments with atmospheric ions. Aerosp Med 36:121–123
Frey AH (1967) Modification of the conditioned emotional response by treatment with small negative air ions. Public Institute for Research, State College, Pennsylvania.
Frey AH, Granada RE (1962) Human reactions to air ions. Part I. Proc Int Conf Ioniz Air pp 1–10
Frey W (1949) Die biologische Wirkung eines elektrischen Feldes. Ann Schweiz Ges Balneol Klimatol 40:29–32
Friedman H, Becker, RO, Bachmann CH (1967) Effect of magnetic fields on reaction time performance. Nature 213:949–950
Friend AW, Jr, Finch ED, Schwan HP (1975) Low frequency electric field induced changes in the shape and motility of amoebas. Science 187:357–358
Frits T, Strauss H, Elges E, Deleanu MC (1963) The action of cholesterol and positive ions on spontaneous mobility. Igiena 12:33–37
Fuerst R (1955) Annual report. M. D. Anderson Hospital and Tumor Institute, University of Texas, p 60
Galejs J (1970) ELF propagation—review paper, 'Sections I' in electromagnetics of the sea. AGARD Conf Proc 77 (Tech Inf Serv Doc AD 716 305)
Gann DS, LaFrance TF (1974) Effect of 60 Hz electric fields on growth and survival of cells in tissue culture. Johns Hopkins University, Baltimore, Maryland
Gardini FJ (1782) De influxu electricitatis atmosphaericae in vegetantia dissertatio ab Academia Luyduimensi. Praemio Donata An Augustae Taurinorum.
Garland GD (1960) Methods and techniques in geophysics. Wiley (Interscience), New York, p 277
Gary C (1976) Effets biologiques d'un champ électrique. Que peut dire l'électricien à ce sujet? RGE-FRA, ISSN 0035-3116, 07. Spec Nr, pp 5–18
Gauquelin M (1961–1962) Die planetare Heredität. Z Parapsychol Grenzgeb Psychol 5 (2/3):168–193
Gauquelin M, Gauquelin F (1975) Review of studies in the USSR on the possible biological effects of solar activity. J Interdiscip Cycle Res 6 (3):249–252
Gavalas RJ, Walter DO, Hamer J, Ross Adey W (1970) Effect of low-level, low frequency electric fields on EEG and behavior in Macaca Nemestrina. Brain Res 18:491–501
Gavalas-Medici R, Sagan PM (1977) Behavioral assays of weak ELF effects: Comments and recommendations. In: Report of the commit-

tee on biosphere effects of extremely-low-frequency radiation. National Academy of Sciences, Washington, DC

Genzel L (1978) MPI für Festkörperphysik, Stuttgart. Vortr Kolloq Münch Phys Germa

Giarola AJ, Krueger WF (1974) Continuous exposure of chicks and rats to electromagnetic fields. IEEE Trans Microwave Theory Tech 22:432–437

Gilbert GO (1973) Effect of negative air ions upon emotionality and brain seroton in levels in isolated rats. Int J Biometeorol 17 (3):267–275

Givoni B (1974) Biometeorological indices. Prog Biometeorol [A] 1 (Part IA):138–145

Gleichmann O (1974a) Vortrag anlässlich der Jahrestagung des Forschungskreises für Geobiologie in Eberbach/Neckar

Gleichmann O (1974b) Personal communication

Glünder G (1976) Einfluss eines elektrostatischen Feldes und eines elektrischen Impulsfeldes auf die humorale Antikörperbildung und die Mastleistung beim Huhn. Dissertation Tierärztl. Hochschule, Hanover

Gnevyshev MN, Novekova KF (1972) The influence of solaractivity on the earth's biosphere. J Interdiscipl Cycle Res 3 (1):99–104

Goodman E, Greenebaum B, Marron M (to be published) Effects of extremly low frequency electromagnetic fields on Physarum polycephalum. Cited in Becker (1975)

Gordon ZV (1964) The effect of microwaves on blood pressure in experiments on animals. The biological action of radio-frequency electromagnetic waves. Moscow, p 57

Gorodetskaya SF (1964) The effect of an SHF electromagnetic field on the reproduction, peripheral blood composition, conditioned-reflex activity, and morphology of the internal organs of white mice. The biological actions of ultra sound and superhigh-frequency electromagnetic Vibrations. Naukova Dumka, Kiev, p 80

Gossel, D (1977) Forschungslaboratorien Philips Hamburg (personal communication)

Grissett JD (1971) Exposure of squirrel monkeys for long periods to extremely low frequency magnetic fields: central nervous system affects as measured by reaction time, NAMRL–1146. Naval Aerospace Medical Research Laboratory, Pensacola, Florida

Grissett JD, Lorge J (1971) Central nervous system affects as measured by reaction time in squirrel monkeys exposed for short periods to extremely low frequency magnetic fields, NAMRL–1137. Naval Aerospace Medical Research Laboratory, Pensacola, Florida

Gross L (1960) An effect of magnetic fields on tumor growth. Biophys Soc Meet Abstr L 8

Grün W (1972) Meine Meinung: Ungesunde Stahlbetonbauten? Baumarkt 29:1266–1267

Gualtierotti R (ed) (1968) Bioclimatology, biometeorology, and aeroionotherapy. Summaries of reports. Carlo Erba Foundation, Milan

Gualtierotti R (1977) Gebirksklima und Gesundheit. Therapeutikhinweise. Kongresshaus Riva del Garda, Italy

Guillerm R, Badre R, Vogt JJ, Hee J (1967) Effets physiologiques chez l'homme d'un séjour de 24 heures en atmosphère chargée en ions positifs ou négatifs. Int J Biometeorol Suppl 11:319

Gulyaev PI et al. (1967) The electric field in the air around excited tissues. Electroauragram paper read to the Leningrad Society of Naturalists, February 13, 1967

Hahn F (1963) Bakterizide und bakteriostatische Wirkung schwacher elektrischer Gleichfelder. Ars Med 6:403–407

Haine E (1961) Nehmen luftelektrische Faktoren Einfluss auf die Aktivitätswechsel kleiner Insekten, insbesondere auf die Häutungs- und Reproduktionszahlen von Blattläusen? Forschungsber Landes Nordrhein-Westfalen 974

Haine E (1969) Beeinflussen luftelektrische Faktoren—insbesondere Ionenkonzentrationswechsel der Luft—Periodizitätserscheinungen im Häuten der Blattläuse? Z Angew Entomol 50(2):222–232

Haine E, König H (1960-1961) Über die Behandlung von Blattläusen (Myous persicae Sulz) mit elektrischen Feldern. Z Angew Entomol 47:459–463

Haine E, König HL, Schmeer H (1964) Aphid moulting under controlled electrical conditions. Int J Biometeorol 7:265–275

Halpern MH (1966a) Effects of reproducible magnetic fields on the growth of cells in culture, NASA CR–75121. National Acronautics and Space Administration, Washington, DC

Halpern MH, van Dyke JH (1966b) Very low magnetic fields: biological effects and their implications for space exploration. Aerosp Med 37:281

Hamburger RJ (1962) On the influence of ar-

tificial ionization of the air on the oxygen uptake during exercise. In: Tromp SW (ed) Biometeorology. Pergamon Press, Oxford
Hamer JR (1965) Biological entrainment of the human brain by low frequency radiation. NSL 65-199. Northrop Space Laboratories
Hamer JR (1969) Effects of low level, low frequency eletric fields on human time judgement. Rep 5th Int Biometeorol Congr
Handa S, Ogawa T, Yasohara M (1971) Damping coefficients of Q-type bursts in the Schumann resonance frequence range. Contrib Geophy Inst Kyoto Univ Nr 11
Hänsche HA (1963) Der Mensch und seine Umwelt. Aerztl Praxis 15:342-346
Hansel C (1966) ESP—A scientific evaluation Scribner's New York
Hansell CW (1960) Atmospheric ionization and its biological effects: Summary of research to 8/1/60. RCA Laboratories, Princeton, New Jersey
Hanselmann JC, Casselman CJ, Tibbais ML, Bickel JE (1964) Field intensity measurement at 10,2 kc/s over reciprocal paths. J Res Nat Bur Stand [D] 68(1)
Hartmann E (1957) Neuartige Therapiemöglichkeiten im Kippschwingungsfeld. Erfahrungsheikunde 6(12)
Hartmann E (1958) Polarität und Rhythmus. Erfahrungsheilkunde. 7(12)
Hartmann E (1967) Krankheit als Standortproblem, Haug Verlag, Heidelberg
Hartmann E (1968) Gibt es eine geopathogene Krebsursache? Wetter Boden Mensch 2:69-78
Hartmann E (1970) 20 Jahre private Krebsforschung. Wetter Boden Mensch 10:517-552
Hartmann E (1972a) Reaktionszeitmessungen über geopathogenen Zone. Wetter Boden Mensch 15:961-963
Hartmann E (1972b) Tumorwachstum bei Ratten in Abhängigkeit von Standort und Milieu. Wetter Boden Mensch 16:988-996
Hartmann E (1976) Krankheit als Standortproblem, 3rd edn. Haug Verlag, Heidelberg
Hartmann E, Brauss FW, König HL, Varga A (1970) Gerät für elektromagnetische Hochfrequenzenergie mit modulierten niederfrequenten Signalen. German Patent Office. Public Register 1, 589, 913, vol 14, no. 5
Hartmann R (1978) Verhaltensphysiologische Untersuchungen zur Wahrnehmung luftelektrischer Felder durch weisse Mäuse (Mus musculus). Graduate work in Biology-Zoology at the University of the Saarland, Saarbrücken, Germany
Hauf G (1974) Untersuchungen über die Wirkung energietechnischer Felder auf den Menschen. Dissertation, Hohen Medizinischen Fakultät, Ludwig-Maximilians-Universität, Munich
Hauf R (1976a) L'influence des champs alternatifs électriques et magnétiques, à 50 Hz, sur les hommes. RGE-FRA, ISSN 0035-3116, 07, Spec Nr, pp 19-26
Hauf R (1976b) Beeinflussen energietechnische Felder den Menschen? 2nd Kolloq Bioklimatol Wirk Luftelektr Faktoren, Munich
Hauf R, Wiesinger J (1973) Biological effects of technical electric and electromagnetic VLF fields. Int J Biometeorol 17(3):213-215
Haufe WO (1976) Development and application of biometeorology. Int J Biometeorol 20(2):92-96
Helson GAH, Penman JER (1970) Biological transformations associated with weather changes. A hypothesis on the flight activity of Wiseana spp. Int J Biometeorol 14:227-247
Henry JP, Ely DL, Stephens PM (1974) The role of psychosocial stimulation in the pathogenesis of hypertension. Verh Dtsch Ges Inn Med 80:1724
Hensel H (1973) In: Precht H, Christophersen J, Hensel H (eds) Temperature and Life. Springer Verlag, New York
Hepburn F (1957) Atmospherics with very low-frequency components below 1 kc/s. J Atmos Terr Phys 10:266
Hepburn F, Pierce E (1953) Atmospherics with very low-frequency components. Nature 171:837 Hepburn F, Pierce E (1954) Philos Mag 45:917
Heppner FH, Haffner J (1973) Electromagnetic fields for communication in bird flocks: A model symposium and workshop on the effects of low frequency magnetic and electric fields on biological communication processes at the 6th annual meeting of the neuroelectric society: Vol. 6
Herrington LP (1935) The influence of ionized air on normal subjects. J Clin Invest 14:70
Herron T (1965) Phase modulation of biomagnetic micropulsations. Nature 207:699
Heydt G (1971) Versuche zur interkontinentalen Lokalisierung von Quellen der Atmospherics-Aktivität durch Bestimmung von

Einfallsrichtungen und Gruppenlaufzeitdifferenzen, Tech Ber Nr. 136. Heinrich-Hertz-Institut für Schwingungsforschung, Berlin-Charlottenburg

Hicks WW (1957) A series of experiments on trees and plants in electrostatic fields. J Franklin Inst 264:1–5

Hildebrandt, G (1974) Chronobiologische Grundlagen der sogenannten Ordnungstherapie. Therapiewoche 24:3883

Hildebrant G (1975a) Institut für Arbeitsphysiologie und Rehabilitations-forschung der Philipps-Universität Marburg. Arbeitsbericht und Verzeichnis der Veröffentlichungen, Marburg/Lahn, West Germany

Hildebrandt G (1975b) Wissenchaftliche Grundlagen der modernen Balneologie. Therapiewoche 25:4122

Hildebrandt G et al. (1972–1976) Veröffentlichungen des Sonderforschungsbereiches Adaptation und Rehabilitation (SFB 122) der Philipps-Univ. Marburg/Lahn, Vol. I

Hoffman JH (1968) AFCRL, University of Texas at Dallas

Holloway RJ (1952) The physiological effects of ionized air upon growth and on the adrenal system of the normal rat. MS thesis, University of San Francisco

Hollwich F (1979) The Influence of Ocular Light Perception on Metabolism in Man and Animals. Springer, New York

Hollwich F, Dieckhues B, Schrameyer B (1977) Die Wirkung des natürlichen and künstlichen Lichtes über das Auge auf den Hormon- und Stoffwechselhaushalt des Menschen. Klin, Monatsbl Augenheilk 171:98–104

Holzer RE, Deal OE, Ruttenberg S (1957) Low Audio frequency natural electromagnetic signals. Symposium on the propagation of VLF waves Paper 45

Hopkins GH, Smith HW (1966) An investigation of the magnetotelluric method for determining subsurface resistivities, Rep Nr 140. Electr Eng Res Lab, University of Texas

Howe NR, Sheikh YM (1975) Anthopleurine: A sea anemone alarm pheromone. Science 189:386–388

Hughes HG (1967a) The directional dependency of slow-tail ELF atmospheric waveforms. J Atmos Terr Phys 29:1629

Hughes HG (1967b) A comparison at extremely low frequencies of positive and negative atmospherics. J Atmos Terr Phys 29:1277

IEEE International Symposium on Electromagnetic Compatibility (1969) Radio noise study weighs major sources of natural and man-made interference. Commun Des Dig 37(7):14–18

Ishida T (1969) Statistical characteristics of atmospheric noise. 16th General Assembly of URSI, Ottawa

Israël H (1957) Atmospheric electricity (Atmosphärische Elektrizität): Vol 1. Akad Verlagsges, Geest & Portikg K-G, Leipzig

Jacobi E, Hagemann G, Kuhnke W (1973) Der Einfluss des Wetters auf die Thrombozytenadhäsivität beim Menschen. Dtsch Med Wochenschr 98:434

Jacobi EG, Krüskemper G (1975) Der Einfluss simulierter Sferics auf die Thrombozytenadhäsivität. Inn Med 2:73–81

Jacobi EG, Krüskemper G (1976) Wirkungen simulierter Sferics auf die Thrombozytenadhäsivität beim Menschen. 2nd Kolloq Bioklimatol Wirk Luftelektr Faktoren, Munich

Jaekkel K-H (1955) An den Grenzen menschlicher Fassungskraft. J.F. Lehmann-Verlag, Munich

Jahn E (1977) Über Einflüsse biophysikalischer und elektrischer Felder auf Lymantria monacha L. Anz. Schädlingskd Pflanz- Umweltschutz 50:101–104, 113–115

Jahn E, Nessler N (1971) Auswirkungen von Bestrahlungen mit langen, sehr langen and insbesondere ultralangen Radiowellen auf Mortalität, weiters auch auf Fruchtbarkeit und Entwicklung der Nonne. Anz Schädlingskd Pflanzenschutz 44(8):113–119

Jahnke H (1961) Über die Verwendbarkeit von objektiv auswertbaren Fragenbogen bei der Prüfung von Arzneimittelwirkungen. Med 5:169–175

Janssen H (1978) Unreifes ZNS provoziert den Krippentod. Selecta 16:1464–1472

Jaski R (1960) Radio waves and life. Radio Electron 31:43

Jendritzky G (1977) Zur Problematik der medizinmeteorologischen Bewertung des thermischen Milieus. Z Angew Bäder- Klimaheilkd 24(4):407–414

Jendritzky G (1978a) Ein Simulationsmodell für den Einfluβ der thermischen Bedingungen in der Klimatherapie. Phys Med 7(1):7–8

Jendritzky G (1978b) Rasche Temperaturschwankungen in Abhängigkeit von synoptischen Parametern.

Jendritzky G, Sönning W, Swantes H (1977) Ein Verfahren zur bioklimatologischen Be-

wertung des thermischen Milieus. Ann Meteorol 12:205–210
Jones DP, O'Connor SA, Collins JV, Watson BW (1976) Effect of long-term ionized air treatment on patients with bronchial asthma. Thorax 31:428–432
Jordon J, Sokoloff B (1959) Air ionization, age, and maze learning of rats. J Geront 14:344–348
Kalmijn AJ (1971) The electric sense of sharks and rays. J Exp Biol 55:371–383
Kamenskii Yu I (1967) The effect of microwaves on the kinetics of nerve impulse parameters. Tr Mosk O-va Isp Prir
Kasnatschej W, Stschurin S, Michailowa L (1973) Photonen—Sprache der Zellen? Bild Wiss 6
Kaufmann W (1968) Der umbaute Raum und seine technische Einrichtung als biologischer Störfaktor. Wetter Boden Mensch 3:103–119
Keefe TJ, Etzold H, Polk C (1973) Detection and processing of ELF (3–30 Hz) natural electromagnetic noise Rep AFCRL-RT-73-0077. University of Rhode Island
Keeton WT (1971) Magnets interfere with pigeon homing. Proc Natl Acad Sci USA 68:102–106
Kellogg et al. (1979) Superoxide involvement in the bactericidal effects of negative air ions on *Staphylococcus albus*. Nature, 281:400–401
Kelvin WT Lord (1859) On the necessity for incessant recording and for simultaneous observations in different localities to investigate atmospheric electricity. Papers on electricity and magnetism. Macmillan, New York, pp 227–229
Kemmer W (1967) W Zoologisches Institut der Universität des Saarlandes, Saarbrücken: Der Einfluss elektrischen Gleichstroms auf im Wasser lebende Tiere. Report, Weinheimer Forschungskreis, Hygieneinstitut, Universität Heidelberg
Kimpara A (1956) The waveform of atmospherics in the day-time and night. Proc Res Inst Atmos Nagoya Univ. 4:1
Kimura N (1967) Universität Kurune. Not Selekta 48:3699
Kimura S, Ashiba M, Matsushima J (1939) Influences of the air lacking light ions and the effect of its artificial ionization upon human beings, in occupied rooms. Jpn J Med Sci 7:1–12
Kingdon KH (1960) Interaction of atmospheric ions with biological material. Phys Med Biol 5:1–10
Kirmaier N, König H-L (1976) Einfluβ von impulsmodulierten elektrischen Feldern auf Probanden im Fahrsimulator. 2nd Kolloq Bioklimatol Wirk Luftelektr Faktoren, Munich
Kirmaier N, Schauerte W, Beierlein HR, Breidenbach H (1978) Probanden im Kraftfahrzeug-Praxistest. Einfluss luftelektrischer Impulsfelder. Münch Med Wochenschr 120(11):367–370
Klingenberg HG, Möse JR, Fischer G, Porta J, Sadjak A (1976a) Changes in osmolality and concentrations of triglycerides and cholesterol during the influences of electrostatic fields. Arch Meteorol Geophys Bioklimatol [B] 24:109–126
Klingenberg HG, Möse JR, Fischer G, Porta J, and Sadjak A (1976b) Metabolic activities of the rat liver during exposition to electrostatic fields and faraday conditions before and after partial hepatectomy. Arch Meteorol Geophys Bioklimatol [B] 24:109–126
Knickerbocker GG, Kouvenhoven WB, Barnes HC (1967) Exposure of mice to a strong AC electric field-An experimental study. IEEE Trans Power Appar Syst 86:26–33
Knoepp L et al. (1962) The effect of low electrical frequencies and various normal and malignant cells. Tex Rep Biol Med 20:623
Knoll M, Kugler J (1959) Subjective light pattern spectroscopy in the electroencephalographic frequency range. Nature 184:1823
Knoll MJ, Rheinstein J, Leonard GF, Highberg PF (1961) Influence of light air ions on human visual reaction time. IRE Trans Biomed Electron 8(4):239–245
Knoll M, Kugler J, Eichmeier J, Höfer O (1962) Note on the spectroscopy of subjective light patterns. Anal Psychol 7(1)
Knoll M, Eichmeier J, Schön RW (1964) Properties, measurements, and bioclimatic action of "small" multimolecular atmospheric ions. Adv Electron Electron Phys 19:177–254
Koch K (1934) Untersuchungen über den Quer- und Längtransport des Wuchsstoffes in pflanzenorganen. Planta 22:190–220
Koe BK, Weissman A (1966) p-Chlorophenylalanine: A specific depletor of brain serotonin. J Pharmacol Exp Ther 154:499–516
Koeppen S (1966) Handb Phys Ther 1:140
Kögl F (1933) Über Auxine. Angew Chem 46:469–473
König HL (1959) Atmospherics geringster Frequenzen. Z Angew Phys 11:264–274

König HL (1962) Über den Einfluss besonders niederfrequenter elektrischer Vorgänge in der Atmosphäre auf die Umwelt. Z Angew Bäder- Klimaheilkd 9(5):481–501

König HL (1968a) Der BIO-Resonator als Höchstfrequenzresonanzspule. Wetter Boden Mensch 3:98–102

König HL (1968b) Hochempfindliche Empfangsspule für ELF-Atmospherics. Z Angew Phys 25(1):14–18

König HL(1976) Zum Problem der Schädlichkeit von Hochspannungsleitungen. 2nd Kolloq Bioklimatol Wirk Luftelektr Faktoren, Munich

König HL (1977) Unsichtbare Umwelt, 2nd enlarged edn. Eigenverlag, Munich

König HL, Ankermüller F (1960) Über den Einfluss besonders niederfrequenter elektrischer Vorgänge in der Atmosphäre auf den Menschen. Naturwissenschaften 47:486–490

König HL, Behringer KH (1970) Simultaneous measurements of ELF-atmospherics by separate stations, AF 61 (052)–836

König HL, Krempl-Lamprecht L (1959) Über die Einwirkung niederfrequenter elektrischer Felder auf das Wachstum pflanzlicher Organismen. Arch Mikrobiol 34:204–210

König HL, Finkle H, Polk C (1964) Measurement of electric fields between 3 cps and 10-000 cps with a vertical antenna over the Atlantic Ocean near the east coast of Florida, Rep 396 (10)/1. Dep Electr Eng, University of Rhode Island

Kop PPAM, Heuts BA (1974) Month of birth and partner choice in marriage. J Interdiscip Cycle Res 5 (1):19–39

Korn JH, Moyer KE (1968) Behavioral effects of isolation in the rat: The role of sex and time on isolation. J Genet Psychol 113:263–273

Kornberg HA (1976) Effets biologiques des champs électriques. RGE-FRA, ISSN 0035-3116, 07, Spec Nr, pp. 51–64

Kornblueh I (1967) Air ions and human health. Seminar on human biometeorology. Public Health Serv Publ 999–AP–25

Kornblueh I, Swope S, Davis FK (1973) Natural ion levels in enclosed spaces. Proc Congr Lacustrine Climatol

Kornblueh JH, Griffin JE (1955) Artificial air ionisation in physical medicine. Am J Phys Med 34:618

Korobkova VP et al. (1971a) Besondere Schutzbestimmungen in der UdSSR bei Arbeiten in Höchstspannungsanlagen. Energetik 11:33–35

Korobkova VP et al. (1971b)

Korobkova VP et al. (1972) Energetik 12:28–30 Besondere Schutzbestimmungen in der UdSSR bei Arbeiten in Höchstspannungsanlagen, Tech commun. Bull Am Soc Eng 63(4):19

Kotaka S, Krueger AP, Andriese PC (1965a) Some observations on air-ion-enhanced iron chlorosis in barley (*Hordeum vulgaris*) seedlings. Plant Cell Physiol 6:233–245

Kotaka S, Krueger AP, Mishizawa K, Ohuchi T, Kogure Y, Takenobu M (1965b) Air ion effects on the iron metabolism of barley (*Hordeum vulgaris*) seedlings. Proc Bot Soc, Jpn [C] 21:43

Kotaka S, Kruegar AP, Nishizawa K, Ohuchi T, Takenobu M, Kogure Y, Andriese PC (1965c) Air ion effects on the oxygen consumption of barley seedlings. Nature 208:1112–1113

Kotaka S, Krueger AP, Nishizawa K, Ohuchi T, Takenobu M, Kogure Y, Andriese PC (1965d) Air ion effects on the oxygen consumption of barley seedlings. Nature 208:1112–1113

Kotaka S, Krueger AP, Andriese PC (1968) The effects of air ions on light-induced swelling and dark-induced shrinking of isolated chloroplasts. Int J Biometeorol 12:85–92

Kracmar F (1962) Die biologischen Grundlagen der Kippschwingungsfeldtherapie. Erfahrungsheilkunde 11(8)

Kraus W (1978) Therapie des Knochens und des Knorpels mit schwacher, langsam schwingender elektromagnetischer Energie. Med-Orthop Tech 98(2):33–43

Kraus W, Lechner F (1973) Heilung im Magnetfeld. Selecta 45:4193–4194

Kretschmar C (1967) Geoelektrische Untersuchungen zur Aufklärung der Häufung von Blitzschlägen in bestimmten Freileitungsabschnitten. Energietechnik 17(6):260–264

Kritzinger HH (1957) Praktische Bioklimatik, insbesondere der Luftelektrizität im freien Raum und im Wohnraum. Fortschr Med 75:469–470

Kritzinger HH (1967) Elektrische Feldwirkungen in ihrer Bedeutung für das Gesundheitswesen. Waerland-Mitt (Bensberg) 23:36–41

Kröling P (1976a) Subjektive Befindenswerte eines Grossraumbürokollektivs unter Elektroklimatisierung. 2nd Kolloq Bioklimatol Wirk Luftelektr Faktoren, Munich

Kröling P (1976b) Das Motilitäts-Tagesprofil

von Labormäusen unter der Einwirkung elektrischer Felder, magnetischer Felder und Luftionen. 2nd Kolloq Bioklimatol Wirk Luftelektr Faktoren, Munich
Krück F (1976) Stress und Hypertension. In: von Eiff AW (ed) Seelische und körperliche Störungen durch Stress. Gustav Fischer Verlag, Stuttgart
Krueger AP (1962) Air ions and physiological function. J Gen Phys 45:233–241
Krueger AP (1969) Preliminary considerations of the biological significance of air ions. Scientia Sept -Oct, pp 1–17
Krueger AP (1970) Personal communication
Kreuger AP (1973) Are negative air ions good for you? N Scientist p 668–670, June 14
Krueger AP (1974) Biological effects of ionization of the air. Prog Biometeorol 1 (Part 1A): 1963–1970
Krueger AP (1975) Are air ions biologically active? Conf Electrostat, June 24-26, Electrostat Soc Am, University of Michigan, Ann Arbor, Michigan
Krueger AP, Kotaka S (1969) The effects of air ions on brain levels of serotonin in mice. Int J Biometeorol 13:25–38
Krueger AP, Levine HB (1967) The effect of unipolar positively ionized air on the course of coccidioidomycosis in mice. Int J Biometeorol 11:279–288
Krueger AP, Reed EJ (1972) Effect of the air ion environment on influenza in the mouse. Int J Biometeorol 16:209–232
Krueger AP, Reed EJ (1975) A study of the biological effects of certain ELF electromagnetic fields. Int J Biometeorol 19:194–201
Krueger AP, Smith R (1960a) The biological mechanisms of air ion action I. 5-hydroxytryptamine as the endogenous mediator of positive air ion effects on the mammalian trachea. J Gen Physiol 43:533–540
Krueger AP, Smith R (1960b) The biological mechanisms of air ion action II. Negative air ion effects on the concentration and metabolism of 5-hydroxytryptamine in the mammalian respiratory tract. J Gen Physiol 44:269–276
Krueger AP, Smith RF (1962) Studies on the effects of gaseous ions on the mammalian trachea. In: Tromp SW (ed) Biometeorology. Pergamon, Oxford
Krueger AP, Smith RF (1959) An enzymatic basis for the acceleration of ciliary activity by negative air ions. Nature 183:1322–1333
Krueger AP, Smith RF, Go IG (1957) The action of air ions on bacteria. I. Protective and lethal effects on suspensions of staphylococci in droplets. J Gen Physiol 41:359–381
Krueger AP, Kotaka S, Andriese PC (1962a) Studies on the effects of gaseous ions on plant growth. I. The influence of positive and negative air ions on the growth of *Avena sativa*. J Gen Physiol 45:879–895
Krueger AP, Kotaka S, Andriese PC (1962b) Some observations on the physiological effects of gaseous ions. Int J Biometeorol 6:33–48
Krueger AP, Andriese PC, Kotaka S (1963a) The biological mechanism of air ion action: The effect of CO_2^+ in inhaled air on the blood level of 5-hydroxytryptamine in mice. Int J Biometeorol 7:3–16
Krueger AP, Kotaka S, Andriese PC (1963b) A study of the mechanism of air-ion-induced growth stimulation in *Hordeum vulgaris*. Int J Biometeorol 7:17–25
Krueger AP, Kotaka S, Andriese PC (1964) Studies on air-ion-enhanced iron chlorosis. I. Active and residual iron. Int J Biometeorol 8:5–16
Krueger AP, Kotaka S, Andriese PC (1965) The effect of abnormally low concentrations of air ions on the growth of *Hordeum vulgaris*. Int J Biometeorol 9:201–209
Krueger AP, Kotaka S, Andriese PC (1966a) Some physiological and pathological effects of gaseous ions, Spec Monogr Ser, 1. Leiden, pp 1–11
Krueger AP, Kotaka S, Nishizawa K, Kogure Y, Takenobu M, Andriese PC (1966b) Air ion effects on growth of the silkworm (*Bombyx Mori L.*). Int J Biometeorol 10:29–38
Krueger AP, Andriese PC, Kotaka S (1968) Small air ions: Their effect on blood levels of serotonin in terms of modern physical theory. Int J Biometeorol 12:225–239
Krueger AP, Kotaka S, Reed EJ, Turner S (1970) The effect of air ions on bacterial and viral pneumonia in mice. Int J Biometeorol 14:247–260
Krueger AP, Kotaka S, Reed EJ (1971) The course of experimental influenza in mice maintained in high concentrations of small negative air ions. Int J Biometeorol 15:5–10
Krueger WF, Giarola AJ, Bradley JW, Daruvalla SR (1972) Influence of low level electric and magnetic fields on the growth of young chickens. Biomed Sci Instrum 9:183–186
Krueger AP, Reed EJ, Day MB, Brook KA (1974) Further observations on the effect of

air ions on influenza in the mouse. Int J Biometeorol 18:45–56

Krueger AP, Reed EJ, Brook KA, Day MB (1975a) Air ion action on bacteria. Int J Biometeorol 19:65–71

Krueger WF, Giarola AJ, Bradley JW, Shrekenhamer A (1975b) Effects of electromagnetic fields on fecundity in the chicken. Ann NY Acad Sci 247:391–4400

Krueger AP, Strubbe AE, Yost MG, Reed EJ (to be published) Electric fields, small air ions and biological effects.

Krueger AP et al. (1978) Electric fields, small air ions and biological effects. Int J Biometeriol 22:202–212

Kuhn T (1962) The structure of scientific revolutions. University of Chicago Press, Chicago

Kuhnke W, Zink O (1950) Erfahrungen mit einer medizinmeteorologischen Vorhersage. Med. Meterol 11–13

Lacey JI (1956) The evolution of autonomic responses: Toward a general solution. Ann NY Acad Sci 67:125–164

Lampert H (1968) Embolieverhütung und Rekanalisierung des Thrombus mittels des galvanischen Stromes. Weserbergland-Klinik (self-published)

Lang S (1971) Änderung des Wasser- und Elektrolythaushaltes bei weissen Mäusen unter Einfluss von Faraday-Bedingungen und eines Rechteckimpulsfeldes der Frequenz 10 Hz. Verh Dtsch Zool Ges Helgol pp 176–179

Lang S (1972a) Stoffwechselphysiologische Auswirkungen der Faradayschen Abschirmung und eines künstlichen luftelektrischen Feldes der Frequenz 10 Hz auf weisse Mäuse. Arch Meteorol Geophys Bioklimatol [B] 20:109–122

Lang S (1972b) Physiologische Untersuchungen an weissen Mäusen im Faradayschen Käfig. Wetter Boden Mensch 14:831–833

Lang S (1973a) Natürliche und künstliche partielle Faraday-Bedingungen bezüglich luftelektrischer Parameter in Wohnsiedlungen. Rep Conf Ges Oekol

Lang S (1973b) Die Eigenschaften moderner Bauten gegenüber atmosphärischen luftelektrischen Faktoren und ihre Auswirkungen auf den Organismus. DAB 8:640–643

Lang S (1975) Influences of an electric field of 10 Hz on the metabolism of lipids and the water electrolyte balance. Int J Biometeorol 19 (Suppl 6):1

Lang S (1976a) Bericht über den 80. Kongress der Deutschen Gesellschaft für Physikalische Medizin. III Medizinische Relevanz des Elektroklimas. Z Angew Bäder-Klimaheilkd 23:58–61

Lang S (1976b) A sensibilisation mechanism for environmental electric fields in mice and rats. Arch Meteorol Geophys Bioklimatol [B] 24:109–126

Lang S (1976c) Frequenzabhängigkeit der biologischen Wirkungen elektrischer Felder im ELF-Bereich? 2nd Kolloq Bioklimatol Wirk Luftelektr Faktoren, Munich

Lang S (1977a) Bericht über das II. Kolloquium Bioklimatische Wirkungen luftelektrischer Faktoren. Technische Universität München, 1976, in Z Angew Bäder- Klimaheilkd

Lang S (1977b) Biologische Wirkungen elektrischer, magnetischer und elektromagnetischer Felder-Stress oder Therapie? Contribution in König (1977)

Lang S, Lehmair M (1975) Einfluss von 10 Hz-Feldern auf die psychische Kondition. 80. Kong Dtsch Ges Phys Med

Lang S, Lehmair M (1977a) Elektrische Umwelt und Leistungsfahigkeit. Klimat Kälte Ing 2:61–66

Lang S, Lehmair M (1977b) Bioklimatische Wirkungen der elektrischen Umwelt am Arbeitsplatz. Dtsch Architektenbl 3:209–211

Lang S, Reuss T (1975) Lipolytische Beeinflussung von Fettgewebe durch atmosphärisch-elektrische Schwingungen. Verh Dtsch Zool Ges pp 281–286

Lang S, Altmann G, Bach W, Lehmair M (1976) Atmospherics und Technics in Arbeitsträumen. 2nd Kolloq Bioklimatol Wirk Luftelektr Faktoren, Munich

Lang S, Altmann G, Lill A, Hartmann R (1977) Verhaltensphysiologische Untersuchungen über die Orientierung von Mäusen in elektrischen Feldern. Z. Z. in Vorbereitung

Lantsman MN (1965) The effect of alternating magnetic field on the phagocytic function of the reticulo-endothelial system in experiments. Questions of hematology, radiobiology, and the biological action of magnetic field. Tomsk, p. 360

Laws CA, Holiday ER (1975) Air ions in physical medicine and environmental hygiene. Symp Soc Environ Eng

Le Bars H (1976) Effets biologiques d'un champ électrique sur le rat et le lapin RGE-FRA, ISSN 0035-3116, 07. Spec Nr, pp 91–97

Lechner F, Ascherl R (1978) Grundlagen und

Klinik der elektrodynamischen Feldtherapie bei Knochenheilungsstörungen. Med-Orthop Tech 98, (2):43–49
Lehmann G, quoted in Brüche (1960)
Lemström S (1902) Elektrokultur Junk, Berlin
Lemström S (1904) Electricity in agriculture and horticulture. London and New York
Lenard P (1892) Über die Elektrizität der Wasserfälle. Ann Phys (Leipzig) 46:584–636
Lenke R, Bonzel J (1975a) Luftelektrische Felder in umbauten Räumen und im Freien. Betontech Ber 11:387–390
Lenke R, Bonzel J (1975b) Betontechnische Berichte 12:425–430
Levengood WC (1966) Cytogenetic variations induced with a magnetic probe. Nature 209:1009–1013
Levengood WC (1967a) Morphogenesis as influenced by locally administered magnetic fields. Biophys J 7:297–307
Levengood WC (1967b) Indicated biological effect from the 9 July 1962, high altitude nuclear test. Int J Biometeorol 11(2):195–199
Leventhal G (1970) Guide for the evaluation of human exposure to whole body vibration. I.S.O. ITC 1081 WG 7 Leventhal G (1971) Chelsea College of Science and Technology, London (personal communication)
Liebermann L (1956) Extremely low-frequency electromagnetic waves. II. Propagation properties. J Appl Phys 27:1477
Lindauer M (1976) Orientierung der Tiere. Verh Dtsch Zool Ges pp 156–183
Lokken JE (1964) Instrumentation for receiving electromagnetic noise below 3000 cps. In: Bleil DF (ed) Natural electromagnetic phenomena below 30 KC/S. Plenum Press, New York, pp 373–428
Lotmar R (1975) Z Phys Med 4:151–154
Lotmar R, Ranscht-Froemsdorff WR, Weise H (1968) Intensität der Gewebeatmung und Wetterfaktoren. Z Angew Bäder- Klimaheilkd 15:1–10
Lotmar R, Ranscht-Froemsdorff WR, Weise H (1969) Dämpfung der Gewebeatmung (QO_2) von Mäuseleber durch künstliche Impulsstrahlung. Int J Biometeorol 13, (3,4):231–238
Lott JR, Linn GH (1975) The effects of an external electric field on action potentials in isolated nerves. Int J Biometeorol 19(Suppl 6):1
Lott JR, McCain HB (1973) Some effects of continous and pulsating electric fields on brain wave activity in rats. Int J Biometeorol 17(3):221–225
Lotz KE (1972) Über die Problematik einer exakten baubiologischen Beurteilung von Baustoffen aufgrund des derzeitigen wissenschaftlichen Kenntnisstandes. Wetter Boden Mensch 16:997–1008
Ludwig HW (1968) A hypothesis concerning the absorption mechanism of atmospherics in the nervous system. Int J Biometeorol 12,(2):93–98
Ludwig HW (1971) Der Einfluss von elektromagnetischen Tiefstfrequenz-Wechselfeldern auf höhere Organismen. Biomed Tech 16(2)
Ludwig HW (1972) Wettereinfluss auf organisches Gewebe. Z Angew Bäder- Klimaheilkd 19:15–17
Ludwig HW (1973a) Wirkung einer nächtlichen Abschirmung der elektrischen Feldstärke bei Rheumatikern. Arch Meteorol Geophys Bioklimatol [B] 21:305–311
Ludwig HW (1973b) Shielding effect of materials in the ULF, ELF and VLF region. Int J Biometeorol 17 (3):207–211
Ludwig H (1976a) Messung elektrischer Felder in Innenräumen. 2nd Kolloq Bioklimatol Wirk Luftelektr Faktoren, Munich
Ludwig HW (1976b) Problems of the shielding and of hospital electro-climatic systems. Arch Meteorol Geophys Bioklimatol [B] 24:109–126
Ludwig HW (1977) Biologische Verträglichkeit elektromagnetischer Felder. Bull SEV/VSE 68(18):941-946
Ludwig HW, Mecke R (1968) Wirkung künstlicher Atmospherics auf Säuger. Arch Meteorol Geophys Bioklimatol [B] 16:251–261
Ludwig HW, Mecke R, Seelewind H (1968) Elektroklimatologie. Arch Meteorol Biophys Bioklimatol [B] 16:237–250
Ludwig HW, Persinger MA, Ossenkopp KP (1973) Physiologische Wirkungen elektromagnetischer Wellen bei tiefen Frequenzen. Arch Meteorol Geophys Bioklimatol [B] 21:99–116
Ludwig HW, Ehrmann W, Sodtke W (1976) Beeinflussungen psychosomatischer Erkrankungen durch magnetische Wechselfelder. 2nd Kolloq Bioklimatol Wirk Luftelektr Faktoren, Munich
Lueders H (1968) Statistischer und objektiver Nachweis elektroklimatischer Wirkungen auf die Abwehrkräfte und das vegetative Nerven-

system des menschlichen Organismus. Wetter Boden Mensch 3:121–140

Lund EJ et al. (1947) Bioelectric fields and growth. University of Texas Press, Austin, Texas

Lyskov Y, Emma Y (1971) Electrical field as a parameter considered in designing electric power transmission of 750–1150 kV (cited in Marino 1975)

Maczynski B, Tyczka S, Marecki B, Córa T (1971) Effect of the presence of man on the air ion density in an office room. Int J Biometeorol 15:11–21

Malakhov AN et al (1965) The electromagnetic hypothesis of biological communication. Bionics, Nauka, Moscow, p 297

Malboysson E (1976) E Surveillance médicale du personnel exposé à l'action des champs électromagnétiques. RGE-FRA, ISSN 0035-3116, 07. Spec Nr, pp. 75–80

Mamontov SG, Ivanova LN (1971) Effect of low frequency electric field on cell division in mouse tissues. Bull Exp Biol Med (Engl Transl) 71:192

Mantell B (1975) Untersuchungen über die Wirkung eines magnetischen Wechselfeldes 50 Hz auf den Menschen. Dissertation, Med Fak Universität Freiburg

Marino AA (1975) Cases 26529 and 26559—Common hearings on health and safety of 765 kV transmission lines. Public Service Commission, New York

Marino AA, Becker RO (1977) Hazard at a distance: Effects of exposure to the electric and magnetic fields of high voltage transmission lines. Med Res Eng 12(N.5):6–9

Marino AA, Becker, RO, Ullrich B (1976a) The effect of continuous exposure to low frequency electric fields on three generations of mice: A pilot study. Experientia 32:565–566

Marino AA, Berger TJ, Austin BP, Becker RO, Hart FX (1976b) Evaluation of electrochemical information transfer system. 1. Effect of electric fields on living organisms. J Electrochem Soc 123:1199–1200

Marr MJ, Rivers WK, Burns CP (1973) ONR final report. Washington, DC

Marsh G (1968) The effect of sixty-cycle AC current on the regeneration axis of *Dugesia*. J Exp Zool 169:65–70

Martin H, Lindauer M (1973) Orientierung im Erdmagnetfeld. Fortschr Zool 21:211–228

Martin HG (1961) Waveguide propagation of VLF radio waves. J Atmos Terr Phys 20:206

Mathewson NS, Oosta GM, Oliva SA, Levin SG, Blasco AP (to be published) Effects of 45 Hz electric field exposure on rats. In: Johnson CC, Shore ML (eds) Biologic effects of electromagnetic waves. US Department of Health, Education and Welfare, Rockville, Maryland

Mattern G (1973) Versuche zur Messung der Radioemission von Blitzen im Frequenzbereich bei 432 MHz. Kleinheubacher-Ber 17:455–460

Maxey ES (1976) Critical aspects of human versus terrestrial electromagnetical symbiosis. 2nd Kolloq Bioklimatol Wirk Luftelektr Faktoren, Munich

Mayasi AM, Terry RA (1969) Effects of direct electric fields, noise, sex and age on maze learning in rats. Int J Biometeorol 13(2):101–111

McDonald RD, Bachman CH, Lorenz PJ (1967) Some psychomotor and physiological tests on humans exposed to air ions. Aerosp Med 38:145–148

McElhaney J, Stalnaker R (1968) Electric fields and bone loss of disease. J Biomech 1:47

McGurk FCJ (1959) Psychological effects of artificially produced air ions. Am J Phys Med 38:136–137

McKinney HE (1973) Physiological and pathological studies of animals exposed to extremely low frequency radiation fields. Bioeffects Proj Resume, MR041.08-0100. Office of Telecommunications Policy, Executive office of the President, Washington, DC

Meda E, Carrescha V, Cappa S (1974) Einfluss elektrischer Felder auf Tiere – Versuchsergebnisse – Bulletin 3/1974 Internat Sektion der IVSS für die Verhütung von Arbeitsunfällen und Berufskrankheiten durch Elektrizität

Medizin-Elektronik (1970) p 447

Merikle R et al. (1964) Plant growth response, in: Biological Effects of Magnetic Fields, vol 1, Plenum Press, New York, p 183

Merkel FW, Fromme HG (1958) Untersuchungen über das Orientierungsvermögen nächtlich ziehender Rotkehlchen. Naturwissenschaften 45:499–500

Miericke J (1972) Recherche über biologische Wirkungen von Magnetfeldern im allgemeinen und speziell beim Menschen. Institut für

WerkstoffwissenschaftenII. Universität Erlangen, Nürnberg
Mikhailova GA (1962) On the spectra of atmospherics and phase velocity of electromagnetic waves at very low frequencies. Geomagn Aeron 2:257
Mikhov D (1976) Sun activity, geomagnetic storms and their biological influence. Dig 4th Int Conf Med Phys
Minch AA (1961) Die Luftionisation als Faktor der Hygiene Zh. Gig Epdimilog Microbiol Immunol 5:409–422
Mittler S (1972) Low frequency electromagnetic radiation and genetic aberrations, Final rep AD 749959. Northern Illinois University, DeKalb, Oregon
Miura T (1972) The biological effect of air ions in room air. Int J Biometeorol 16(3):304–305
Monaco P, Acker CW (1963) Psychophysiological effects of ionized air on psychiatric patients. Newsl Res Psych 5:23–25
Moos WS, Schmitz R, Clark RK(1967) The effects of electric fields on mouse weights. Int J Biometeorol [Suppl] 11:321
Mörner N-A (1973) Climatic cycles during the last 35 000 years. J Interdiscipl Cycle Res 4(2):189–192
Möse JR, Fischer G (1970) Zur Wirkung elektrostatischer Gleichfelder, weitere tierexperimentelle Ergebnisse. Arch Hyg Bakteriol 154(4):378–386
Möse JR, Fischer G (1976) the electrostatic field as bioclimatological significant value. Arch Meteorol Geophys Bioklimatol [B] 24:109–126
Möse JR, Fischer G, Fischer M (1969) Einfluss des elektrischen Gleichfeldes auf die Wirkung einiger die glatte Muskulatur stimulierender Pharmaka. Z Biol pp 344–363
Möse JR, Schuy S, Fischer G (1972) Versuchsanlage zum Studium der Wirkungen von elektrostatischen Gleichfeldern an kleinen Laboratoriumstieren und die damit erzielten Ergeb Biomed Tech 17(2):65–70
Möse JR, Fischer G, Strampfer H (1973) Immunbiologische Reaktion im elektrostatischen Gleichfeld und Faraday-Käfig. Z Immunitäts forsch Exp Klin Immunol 145:404–412
Mühleisen R (1974) The global circuit and its parameter 5th Int Conf Atmos Electr
Müller-Velten H (1966) Über den Angstgeruch bei der Hausmaus. Z Vergl Physiol 52:401–429
Murr LE (1963) Plant growth response in a simulated electric field environment. Nature 200:490–491
Murr LE (1964) Mechanism of plant cell damage in an electrostatic field. Nature 201:1305–1306
Murr LE (1966a) The biophysics of plant growth in a reversed electrostatic field: A comparison with the conventional electrostatic and electrokinetic field growth responses. Int J Biometeorol 10(2):135–146
Murr LE (1966b) Physiological stimulation of plants using delayed and regulated electric field environments. Int J Biometeorol 10:147–153
Nazzaro JR, Jackson DE, Perkins LE (1967) Effects of ionized air on stress behavior. Med Res Eng 6:25–28
Nefedov YG, Anisimov BV et al. (1966) The aero-ionic composition of pressurized cabin air and its influence on the human body. Sov Conf. Space Biol Med pp 32–46
Ng W, Piekarsky K (1975) The effect of an electrostatic field on the mitosis of cells. Med Biol Eng pp 107–111
Nielsen CB, Harper HA (1954) Effect of air ions on succinoxidase activity of the rat adrenal gland. Proc Soc Exp Biol Med 86:753–756
Nikonowa KV (1964) Effect of a high frequency electromagnetic field on the blood pressure and body temperature of experimental animals. The biological action of radiofrequency electromagnetic fields. Moscow, p 61
Norton L (1974) In vivo bone growth in a controlled electric field. Ann NY Acad Sci 238:466
Noval JJ, Sohler A, Reisberg RB, Coyne H, Straub KO, McKinney H (1976) Extremely low frequency electric field induced changes in rate of growth and brain and liver enzymes of rats, Final rep Naval Air Development Center, Johnsville, Pennsylvania
Novitskii Yu I (1966a) The action of a constant magnetic field on gels of substances of plant origin. Proc Conf Eff Magn Fields Biol Obj, Moscow, p 50
Novitskii Yu I (1966b) Effect of a magnetic field on the dry seeds of some cereals. Proc Conf Eff Magn Fields Biol Obj Moscow, p 52
Novitskii Yu I et al. (1965) A further study of the effect of a constant magnetic field on plants. In: Questions of hematology, radiobi-

ology and the biological action of magnetic fields. Tomsk, p 320

Novitskii Yu I et al. (1966) Effect of a weak magnetic field on the movement of chloroplasts in Elodea. Proc Conf Eff Magn Fields Biol Obj, Moscow, p 53

Odintsov Yu N (1965) Effect of an alternating magnetic field on some immunological indices in experimental listeriosis. Questions of hematology, radiobiology and the biological action of magnetic fields. Tomsk, p 382

Oehrl W, König HL (1968) Messung und Deutung elektromagnetischer Oszillationen natürlichen Ursprungs im Frequenzbereich unter 1 Hz. Z Angew Phys 25(1):6–14

Ogawa T (1973) Analyses of measurement techniques of electric fields and currents in the atmosphere. Contrib Geophys Inst Kyoto Univ Nr 13

Ogawa T, Murakami Y (1973) Schumann resonance frequencies and the conductivity profiles in the atmosphere. Contrib Geophys Inst Kyoto Univ Nr 13

Ogawa T, Tanaka Y (1970) Q-factors of the Schumann resonances and solar activity. Spec Contrib Geophys Inst Kyoto Univ Nr 10

Ogawa T, Tanaka Y, Yasuhara M (1969) Schumann resonances and worldwide thunderstorm activity. J Geomagn Geoelectr 21(1):447–452

Olivereau JM (1971a) Incidences psychophysiologiques des facteurs climatiques de l'environment. Bull Psychol (Paris) 24:597–606

Olivereau JM (1971b) Incidences psychophysiologiques de l'ionization atmosphérique. PhD thesis, University of Paris.

Olivereau JM (1973) Influence des ions atmosphériques négatifs sur l'adaptation à une situation anxiogène chez le rat. Int J Biometeorol 17:277–284

Ossenkopp KP (1972) Maturation and open field behavior in rats exposed prenatally to an ELF low intensity rotating magnetic field. Psychol Rep 30:371–374

Ossenkopp K-P, Ossenkopp MD (1975) Open-field behaviour in juvenile rats exposed to an ELF rotating magnetic field. Differential sex effects. Int J Biometeorol 19 (Suppl 6):1

Ossenkopp KP, Koltek WT, Persinger MA (1972) Prenatal exposure to an extremely low frequency, low intensity, rotating magnetic field and increase in thyroid and testicle weights in rats. Dev Psychobiol 5:275–285

Palm H (1968) Das gesunde Haus. Eigenverlag,

Palti Y, DeNour E, Abrahamov A (1966) The effect of atmospheric ions on the respiratory system of infants. Pediatrics 38:405–411

Pautrizel R, Rivière M, Prioré A, Berlureau F (1966) Influences d'ondes électromagnétiques et de champs magnétiques associés sur l'immunité de la souris infestée par trypanosome équiperdum. CR Acad Sci (Paris) 236:579–682

Pautrizel R, Prioré A, Berlureau F, Pautrizel AN (1969) Stimulation par des moyens physiques des défenses de la souris et du rat contre la trypanosome expérimentale. CR Acad Sci (Paris)268:1889–

Pautrizel R, Prioré A, Berlureau F, Pautrizel AN (1970) Action de champs magnétiques combinés à des ondes électromagnétiques sur la trypanosomose expérimentale du lapin. CR Acad Sci (Paris)271:877–880

Pautrizel R, Prioré A, Dallachio M, Crockett R (1972) Action d'ondes électromagnétiques et de champs magnétiques sur les modifications lipidiques provoquées chez le lapin par l'administration d'un régime alimentaire hypercholesterole. CR Acad Sci (Paris) 274:488–491

Pautrizel R, Prioré A, Mattern P, Pautrizel AN (1975) Stimulation des défenses de la souris trypanosomées par l'action d'un rayonnement associant champs magnétiques et ondes électromagnétiques. CR Acad Sci (Paris) 280:1915–1918

Persinger MA (1969) Open field behavior in rats exposed prenatally to a low intensity frequency rotating magnetic field. Dev Psychobiol 2:168–171

Persinger MA (1973) Possible cardiac driving by an external rotating magnetic field. Int J Biometeorol 17(3):263–266

Persinger MA (1974) ELF and VLF electromagnetic field effects. Plenum Press, New York

Persinger MA (1975a) Comments on transient seismo-electricmagnetic fields and proximal human behaviour. Int J Biometeorol 19(Suppl 6):1

Persinger MA (1975b) Lag responses in mood reports to changes in the weather matrix. Int J Biometeorol 19(Suppl 6):1

Persinger MA (1976) Day time wheel running activity in laboratory rats following geomagnetic event of 5–6 July 1974. Int J Biometeorol. 20(1):19–22

Persinger MA, Janes JT (1975) Significant correlations between human anxiety scores and perinatal geomagnetic activity. Int J Biometeorol 19(Suppl 6):1

Persinger MA, Lafrenière GF (1975) Relative hypertrophy of rat thyroid following ten day exposures to an ELF magnetic field: Determing intensity thresholds. Int J Biometeorol 19(Suppl 6):1

Persinger MA, Pear JJ (1972) Prenatal exposure to an ELF rotating magnetic field and subsequent increase in conditioned suppression. Dev Psychobiol 5:269–274

Persinger MA, Glavin GB, Ossenkopp KP (1972) Physiological changes in adult rats exposed to an ELF rotating magnetic field. Int J Biometeorol 16, (2):163–172

Persinger MA, Lafrenière GF, Mainprize DN (1975) Human reaction time variability changes from low intensity 3 Hz and 10 Hz electric fields: Interaction with stimulus pattern, sex and field intensities. Int J Biometeorol 19, (1):46–54

Persinger MA, Lafrenière GF, Carrey NJ (1978) Thyroid morphology and wet organ weight changes in rats exposed to different low intensity 0.5 Hz magnetic fields and preexperimental caging conditions. Int J Biometeorol 22(1):67–73

Peschka W (1974a) Dynamische Effekte an Proben bestehend aus Hochfrequenzschwingkreisen bzw Leitungselementen unter Verwendung von Wasser als Dielektrikum. Wetter Boden Mensch 19:1200–1213

Peschka W (1974b) Wasser im Hochfrequenzfeld. Südd Forsch Tech 181:22

Petersen WF (1934-1938) The patient and the weather: Vols I-IV. Edwards, Ann Arbor, Michigan

Petersen WF (1946) Hippocratic wisdom. Thomas, Springfield, Illinois

Petrov FP (1935) Effect of an electromagnetic field on isolated organs. In: Physico chemical bases of higher nervous activity. Leningrad

Petrov FP (1952) Effect of a low-frequency electromagnetic field on higher nervous activity. Tr Inst Fiziol Akad Nauk SSSR 1:369

Petschke H (1953a) Über Beziehung zwischen der Blutkörperchensenkungsreaktion radioästhetischen Befunden und meteorologischen Vorgängen. Medizinische 39:1263

Petschke H (1953b) Über Beziehung zwischen der Blutkörperchensenkungsreaktion radioästhetischen Befunden und meteorologischen Vorgängen. Medizinische 52:1759–1761

Phillips GB, Harris GJ, Jones MW (1964) Effect of air ions on bacterial aerosols. Int J Biometeorol 8:27–37

Piccardi G (1970) The chemical basis of medical climatology. Thomas, Springfield. Illinois

Pirenne MH (1958) Photobiology group meeting, Oxford University. Quoted in Office of Naval Research European Scientific Notes 12:11

Pischinger A (1969) Über das vegetative Grundsystem. Phys Med Rehabil 10(3):53–57

Pittmann UZ (1964) Magnetism and plant growth. II. Effect on root growth of cereals. Can J Plant Sci 44:283–294

Pittmann U (1967) Gemäss Not Selecta 48:3716

Plaksin IN et al. (1966) Effect of frequency of electric field on optical properties of water. Dokl Akad Nauk SSSR 168:1

Plekhanov GF, Vedyushkina VV (1966) Elaboration of a vascular conditioned reflex in man to change in strength of a high-frequency electromagnetic field. Zh Vyss Nervn Deyat IP Pavlova. 16:34

Polk C (1969) Relation of ELF noise and Schumann resonances to thunderstorm activity. In: Coroniti FV, Hughes HG (eds) Planetary Electrodynamics: Vol 2 Gordon & Breach, New York, pp 55–83

Polk C (1973) Source, propagation, amplitude and temporal variation of extremely low frequency (0–100 Hz) electromagnetic fields. Symposion and workshop on the effects of low-frequency magnetic and electric fields on biological communication processes

Popp FA, Ruth B (1977) Arzneim-Forsch 27:933–940

Pratt R, Barnard RW (1960) Some effects of ionized air on Penicillum notatum. J Pharm Assoc [Sci Ed] 49:643–649

Presman AS (1970) Electromagnetic fields and life (transl from Russian into English by FL Sinclair). Plenum Press, New York

Presman AS (1972) Akzeleration und elektromagnetische Felder der Biosphäre. Mod Med 2(4):224–228

Proceedings of the *Ad Hoc* Committee for the Review of Biomedical and Ecological Effects of ELF Radiation (1973) Bureau of Medicine and Surgery, Department of the Navy, Washington, DC

Rager GR (1970) Experimental bases of negative air-ionisation (one paragraph summary only, in English). Aggressologie 11:105–110

Rajewsky B (1956) Strahlendosis und Strahlenwirkung. Stuttgart

Rambeau V (1969) Besteht ein Zusammenhang zwischen der Tektonik der Erde und dem Krankheitsproblem? Wetter Boden Mensch 7:341–354

Ranscht-Froemsdorff WR (1966a) Krankheitshäufungen bei Wettervorgang Null. Selecta 11:630–640

Ranscht-Froemsdorff WR (1966b) Selecta 48:2866

Ranscht-Froemsdorff W (1969) Dämpfung der Zellatmung durch Elektroklima. Umschau 24:803–804

Ranscht-Froemsdorff W (1976) Diagnostik von "Wetterfühligkeit" und "Wetterschmerz" Z Allg Med, 52 (5):228–236

Ranscht-Froemsdorff WR, Rinck O (1972) Elektroklimatische Erscheinungen des Föhns (Korrelationen von Blutgerinnung und simulierten Sferics-Programmen). Z Angew Baeder- Klimaheilkd 19:169–176

Ranscht-Froemsdorff WR, Weise H (1969) Sferics-Koinzidenzmessungen im Gelände mit Schmalbandempfängern. Kleinheubacher-Ber 13:73

Reinders H (1964) Ein Beitrag zur Klärung der Einflüsse elektrischer Gleichstromfelder auf die Gestaltung der Raumluft in Aufenthaltsräumen. Kaarst bei Neuss, Niederrhein-Verlag

Reinke R (1977) Der medizinmeteorologische Dienst – eine zusätzliche Hilfe für den behandelnden Arzt. Dtsch Aerzt 27 (1,2):67–76, 90–98

Reis A (1974) Biomedizinische Technik. Ingenieuraufgaben in der Medizin. VDI-Verlag, Düsseldorf

Reiter R (1953) Neuere Untersuchungen zum Problem der Wetterabhängigkeit des Menschen, ausgeführt unter Verwendung biometeorologischer Indikatoren. Arch Meteorol Geophys Bioklimatol [B]4(3):327–377

Reiter R (1954) Umwelteinflüsse auf die Reaktionszeit des gesunden Menschen. Münch Med Wochenschr 96 (17,18):(17):479–481, 526–529

Reiter R (1960) Meteorologie und Elektrizität der Atmosphäre. Akad. Verlagsges, Geest & Portig K-G, Leipzig

Reiter R (1962) Welche atmosphärisch-elektrischen Elemente können auf den Organismus einwirken? Tag Med Meteorol, Timmendorfer Strand

Reiter R (1970) Sind luftelektrsche Grössen als Komponenten des Bioklim in Betracht zu ziehen? Heiz-Lüft-Haustech, 21(8):258–262, 279–285

Reiter R (1973) Introductory remarkes. Int 7 Biometeorol. 17(3):205–206

Report of the Committee on Biosphere Effects of Extremely-Low-Frequency Radiation (1977) Biologic effects of electric and magnetic fields associated with proposed project Seafarer, Contract Nr N00014-75-C-0258. Office of Naval Research of the Department of the Navy, National Academy of Sciences, Washington, DC

Rheinstein J (1960) Der Einfluss von künstlich erzeugten atmosphärischen Ionen auf die einfache Reaktionszeit und auf den optischen Moment. Dissertation, Technischen Hochschule, Munich

Rheinstein J (1961) Statistical evaluation of the influence of atmospheric ions on the simple reaction time and on the optical moment. Proc Int Conf Air Ioniz 2(19)

Rhoads FJ, Garner WE (1967) An investigation of the modal interference of VLF radio waves. Radio Sci 2:539

Riesen W, Aranyi C, Kyle J, Valentino A, Miller D (1971) A pilot study of the interaction of extremely low frequency electromagnetic fields with brain organelles, Nr 3, IITRI Proj E 6185 IIT Res Inst, Chicago

Rinck O (1972) Freiburger Dissertation (in preparation). Freiburg, Germany

Rinfret A, Wexler R (1953) Effect of ionized air and tobacco smoke on the adrenal lipid content in rats. Stanford Med Bull 2:125

Rivière J (1976) Action des champs magnétique et électrique sur la croissance et le taux des mutations de divers microorganismes. RGE-FRA, ISSN 0035-3116, 07. Spec Nr, pp 98–101

Rivière MR, Prioré A, Berlureau F, Fournier M, Guérin M (1964) Action de champs électromagnétiques sur les greffes de le tumeur T 8 chez la rat. CR Acad Sci (Paris) 259:4895–4897

Rivière MR, Prioré A, Berlureau F, Fournier M, Guérin M (1965) Phénomènes de régression observés sur les greffes d'un lymphosarcome chez des souris exposées à des champs électromagnétiques. CR Acad Sci (Paris) 260:2639–2102

Robertson K (1977) Personal communication, October

Rocard Y (1964) Actions of a very weak magnetic gradient: The reflex of the dowser. In: Barnothy MF (ed) Biological effects of magnetic fields. Plenum Press, New York, Part IV, Chap. 2

Rohracher H (1949) Mechanische Mikroschwingungen des menschlichen Körpers. Urban & Schwarzenberg, Vienna

Rohracher H (1952) Neue Untersuchungen über biologische Mikroschwingungen. Anzeiger des phil hist. Kl Oest Akad Wiss Nr 11

Rohracher H, Inaga K (1969) Die Mikrovibration. Huber, Bern

Rosenberg BL (1972) A study of atmospheric ionization: Measurement of the ion conditions in an ATC laboratory and a review of the literature of ion effects on performance. National Aviation Facilities Experimental Center.

Rosenthal GM, Jr (1975) Germination and early growth of sunflowers in weak ELF electromagnetic fields, Final rep. IIT Research Institute, Chicago

Ross Adey W (1974) Gehirn reagiert auf Potentialschwankungen. Südd. Forsch Wiss Tech 106:34

Rupilius JP (1976) Untersuchungen über die Wirkung eines elektrischen und magnetischen 50 Hz-Wechselfeldes auf den Menschen. Dissertation, Med Fak, Universität Freiburg

Ruth B, Popp FA (1976) Z Naturforsch [C]31:741–745

Rycroft MJ (1965) Resonances of the earth-ionosphere cavity observed at Cambridge, England. J Res Natl Bur Stand [D] 69(8)

Sale AJH, Hamilton WA(1967) Effects of high electric fields on microorganisms. I. Killing of bacteria and yeasts. II. Mechanism of action of the lethal effect. Biochim Biophys Acta 148:781–800

Sao K, Jindoh H (1974) Real time location of atmospherics by single station techniques and primelary results. Atmos Terr Phys 36:261–266

Sao K, Yamashita M, Tanahashi S, Jindoh H, Ohta K (1973) Experimental investigations of Schumann resonance frequencies. Atmos Terr Phys 35:2047–2053

Sazonova T (1967) A physiological assessment of the work conditions in 400 kV and 500 kV open switchyards. Scientific publications of the institutes of labor protection of the All-Union Central Council of Trade Unions. Profizdat, Issue 46.

Schaefer H, Silny J (1977) Plötzlicher und unerwarteter Tod im Kleinkindesalter und elektromagnetische Felder. Med Klin 72(19):871–874

Scheller EF (1969) Strahleneinwirkung im Blut und Krebs. Wetter Boden Mensch, 6:269–272

Schmid A (1936) Biologische Wirkungen der Luftelektrizität. Bern-Leipzig

Schneider F (1974) Devitationsrhythmen in Bezug auf künstliche magnetische Felder. Mitt Schweiz Entomol Ges 47(1-2)

Schneider K-H, Studinger H, Weck K-H, Steinbigler H, Utmischi D, Wiesinger J (1974) Displacement currents to the human body caused by the dielectric field under overhead lines. CIGRE-Ber 36–04, Session

Schneider M (1964) Einführung in die Physiologie des Menschen. Springer-Verlag Berlin, Göttingen, Heidelberg

Schneider R (1973) Über neuere Untersuchungen und Ergebnisse zum Problem der 'Wünschelrute.' Wetter Boden Mensch 17:1093–1095

Schneider R (1978) Leitfaden und Lehrkurs der Ruten und Pendelkunst. Eigenverlag Forschungskreis für Geobiologie, Germany

Schüa L (1953) Die Fluchtreaktion von Goldhamstern aus elektrischen Feldern. Naturwissenschaften 40:514

Schuldt H (1974a) Dr. med., Dipl.-Ing., 2 Hamburg 52, Up de Schanz 60 (personal communication)

Schuldt H (1974b) Wetterfühligkeit des Menschen aus neuer medizinisch-meteorologischer Sicht. Hamburg Äerztebl pp 143–144

Schulz H (1970) Über allgemeine und persönlichkeitsspezifische Wirkungen stationärer elektrischer Felder auf Leistung und Befindlichkeit. Dissertation, Math-Naturwiss, Fak, Universität Düsseldorf

Schulz K-H (1965) Die Bedeutung der Luftelektrizität, der Elektro-Aerosol-Therapie, der direkten elektrostatischen Aufladungstherapie nack Takata und ihre biologische Wirkung auf das vegetative Nervensystem unter

Berücksichtigung des Begetonogramms. Z Aerosol-Forsch 12:206–212, 455–466
Schulz K-H (167) Untersuchungen der biologischen Wirkung der Eletro-Aerosole und der direkten Aufladungstherapie auf das vegetative Nervensystem. Lecture at the conference of the Weinheimer research group at the Institute for Hygiene of Heidelberg University
Schulze R (1970) Strahlenklima der Erde. Dr. Dietrich Steinkopff Verlag, Darmstadt
Schumann WO (1952) Über die strahlungslosen Eigenschwingungen einer leitenden Kugel, die von einer Luftschicht und einer Ionensphärenhülle umgeben ist. Z Naturforsch [A] 7:149–154
Schumann WO, König HL(1954) Atmospherics geringster Frequenzen. Naturwissenschaften 41:183–184
Schumann WO, Rohrer L, König HL (1966) Experimentelle Untersuchungen elektromagnetischer Wellen in der Atmosphäre mit 4–40 sec Periodendauer. Naturwissenschaften 3:79
Schwamm E (1957) Ultrarot-Körpermessungen im Kippschwingungsfeld. Erfahrungsheilkunde 6(12)
Searle AG (1965) Science News Lett, 88:217; cited in Selecta 14:832 (1966)
Seeger PG (1968) Die Erfolge der Vitanova Feldtherapie bei akuten und chronischen Erkrankungen in biochemischer Sicht. Biologisch physikalische Forschungsgesellschaft, Oberjesingen
Seidel D, Knoll M, Eichmeier J (1968) Anregung von subjektiven Lichterscheinungen (Phosphenen) beim Menschen durch magnetische Sinusfelder. Pfluegers Arch 299:11 bis 18
Selecta (1976) Medizinisches Forum: Was ist Elektro-Akupunktur? Selecta 46:4367–4372
Selye H (1950) Stress. Acta Inc., Med Publ, Montreal
Selzer E (1970) Oscillations de Schumann. Handb Phys 49/4(Sect. 55):320–330
Senetra D, Perego A, Giubilaro G (1978) Biological effects of low-level, very low-frequency (VLF) electric fields on the blood sedimentation rate. Int J Biometeorol 22 (1):59–66
Sharp EL (1967) Atmospheric ions and germination of uredospores of *Puccinia striiformis*. Science 156:1359–1360
Sheppard AR, Eisenbud M (1977) Biological effects of electric and magnetic fields of extremely low frequency. New York University Press, New York
Shewchuk SR, Iribarne JV (1970) Electrification associated with drop splashing. Trans Faraday Soc 66:2092–2105
Sidaway GH (1966) Influences of electrostatic fields on seed germination. Nature 211:303
Sidaway GH (1975) Some early experiments in electro-culture. J Electrostatics pp 389–393
Siebert M (1964a) Erdmagnetische Pulsationen. Umschau Wissenschaft Technik. 4:110–113
Siebert M (1964b) Erdmagnetische Pulsationen. Umsch Wiss Tech 6:182–184
Silny J (1976a) Einwirkungen der elektrischen 50 Hz-Felder hoher Feldstärken auf den Organismus eines Warmblüters. 2nd Kolloq Bioklimatol Wirk Luftelektr Faktoren, Munich
Silny J (1976b) Effet d'influence d'un champ électrique à 50 Hz sur l'organisme. RGE-FRA, ISSN 0035-3116, 07. Spec Nr, pp 81–90
Silverman D, Kornblueh I (1957) Effect of artificial ionization of the air on the electroencephalogram. Am J Phys Med 36:352–358
Sinz R (1978) (personal communication)
Skorobogatova AM (1962) On the mechanism of action of ionized air on the organism. In: Tromp SW (ed) Biometeorology. Macmillan New York, pp 507–509
Slote L (1962) An experimental evaluation of man's reaction to an ionized air environment. Proc Int Conf on Ioniz 2:1–22
Snorrason E (1964) Early history of medical climatology. Med Climatol vol 8
Soderberg EE, Finkle M (1970) A Comparsion of ELF-atmospherics noise spectra measured above and in the sea. Navy Underwater Sound Laboratory, New London, Connecticut
Solov'ev NA (1963) Action of a high-voltage electric field of 50–2000 Hz on white mice and drosophila. Proceedings of conference on labor hygiene and the biological action of radio-frequency electro-magnetic fields. Moscow, p 91
Solov'ev NA (1967) Experimental study of the biological action of a low frequency electrical field. Nov Medit Driborostr 3:101–107
Sönning W (1974) Die nächtliche relative Luftfeuchtigkeit am Boden als Indikator für Wettervorgänge in höheren troposphärischen Schichten. Meteorol Rundsch 27:142–148
Sönning W (1976) Die Biotropie des Wetters

und der Witterung und ihre analytische Darstellung in der Biosynoptik. 2nd Kolloq Bioklimatol Wirk Luftelektr Faktoren, Munich
Southern WE (1969) Orientation behavior of ring-billed gull chicks and fledglings. Condor 71:418–425
Southern WE (1971) Gull orientation by magnetic cues: A hypothesis revisited. Ann NY Acad Sci 188:295–311
Southern WE (1972) Magnets disrupt the orientation of juvenile ring-billed gulls. BioScience 22:476–479
Southern WE (1975) Influence of disturbances in the earth's magnetic field on ring-billed gull orientation. Condor 74:102–105
Spittka O, Taege M, Tembrock G (1969) Experimentelle Untersuchungen zum operanten Trinkverhalten von Ratten im 50 Hz-Hochspannungswechselfeld. Biol Zentralbl 88:273–282
Stängle JWF (1973) Grundstrahlungsmessungen über geopathischen Reizstreifen. Wetter Boden Mensch 18:1146–1153
Stein L (1971) Neurochemistry of reward and punishment: Some implications for the etiology of schizophrenia. J Psychiatr Res 8:345–361
Stetson HT (1957) Note on possible effect of electric field on the growth of plants. J Franklin Inst 264(3):169–180
Stevens D, Fechter D (1969) The effects of p-chlorophenylalanine, a depletor of brain serotonin, on behavior. II. Retardation of passive avoidance learning. Life Sci 8:379–385
Storey LR (1953) An investigation of whistling atmospherics, Phils Trans R Soc Lond [Math Phys Sci] 246(905)
Strampfer H, Geyer N (1974) Ersterhebung über die Wirkung eines elektrostatischen Gleichfeldes und des 10 Hz-Impulsfeldes auf das psychische Verhalten 1st Kolloq Bioklimatol Wirk Luftelektr Faktoren, Graz
Strumza M (1970) Influence sur la santé humaine de la proximité des conducteurs d'électricité à haute tension. Arch Mal Prof Med Trav Secur Soc 31:269–276
Sulman FG (1971a) Serotonin migraine in climatic heat stress, its prophylaxis and treatment. Proc Int Headache Symp pp 205–210
Sulman FG (1971b) Meteorologische Frontverschiebung und Wetterfühligkeit. Aerztl Praxis 23(17):998–999
Sulman FG (1972) Urinalysis and treatment of patients suffering from climatic heat stress. In Bar-Or O (ed) Proceedings of the Fourth International Symposium Wingate Institute, Israel, Lect Nr 31
Sulman FG, Danon A, Pfeifer Y, Tal E, Weller CP (1970) Urinalysis of patients suffering from climatic heat stress (Sharav). Int J Biometeorol 14:45–53
Sulman FG, Levy P, Levy A, Pfeifer Y, Superstine E, Tal E (1974) Air-ionometry of hot, dry desert winds (Sharav) and treatment with air ions of weather-sensitive subjects. Int J Biometeorol 18:313–318
Sulman FG, Levy D, Pfeifer Y, Superstine E, Tal E (1975) Effects of the Sharav and Bora on urinary neurohormone excretion in 500 weather sensitive females. Int J Biometeorol 19:202–209
Sulman FG, Levy D, Lunkan L, Pfeifer Y, Tal E (1978) Absence of harmful effects of protracted negative air ionisation. Int J Biometeorol 22(1):53–58
Swantes HJ (1977) Staubbeimengungen der atmosphärischen Luft und ihr Einfluss auf den pH-Wert der menschlichen Haut. Z Hautkr 52(9):520
Swider W (1968) AFCRL, Bedford, Massachusetts
Takata M, Dohmoto S (1936) Tohoku J Exp Med 28:522
Takata M, Dohmoto S (1936–1937) Tohoku J Exp Med 30:219–250
Tal E, Pfeifer Y, Sulman FG (1976) Effect of air ionization on blood serotonin in vitro. (Specialia.) Experientia 15(3):326–327
Tanner JA (1966) Effect of microwave radiation on birds. Nature 210:636 only
Tanner JA, Romero-Sierra C, Davie SJ (1967) Non-thermal effects of microwave radiation on birds. Nature 216:1139
Taylor WL (1960a) Daytime attenuation rates in the VLF band using atmospherics. J Res Nat Bur Stand [D]64:349
Taylor WL (1960b) VLF attenuation for east-west and west-east daytime propagation using atmospherics. J Geophys Res 65 (7)
Tchijevsky AL (1933) Transaction of central laboratory scientific research on ionification: 3 vols. The Commune, Voronej.
Tchijevsky AL (1960) Air ionification, its role in the national economy. Publishing House of the State Planning Commission of the USSR.
Tegenkamp TR (1969) Mutagenic effects of

magnetic fields on *Drosophila melanogaster* In: Barnothy MF (ed) Biological effects of magnetic fields: Vol 2. Plenum Press, New York, pp 189–206

Teilhard de Chardin P (1959) The phenomenon of man. Harper & Row, New York.

Tenen S (1967) The effects of p-chlorophenylalanine, a serotonin depletor, on avoidance acquisition, pain sensitivity, and related behavior in the rat. Psychopharmacologia 10:204–219

Tepley L (1959) A comparison of spherics as observed in the VLF and ELF bands. J Geophys Res 64:2315

Terry RA, Harden DG, Mayyasi AM (1969) Effects of negative air ions, noise, sex and age on maze learning in rats. Int J Biometeorol 13:39–49

Thompson SP (1910) The effects of magnetic fields. Proc R Soc Lond [Biol] 82:996

Thomson JJ (1898) On the charge of electricity carried by the ions produced by Röntgen rays. Philos Mag 46:528–545

Thudichum JWL (1884) A treatise on the chemical constitution of the brain. Balliere, Tindall and Cox, London

Toomey J, Polk C (1970) Research on extremely low frequency propagation with particular emphasis on Schumann resonance and related phenomena. Contract Nr AF 19 (628)–4950. University of Rhode Island

Tran A, Polk C (1972) Propagation and resonance parameters of the earth-ionosphere cavity-spherical versus planar stratification of the ionosphere, Report AFCRL-72-0682. Department of Electrical Engineering, University of Rhode Island

Tran A, Polk C (1973) Electrical conductivity of the mesophere (40 km to 100 km) from ELF spectra. F 19628-70-C-0090, AFCRL-TR-73-0168

Tromp SW (1954). Cited in Wüst and Petschke (1954)

Tromp SW (1963) Medical biometeorology. Elsevier, Amsterdam

Tromp SW (1968) A survey of recent advances in electro-biometeorology, Rep Nr 10. Biometeorol Res Cent Leiden, The Netherlands

Tromp SW (1970) Seasonal and yearly fluctuations in meteorologically induced electromagnetic wave patterns in the atmosphere (period 1956–1968) and their possible biological significance. J Interdiscipl Cycle Res 1(2):193–199

Tromp SW (1972) Waterdivining (dowsing). Encyclopedia of Geochemistries and Earth Science, New York, pp 1252–1258

Tromp SW (ed) (1974) Definition and classification of biometeorology. Prog Biometeorol [A] (Part 1):1–

Tylor PE (1975) Biologic effects of non-ionizing radiation. Ann NY Acad Sci 247:1–545

Ucha Udabe R (1963) Significado biologico de la aeroionizacion, su relacion con las neurohormonas. Semana Med 122:1399–1403

Ucha Udabe R, Kertesz R, Franceschetti L (1968) Etudes sur l'utilisation des ions négatifs dans les maladies du système nerveux. In: Gualtierotti R, Kornblueh IH, Sirtoni C (eds) Bioclimatology, biometeorology, and aeroionotherapy. Carlo Erba Foundation, Milan.

Undt W (1962) 4. Medizin-Meteorologische Arbeitstagung in Timmendorfer Strand vom 12.–14. April 1962. Wetter Leben 14:128–134

Undt W (1976) Ergebnisse von Versuchen mit künstlichen elektrischen Feldern. 2nd Kolloq Bioklimatol Wirk Luftelektr Faktoren, Munich

Ungeheuer H (1955) Ein meteorologischer Beitrag zu Grundproblemen der Medizinmeteorologie. Ber Dtsch Wetterdienstes 16:1, and (1956) Medizin-Meteorolog Nr. 2:26–29

Ungeheuer H, Brezowsky H (1965) Lufttemperatur und Luftfeuchtigkeit als Indikatoren biosphärischer Akkordschwankungen. Meteorol Rundsch 18(4):120–123

Ungeheuer H, Kügler H (1957) Meteorologie - Biologie-Medizin. Arzneim Forsch 7(6):370 –380

URSI (1975) United States National Committee, Institute of Electrical and Electronics Engineers, University of Colorado (ann meet)

Usemann K, Vogel W (1976) Fachbereich Architektur und Bautechnik der Universität Kaiserslautern (personal communication)

Valdes-Dapena MA (1967) Sudden and unexpected death in infancy: A review of world literature 1954-1966. J Pediatr 39:123–138

Valenstein ES, Cox VC, Kakolewski JW (1970) A reexamination of the role of the hypothalamus in motivation. Psychol Rev 77:16–31

Valfre F et al. (1964) La sensibilità degli organismi animali alle variabili cosmiche. Prove effectuate con acqua normale e con acqua fisicamente attivata. Geophys Meteorol 13:76

van Dyke JH, Halpern MH (1965) Observa-

tions on selected life processes in null magnetic fields. Anat Rec 151:480
van Everdingen W (1940) Moleculare veranderingen tengevolge van bestraling met golvan van hertz met cen frequentie van 1875 MHz. Ned Tijdschr Geneesk 84:4370
Varga A (1968a) Wirkung von Luftionen auf die Herzfrequenz. Bericht Weinheimer Forschungskreis für Bio-Elektroklimatologie, Heidelberg
Varga A (1968b) Wirkung von Luftionen auf die Herzfrequenz. Umschau 68:152
Varga A (1972a) Hygiene-Institut der Universität Heidelberg, Elektro-Bioklimatische Forschungsstelle (personal communication)
Varga A (1972b) Forschungsbericht über die physiologische Wirkung von Luftionen und deren Bedeutung als Umweltfaktoren. Elektro-Bioklimatische Forschungsstelle des Hygiene-Instituts der Universität Heidelberg
Varga A (1973) Einfluss von Magnetfeldern auf das Wachstum von Mikroorganismen am Beispiel von Escherichia Coli und Bacillus Subtilis. Dissertation, Fak Bio- Geowiss, Technischen Universität, Karlsruhe
Vasiliev IL (1960) The physiological mechanisms of aeroions. Am J Phys Med 39:124–128
Vasiliev LL (1966) Theory and practice of ionized-air therapy. Foreign Technology Division, Wright-Patterson Air Force Base, Ohio
Verheijen FJ (1958) The mechanisms of the trapping effect of artificial light sources upon animals. Dissertation, Utrecht
Vill H, Jahnke H (1975) Forschungsgemeinschaft für bioelektronische Funktionsdiagnostik und Therapie e. V., Nürnberg
Voll R (1966) Gelöste und ungelöste Probleme der Elektroakupunktur-Diagnostik und-Therapie. Schriftenreihe des Zentralverbandes der Ärzte für Naturheilverfahren e. V:Vol 18. Sonderheft, Medizinisch-Literarischer Verlag, Dr. Bume & Co., Ülzen, p 5
von Leitner H, Ludwig HW (1975) Therapy with 10 cps pulses. Int J Biometeorol 19(6):1
von Philipsborn E (1949) Bioklimatologie im Alten Griechenland. Meteorol Rundsch 2(3/4):100–101
von Pohl, Baron, cited in Hartmann (1976) and Stängle (1973)
Wageneder FM et al. (1973) Gehirn mit Strom durchflutet. Selecta 41:3722–3725
Waibel R (1975) Der Einfluss niederfrequenter elektrischer Felder auf Lebewesen. Dissertation, Universität Graz
Wait GR (1934) Ions in the air. Carnegie Inst Wash New Serv Bull 111(12):87–91
Warnke U (1973) Physikalisch-physiologische, Grundlagen zur luftelektrisch bedingten "Wetterfühligkeit" der Honigbiene (Apis mellifica). Dissertation, Universität des Saarlandes, Saarbrücken
Warnke U (1975a) Insekten und Vögel erzeugen elektrische Felder. Umschau 15:479
Warnke U (1975b) Bienen unter Hochspannung. Umschau 13:416
Warnke U (1976a) Bioelektrische und biomagnetische Eigenschaften der Körperoberflächen von Tieren im Einfluss meteorologischer Faktoren. 2nd Kolloq Bioklimatol Wirk Luftelektr Faktoren, Munich
Warnke U (1976b) Effects of electric charges on honey bees. Bee World 57(2,5):50–56
Warnke U (1976c) Die Wirkung von Hochspannungswechselfeldern auf das Verhalten von Bienensozietäten. Z Angew Entomol 82:88
Warnke U (1978) Aspekte zur magnetischen Kraftwirkung auf biologische Systeme. Heilkunst 91(1):1–12
Warnke U, Paul R (1976) Das Verhalten von Bienen unter Hochspannungsleitungen. Dokumentarischer Super-8-Film, Fachbereich Biologie der Universität des Saarlandes, Saarbrücken
Warnke U, Altmann G, Paul R (1975) Der Temperaturgang im Bienenvolk als Störungsindikator. Z Angew Entomol 78:150–159
Wassermann G (1956) An outline of a field theory of organismic form and behaviour. Churchill, London
Watson J, De Haas WG, Hauser SS (1975) Effect of electric field on growth rate of embryonic chick tibiae in vitro. Nature 254:331
Watt AD, Groghan RD (1964) Comparison of observed VLF attenuation rates and excitation factors with theory. J Res Natl Bur Stand[D]68(1)
Watt AD, Maxwell EL (1957) Characteristic of atmospheric noise from 1 to 100 Kc/s. Symp Propagation VFL Waves Paper 35
Wedler E (1960) Cited in Undt (1962)
Wedler E (1970) Erfahrungen aus einem Medizin-Meteorologischen Testjahr. Schriftenr Ver Wasser- Boden- Lufthyg Berlin-Dahlem 30:53–88

Wehner AP (1962) Electro-aerosol therapy. Am J Phys Med 41:23–40

Wehner AP (1964) Growth of microorganism in electrostatic fields. Int J Biometeorol 2:277–282

Wehner AP (1969) Electro aerosols, air ions, and physical medicine. Am J Phys Med 48:119–149

Weihe NH (1976) The application of meteorology in medical sciences. Int J Biometeorol 20(2):157–165

Weihe WH, Brezowsky H (1963) Die wissenschaftlichen Veröffentlichungen der medizinmeteorologischen Forschungsstellen des Deutschen Wetterdienstes in der BRD. Int J Biometeorol 6:113–128

Weissenborn G (1968) Das längst bekannte Magnetfeld - seine unbekannten Wirkungen auf den menschlichen Kreislauf. Medizin 17a(9):270–274

Weissman A (1973) Behavioral pharmacology of p-chlorophenylalanine (p-CPA). In: Barchas JD, Usdin E (eds) Serotonin and behavior. Academic Press, New York

Wellenstein G (1973) Der Einfluss von Hochspannungsleitungen auf Bienenvölker. Z Angew Entomol 74(1):86–94

Went FW (1932) Eine botanische Polaritatstheorie. Jahrb Wiss Bot 76:528–557

Wenzel HG (1975) Physiologically equivalent combinations of elevated environmental temperature and air humidity. Proc Int Congr Mine Vent

Wessel W (1939) Über den Durchgang elektrischer Wellen durch Drahtgitter. Z Hochfrequenztech Elektroakust 54:62

Wever R (1967) Über die Beeinflussung der circadianen Periodik des Menschen durch schwache elektromagnetische Felder. Z Vergl Physiol 56:111–128

Wever R (1968) Einfluss schwacher elektromagnetischer Felder auf die circadiane Periodik des Menschen. Naturwissenschaften 55(1): 29–32

Wever R (1973) Human circadian rhythms under the influence of weak electric fields and the different aspects of these studies. Int J Biometeorol 17(3):227–232

Wever R (1976) Effects of weak electric 10 Hz fields on separated vegetative rhythms involved in the human circadian multi-oscillator system. Arch Meteorol Geophys Bioklimatol [B]24:109–126

Wilder J (1957) The law of initial value in neurology and psychiatry. J Nerv Ment Dis 125:73–86

Wilhelmi T (1962) Der Einfluss von Gewittern auf das Radialwachstum unserer Waldbäume. Umschau 22:705–706

Williams TC, Williams JM, Larkin RP, Sutherland PJ, Cohen B (1977) A radar investigation of the effects of extremely low frequency electromagnetic fields on free flying migrant birds. Final rep., Contract N00014-75- - -341. US Navy, Office of Naval Research, Washington, DC

Williams WJ, Lorenz PL (1968) World oil. April 1957, cited in Wetter Boden Mensch 3:97

Wiltschko W (1973) Kompasssysteme in der Orientierung von Zugvögeln. Akad Wiss Lit Mainz 2

Wiltschko W, Fleissner G (1976) Die Orientierung von Rotkehlchen in magnetischen Wechselfeldern. Verh Dtsch Zool Ges 265:1976

Winsor T, Beckett JC (1958) Biological effects of ionized air in man. Am J Phys Med 37:83–89

Wise C, Berger B, Stein L (1970) Serotonin: A possible mediator of behavioral suppression induced by anxiety. Dis Nerv Syst 31:34–37

Wofford JC (1966) Negative ionization: An investigation of behavioral effects. J Exp Psychol 71:608–611

Woolley DW (1967) Involvement of the hormone serotonin in emotion and mind. In: Glass D (ed) Neurophysiology and emotion. Rockefeller University Press, New York, pp 108–116

Worden JL (1961) Some biological effects of air ionization. Proc Int Conf Ioniz Air 1:1–24

Wüst J (1953) Über physikalische Nachweismethoden der sog. 'Erdstrahlen' 6. Beiheft zur Zeitschrift 'Erfahrungsheilkunde' Geopathie, Haug Verlag, Ulm

Wüst J (1956) Dtsch Z Akupunktur 4(7/8)

Wüst J (1959) Rutengänger und ultrakurze Wellen. Z Radioästh 11(1/2)

Wüst J, Petschke H (1954) Zur gegenwärtigen Situation der Geopathie. Erfahrungsheilkunde 111(12)

Yaglou CP, Benjamin LC, Brandt A (1933) Observations on a group of subjects before, during and after exposure to ionized air. J Ind Hyg 15:8–17

Zahner R (1964) Zur Wirkung des elektrischen Feldes auf das Verhalten des Goldhamsters. Z Vergl Physiol 49:172–190

Zhurbitskii ZI (1958) Effect of the electrical fields of the atmosphere on the assimilation of nutrient elements by plants. Fiziol Rast 5:300–305

Zhurbitskii ZI, Shidlovskaya IL (1967) The action of an electrical field and ionized air on the absorption of mineral ions by wheat sprouts. (abstr). Biol Abstr p 100181 (1969)

Subject Index

B

C

GPSR Compliance

The European Union's (EU) General Product Safety Regulation (GPSR) is a set of rules that requires consumer products to be safe and our obligations to ensure this.

If you have any concerns about our products, you can contact us on ProductSafety@springernature.com

In case Publisher is established outside the EU, the EU authorized representative is:

Springer Nature Customer Service Center GmbH
Europaplatz 3
69115 Heidelberg, Germany

Batch number: 10762209

Printed by Printforce, the Netherlands